Recent Titles in This Series

163 **L. A. Bokut′, M. Hazewinkel, and Yu. G. Reshetnyak,** Third Siberian School "Algebra and Analysis"
162 **S. G. Gindikin, Editor,** Applied Problems of Radon Transform
161 **Katsumi Nomizu, Editor,** Selected Papers on Analysis, Probability, and Statistics
160 **K. Nomizu, Editor,** Selected Papers on Number Theory, Algebraic Geometry, and Differential Geometry
159 **O. A. Ladyzhenskaya, Editor,** Proceedings of the St. Petersburg Mathematical Society, Volume II
158 **A. K. Kelmans, Editor,** Selected Topics in Discrete Mathematics: Proceedings of the Moscow Discrete Mathematics Seminar, 1972–1990
157 **M. Sh. Birman, Editor,** Wave Propagation. Scattering Theory
156 **V. N. Gerasimov, N. G. Nesterenko, and A. I. Valitskas,** Three Papers on Algebras and Their Representations
155 **O. A. Ladyzhenskaya and A. M. Vershik, Editors,** Proceedings of the St. Petersburg Mathematical Society, Volume I
154 **V. A. Artamonov et al.,** Selected Papers in K-Theory
153 **S. G. Gindikin, Editor,** Singularity Theory and Some Problems of Functional Analysis
152 **H. Draškovičová et al.,** Ordered Sets and Lattices II
151 **I. A. Aleksandrov, L. A. Bokut′, and Yu. G. Reshetnyak, Editors,** Second Siberian Winter School "Algebra and Analysis"
150 **S. G. Gindikin, Editor,** Spectral Theory of Operators
149 **V. S. Afraĭmovich et al.,** Thirteen Papers in Algebra, Functional Analysis, Topology, and Probability, Translated from the Russian
148 **A. D. Aleksandrov, O. V. Belegradek, L. A. Bokut′, and Yu. L. Ershov, Editors,** First Siberian Winter School "Algebra and Analysis"
147 **I. G. Bashmakova et al.,** Nine Papers from the International Congress of Mathematicians, 1986
146 **L. A. Aĭzenberg et al.,** Fifteen Papers in Complex Analysis
145 **S. G. Dalalyan et al.,** Eight Papers Translated from the Russian
144 **S. D. Berman et al.,** Thirteen Papers Translated from the Russian
143 **V. A. Belonogov et al.,** Eight Papers Translated from the Russian
142 **M. B. Abalovich et al.,** Ten Papers Translated from the Russian
141 **H. Draškovičová et al.,** Ordered Sets and Lattices
140 **V. I. Bernik et al.,** Eleven Papers Translated from the Russian
139 **A. Ya. Aĭzenshtat et al.,** Nineteen Papers on Algebraic Semigroups
138 **I. V. Kovalishina and V. P. Potapov,** Seven Papers Translated from the Russian
137 **V. I. Arnol′d et al.,** Fourteen Papers Translated from the Russian
136 **L. A. Aksent′ev et al.,** Fourteen Papers Translated from the Russian
135 **S. N. Artemov et al.,** Six Papers in Logic
134 **A. Ya. Aĭzenshtat et al.,** Fourteen Papers Translated from the Russian
133 **R. R. Suncheleev et al.,** Thirteen Papers in Analysis
132 **I. G. Dmitriev et al.,** Thirteen Papers in Algebra
131 **V. A. Zmorovich et al.,** Ten Papers in Analysis
130 **M. M. Lavrent′ev, K. G. Reznitskaya, and V. G. Yakhno,** One-dimensional Inverse Problems of Mathematical Physics
129 **S. Ya. Khavinson,** Two Papers on Extremal Problems in Complex Analysis
128 **I. K. Zhuk et al.,** Thirteen Papers in Algebra and Number Theory
127 **P. L. Shabalin et al.,** Eleven Papers in Analysis
126 **S. A. Akhmedov et al.,** Eleven Papers on Differential Equations
125 **D. V. Anosov et al.,** Seven Papers in Applied Mathematics

(Continued in the back of this publication)

American Mathematical Society

TRANSLATIONS

Series 2 • Volume 163

Third Siberian School

Algebra and Analysis

Proceedings of the Third Siberian School
Irkutsk State University, Irkutsk
1989

L. A. Bokut'
M. Hazewinkel
Yu. G. Reshetnyak
Editors

American Mathematical Society
Providence, Rhode Island

Translation edited by DAUSA ZVIRENAITE and SIMEON IVANOV

Library of Congress Cataloging-in-Publication Data

Siberian School "Algebra and Analysis" (3rd : 1989 : Irkutsk State University).

[Algebra i analiz. English]

Third Siberian School "Algebra and Analysis" : proceedings of the Third Siberian School, Irkutsk State University, Irkutsk, 1989 / L. A. Bokut′, M. Hazewinkel, Yu. G. Reshetnyak, editors; [translation edited by Dausa Zvirenaite and Simeon Ivanov].

p. cm. — (American Mathematical Society translations, ISSN 0065-9290; ser. 2, v. 163.)

Meeting held Aug. 30–Sept. 4, 1989.

Includes bibliographical references.

ISBN 0-8218-0286-0 (acid-free)

1. Algebra—Congresses. 2. Mathematical analysis—Congresses. I. Bokut, L. A. (Leonid A.), 1937– . II. Hazewinkel, Michiel. III. Reshetni͡ak, I͡Uriĭ Grigor′evich. IV. Zvirenaite, Dausa. V. Ivanov, Simeon. VI. Title. VII. Series: American Mathematical Society translations; ser. 2, v. 163. VIII. Series

QA3.A572 ser. 2, vol. 163

[QA150]

510 s—dc20

[512] 94-40144

CIP

The paper used in this book is acid-free and falls within the guidelines
established to ensure permanence and durability.

Printed on recycled paper.

This volume was typeset using $\mathcal{A}_{\mathcal{M}}\mathcal{S}$-TEX,
the American Mathematical Society's TEX macro system.

10 9 8 7 6 5 4 3 2 1 99 98 97 96 95

Contents

Amer. Math. Soc. Transl.
(2) Vol. **163**, 1995

On the III Siberian School "Algebra and Analysis", Irkutsk, August 30–September 4, 1989

Yu. G. Reshetnyak and L. A. Bokut′

The State University of Irkutsk held its III Siberian School "Algebra and Analysis" from August 30 to September 4, 1989. The chairman of the organizing committee was academician Yu. G. Reshetnyak. There were 130 participants in the school; ten of these were from Irkutsk itself. More than 30 participants represented Novosibirsk; 15, Moscow; ten, Leningrad. Others were from Omsk, Barnaul, Tomsk, Krasnoyarsk, Tbilisi, Tartu, Yaroslavl′, Sverdlovsk, Chita, Khabarovsk, Tashkent, Alma-Ata, and other areas. Also eight foreign scientists from the International Conference on Algebra in memory of A.I. Mal′cev attended this school. More than 100 participants in the school held the doctor or the candidate of science degrees.

The work of the school was organized in the following way: in the morning, three one-hour lectures were given, and in the afternoon three or four reports of 45 minutes' length were read. The evenings were filled with informal discussions on mathematical problems. The "Charter" for a Mathematical Association of the Soviet Union (a kind of Soviet mathematical society) was drafted and accepted.

Traditionally, the school has two main goals. First of all it should acquaint Siberian and other mathematicians with the newest achievements in mathematics. Therefore, leading mathematicians from Moscow, Leningrad, and so forth were invited as lecturers. The second objective is to give an opportunity for young scientists, especially those from Siberia, to present results obtained after their (candidate's) dissertations, and in this way to help them in their work toward a doctor of science degree. Some success was achieved in both directions.

The program of lectures with some additional comments is detailed below.

A. N. Grishkov (Omsk). *Modular Lie algebras.* This paper deals with a new approach to a classical problem, viz., the classification of simple Lie algebras. The approach does not use Dynkin diagrams, but it is related to completely different combinatorial objects. This method can also be applied to the case of Lie subalgebras.

A. N. Varchenko (Moscow). *The Aomoto dilogarithm and algebraic K-theory.* The author presented a detailed discussion of the conjecture that the cohomology groups of standard complexes constructed by means of the comultiplication in the Hopf algebra $A(k) + A_0 + A_1 + \cdots + A_n + \cdots$, where A_n is the group of pairs of n-dimensional simplexes in n-dimensional projective space $P^n(k)$ over a field k, coincide with the Quillen K-groups of the field k.

M. A. Shubin (Moscow). *Lefschetz type theorems (analytical aspects).* The author describes a Lefschetz type theorem on fixed points, proved in collaboration with S. Sei-

0065-9290/95/$1.00 + $.25 per page

farth, which uses L^2-cohomology on manifolds with cylindric ends. An interesting feature of this theorem is that in this case the Lefschetz number in L^2-cohomology is expressed not only through the contributions of the fixed points, but the formula also contains global invariants of the degree of a mapping type.

Hèléne Esnault (Paris). *Effective bounds for positive bundles.* A result obtained together with E. Vieweg is presented here: an effective bound in the Manin theorem on the Mordell conjecture is found.

G. A. Margulis (Moscow). *Values of indeterminate quadratic forms and unitary flows.* An outline of the proof of the following theorem, which represents the solution of the Davenport problem (1940) is given:

Let B be an indeterminate real quadratic form in n variables, $n \geq 3$, that is not proportional to a form with rational coefficients. Then $\inf_{x\in\mathbf{Z}^n, x\neq 0} |B(x)| = 0$.

I. L. Kantor (Moscow). *A connection between Poisson brackets and Jordan and Lie superalgebras.* The Cantor functor which connects simple Poisson superalgebras and simple Jordan and Lie superalgebras is considered.

A. A. Kirillov (Moscow). *Introduction to the Neretin program.* Recently a new and interesting direction in representation theory was noticed: a hidden symmetry in the representations of "large" groups was discovered (of groups such as the group of operators on an infinite-dimensional linear space, the group of diffeomorphisms of a circle, the gauge groups of field theory). Namely, unitary representations of the "large" groups automatically extend to the representations of certain compact semigroups which contain the initial group as a skeleton. Yu. A. Neretin was the first to construct such a semigroup for a group of diffeomorphisms.

E. B. Vinberg (Moscow). *On some commutative subalgebras of universal enveloping algebras.* This paper considers commutative subalgebras that are connected with the Gel′fand-Tsetlin basis and it is shown that the graded subalgebras of the Poisson-Lie-Berezin algebra, corresponding to such subalgebras, are limits of commutative subalgebras that are obtainable by shifting invariants.

A. G. Khovanskiĭ (Moscow). *Differences of convex polyhedrons and integrals over the Euler characteristic.* Results obtained together with A.V. Pukhlikov concerning operations on convex polyhedrons in the space $\mathbb{R}^n$, e.g., Minkowski summation, "sectioning", and product of polyhedrons, are presented. As a special case of these results there appear the Minkowski theorems on the polynomiality of the extension of a linear combination of convex polyhedrons, the theorem on the polynomiality of the number of integers in an integral linear combination of integral convex polyhedrons, and the duality law for such polynomials.

A. N. Parshin (Moscow). *Geometry of arithmetic surfaces.* In this lecture attention was paid mostly to Arakelov's intersection theory and the formulation of an analogue of the Van der Ven–Bogomolov–Yau inequality for arithmetic surfaces. From this inequality a positive solution of the Fermat problem for sufficiently large n should follow.

Yu. A. Neretin (Moscow). *Classification of representations of categories A, B, C, D.* Continuation of A. A. Kirillov's lecture.

R. I. Grigorchuk (Moscow). *On degrees of growth of groups and applications to analysis.* A survey where particular attention is paid to groups of intermediate growth is presented. These groups can be used in the theory of invariant means, the theory of Banach algebras, the spectral theory of operators, Riemannian geometry, the theory of random walks, and the theory of finite automata.

V. L. Popov (Moscow). *Automorphisms of a ring of polynomials.* This is a survey which, in particular, presents the connections between several well-known problems of an affine algebraic geometry: the Jacobi problem, the Zariski (cancellation) problem, and ruled varieties.

M. E. Kapovich (Khabarovsk). *Uniformization of the three-dimensional manifolds.* In his lecture the author considered the following problems related to the three types of geometrical structures (hyperbolic, plane conformal, and birational) on three-dimensional manifolds: (a) existence of the structures (V. Tersten, M. Kapovich), (b) description of the structures (R. Mostow, R. Schoen, and S.-T. Yau), (c) realization of the finite groups of diffeomorphisms as automorphisms of structures (M. Kapovich).

Yu. S. Ilyashenko (Moscow). *Stokes phenomenon in nonlinear analysis.* The Stokes phenomenon lies in the fact that for resonating vector fields and mappings the local dynamics determines not a map but an atlas of maps with nontrivial transfer functions containing significant information on the dynamics. Results obtained by P. M. Elizarov and S. I. Trifonov consider the computation of these transfer functions.

A. M. Vershik (Leningrad). *Topology of spaces of the configuration of convex polyhedrons and the theorem of universality.* A profound theorem by N. E. Mnev states that any semialgebraic set up to stabilization can be the space of configurations of a given combinatorial type. This fact leads to the "configuration topology". Another program which is connected with the one mentioned relates to the theory of real complexity (the real analogue of the problem $P = NP$).

V. Kaup (Tubingen). *Boundeded symmetric domains and Jordan algebras.* The lecture presents contemporary extensions in the infinite-dimensional case of results by Cartan on the classification of symmetric domains (these generalizations are due to Kaup and his student Upmeier).

G. I. Ol′shansky (Moscow). *Brauer-Weyl duality and representations of an infinite symmetric group.* In recent years the theory of unitary representations of "large" (i.e., not locally compact) groups and subgroups related to them has been actively developed. Results of the author on certain large groups which are of type I in the sense of von Neumann, and certain related finite Brauer semigroups which are related to the groups mentioned are considered in this lecture. Objects of this kind are interesting at the moment in view of their connections with knots and quantum groups.

V. V. Bludov (Irkutsk). *The Todd-Coxeter method and its realization on a computer.* Results by the author in computer algebra were presented in the lecture.

In addition to the lectures the following reports were delivered:

A. B. Zhubr (Sytyvkar). On the classification of the smooth manifolds. I.

N. Yu. Netsvetaev (Leningrad). On the classification of the smooth manifolds II.

M. Hazewinkel (Amsterdam). Quantum groups.

V. K. Ionin (Novosibirsk). Transformations of geometric structures.

V. P. Golubyatnikov (Novosibirsk). Generalized convexity and applications.

A. P. Yuzhakov (Krasnoyarsk). Multidimensional residues and applications.

Yu. B. Rudyak (Moscow). Some problems in algebraic topology (realization of homology classes).

R. G. Nadiradze (Tbilisi). Classification of formal groups of the form $f(u,v) = [uR(v) + vR(u)]\psi(uv)$.

M. A. Nazarov (Moscow). Symplectic categories over p-adic fields and their Weil representations.

Z. I. Borevich (Leningrad). On the relation of normality in the group $GL(2,\mathbb{Q})$.

A. V. Yakovlev (Leningrad). Representations of lattices and integral representations.

V. P. Gerdt (Dubna). Computer algebra systems.

E. Yu. Eroshkina (Moscow). Computer algebra for differential equations.

A. S. Dzhumadildaev (Alma-Ata). Cohomology of Lie algebras of characteristic p, their applications and analogs in characteristic zero.

O. I. Tavgen (Minsk). Finite width of the Chevalley groups over the rings of S-integers.

A. A. Premet (Moscow). Invariant functions and nilpotent elements.

G. A. Soifer (Kemerovo). Affine crystallographic groups (on the L. Auslander problem).

Yu. B. Khakimzhanov (Tashkent). Manifolds of left structures and characteristically nilpotent Lie algebras.

I. L. Panin (Leningrad). The role of algebraic K-theory in the solution of classical problems.

A. V. Uglanov (Yaroslavl). Fubini theorem for vector measures.

O. I. Ivanov, N. Yu. Netsvetaev (Leningrad). Quadratic form of a sum of manifolds.

The lectures and reports were all at a level one expects at international conferences. The school illustrated the great value of bringing together representatives of the Siberian school (in Novosibirsk, Omsk, Barnaul, Kemerovo, the schools of Mal′cev, Shirikov, Sobolev, Kantorovich, also A. D. Aleksandrov, B. G. Reshetnyak, M. M. Lavrent′ev), the Moscow school (of Kurosh, Kolmogorov, also I. R. Shafarevich, I. M. Gel′fand), the Leningrad school (of Rokhlin, Kantorovich, also D. K. Faddeev, L. D. Faddeev), the Minsk school (of V. P. Platonov), etc. The proceedings of the first school (1987) are deposited in VINITI and are published by the American Mathematical Society, the proceedings of the second school (1988) likewise.

Yu. G. Reshetnyak
Chairman of the Organizing Committee
Academician

L. A. Bokut′
Vice-Chairman of the Organizing Committee
Professor

Amer. Math. Soc. Transl.
(2) Vol. **163**, 1995

Quasiderivations in Diagonal Matrix Algebras

A. Z. Anan′in

Let A be the algebra of all diagonal matrices over a field k of characteristic zero.

THEOREM. *The algebra A has no nonzero quasiderivations.*

This answers one of the questions that E. B. Vinberg asked the participants of the third school "Algebra and Analysis".

It is clear that the field k may be assumed algebraically closed and that we can use induction on the matrix size.

LEMMA 1. *Let $I \subset A$ be a proper ideal and $D(I) \subseteq I$; then $D = 0$* (D denotes a quasiderivation in A).

PROOF. D induces a quasiderivation $\overline{D}$ in the algebra A/I and a quasiderivation $D|_I$ in the algebra I. By the induction assumption $\overline{D} = 0$ and $D|_I = 0$; that is, $D(A) \subseteq I$ and $D(I) = 0$. There is the following identity for a quasiderivation (see formula (4) in the paper by Vinberg in this volume):

$$D^2(xy) - 2D\big((Dx)y + x(Dy)\big) + \big(D^2x\big)y + x\big(D^2y\big) + 2(Dx)(Dy) = 0. \tag{1}$$

Now take $x = y \in A$ and use $D\big(AD(A)\big) = D^2(A) = 0$ to conclude that $2(Dx)^2 = 0$. The fact that the field k is of characteristic zero and the absence of nilpotent elements in the algebra A complete the proof. □

LEMMA 2. *D is a nilpotent k-linear transformation of A.*

PROOF. If D is not nilpotent, then there exist $0 \neq \alpha \in k$ and $0 \neq a \in A$ such that $Da = \alpha a$. Substituting $x = y = a$ in (1) we obtain $(D - 2\alpha Id)^2\big(a^2\big) = 0$. Now there exists $0 \neq b \in A$ such that $Db = 2\alpha b$. This contradicts the fact that k is of characteristic zero. □

LEMMA 3. $D(1) = 0$.

PROOF. Equation (1) takes the form $xD^2(1) + 2(Dx)(D1) - 2D\big(x(D1)\big) = 0$ for $y = 1$ and $D^2(1) = 2(D1)^2$ for $x = 1$. This means that

$$D\big(x(D1)\big) = (Dx)(D1) + x(D1)^2. \tag{2}$$

Because of Lemma 2 we need only to prove that $D^i(1) = i!\,(D1)^i$ for $1 \leq i \leq s$ using

1991 *Mathematics Subject Classification*. Primary 16W25.

induction (the statement is already proved for $s = 2$). By the induction assumption $D^s(1) = s!\,(D1)^s$ and $D^{s-1}(1) = (s-1)!\,(D1)^{s-1}$. Hence,

$$D\big((D1)^{s-1}\big) = \frac{1}{(s-1)!}D^s(1) = s(D1)^s$$

and using (2) we obtain

$$\begin{aligned} D^{s+1}(1) &= D\big(D^s(1)\big) = D\big(s!\,(D1)^{s-1}(D1)\big) \\ &= s!\,D\big((D1)^{s-1}\big)(D1) + s!\,(D1)^{s-1}(D1)^2 \\ &= s!\,(s+1)(D1)^{s+1} = (s+1)!\,(D1)^{s+1}. \end{aligned}$$

The lemma is proved. □

We may use the lemma above as a start for induction on the matrix size.

We have $A = ke_1 \oplus \ldots \oplus ke_n$, where $e_1, \ldots, e_n$ is a system of orthogonal idempotents may be regarded as the set of vertices V of a graph G. Two vertices e_i and e_j are connected by an edge with the source e_i and the target e_j if and only if (this is a definition) $i \neq j$ and $e_j D(e_i) \neq 0$. An arbitrary graph G is said to be reducible if and only if for its set of vertices V there exist V_1 and V_2 such that $V = V_1 \cup V_2$, $V_1 \cap V_2 = \varnothing \neq V_1, V_2$, and there is no edge from V_1 to V_2. If our graph G is reducible, then $D = 0$ by Lemma 1. We will connect e_i and e_j by an unoriented edge instead of two edges both ways from e_i to e_j and back in order to simplify the pictures. To analyze the graph G we write $De_i = \sum_{1 \le j \le n} d_i^j e_j$ for suitable $d_i^j \subset k$. Substituting $x = e_i$, $y = e_i - e_j$ and multiplying (1) by e_j, $i \neq j$, we get

$$\begin{aligned} 0 &= e_j D^2 e_i - 2e_j D\big((De_i)(e_i - e_j)\big) \\ &\quad - 2e_j D\big(e_i D(e_i - e_j)\big) - e_j D^2 e_i + 2e_j\big(De_i\big)D(e_i - e_j) \\ &= -2e_j D\big(e_i De_i\big) + 2e_j(De_i)De_j - 2e_j D\big(e_i De_i\big) \\ &\quad + 2e_j D\big(e_i De_j\big) + 2e_j\big(De_i\big)De_i - 2e_j\big(De_i\big)De_j \\ &= 2e_j D\big(e_i De_j\big) + 2e_j\big(De_i\big)De_i - 4e_j D\big(e_i De_i\big) \\ &= 2e_j\big(d_i^j d_j^i + d_i^j d_i^j - 2d_i^j d_i^i\big) = 2e_j d_i^j\big(d_i^j + d_j^i - 2d_i^i\big); \end{aligned}$$

here we have used the equality $e_j D\big((De_i)e_j\big) = e_j(De_i)De_j$. Hence

$$d_i^j\big(d_i^j + d_j^i - 2d_i^i\big) = 0 \qquad \text{for } i \neq j. \tag{3}$$

Substituting $x = e_i$, $y = e_j$ in (1) and multiplying (1) by e_k, we get

$$\begin{aligned} 0 &= -2e_k D\big(e_j De_i\big) - 2e_k D\big(e_j De_j\big) + 2e_k\big(De_i\big)De_j \\ &= 2e_k\big(d_i^k d_j^k - d_i^j d_j^k - d_j^i d_i^k\big) \end{aligned}$$

for pairwise distinct e_i, e_j, e_k. Hence,

$$d_i^k d_j^k = d_i^j d_j^k + d_j^i d_i^k \qquad \text{for pairwise distinct } i, j, k. \tag{4}$$

Lemma 3 means that for every i

$$\sum_k d_k^i = 0. \tag{5}$$

The graph G may be complete; i.e., every two vertices $e_i, e_j \in V$, $i \neq j$, are connected by an nonoriented edge. To study the question we need the elementary

LEMMA 4. *Let $a_2, \dots, a_n \in k$ be pairwise distinct. Then*

$$\sum_{\substack{1<j,k\leq n\\ j\neq k}} \frac{a_k}{a_k - a_j} = \frac{(n-1)(n-2)}{2} \quad \textit{for } n \geq 2.$$

PROOF. We have

$$\begin{aligned}\sum_{\substack{1<j,k\leq n\\ j\neq k}} \frac{a_k}{a_k - a_j} &= \sum_{1<j<k\leq n} \frac{a_k}{a_k - a_j} + \sum_{1<k<j\leq n} \frac{a_k - a_j + a_i}{a_k - a_j} \\ &= \sum_{1<k<j\leq n} 1 + \sum_{1<j<k\leq n} \frac{a_k}{a_k - a_j} - \sum_{1<k<j\leq n} \frac{a_j}{a_j - a_k} \\ &= \frac{(n-1)(n-2)}{2},\end{aligned}$$

proving the lemma. □

Suppose that the graph G is complete. This means that $d_i^j \neq 0$ for every i, j, $i \neq j$. Using (3) we get $d_i^j + d_j^i = 2d_i^i$; therefore $d = d_i^i$ does not depend on i. The trace of D (which is equal to zero by Lemma 2) is equal to nd; hence $d = 0$ and $d_i^j = -d_j^i$ for any i, j. Write $a_i = d_1^i$ for $1 < i \leq n$. Note that $a_2, \dots, a_n \in k$ are pairwise distinct. Indeed, if $a_i = a_j$ for $1 < i, j \leq n$, $i \neq j$, then using (4) for $k = 1$ we get $d_i^1 d_j^1 = d_i^j d_j^1 + d_j^i d_i^1$; that is, $d_i^1 d_j^1 = d_i^j(-a_j + a_i) = 0$. This contradicts the completeness of G. It follows from the equality above that

$$d_i^j = \frac{a_i a_j}{a_i - a_j} \quad \text{for } i \neq j,\ \ 1 < i, j \leq n.$$

In view of (5) and Lemma 7 we have

$$\begin{aligned} 0 &= \sum_{1<i\leq n} \frac{1}{a_i} \sum_{1\leq k\leq n} d_k^i = \sum_{1<i\leq n} \frac{1}{a_i}\Big(a_i + \sum_{1<k\leq n} d_k^i\Big) \\ &= (n-1) + \sum_{1<i\leq n} \sum_{\substack{1<k\leq n\\ k\neq i}} \frac{a_k}{a_k - a_i} = (n-1) + \sum_{\substack{1<j,k\leq n\\ j\neq k}} \frac{a_k}{a_k - a_j} \\ &= (n-1) + \frac{(n-1)(n-2)}{2} = \frac{n(n-1)}{2}.\end{aligned}$$

This is not equal to zero when $n > 1$. We have obtained

LEMMA 5. *The graph G is not complete.*

Consider the graph G in detail. Three pairwise distinct vertices of the graph taken together with edges between them may look as shown in Figure 1 (next page).

The second and the fourth rows contain all eight types of graph fragments which are not forbidden by (4). Their vertices are indexed to simplify verification. We shall

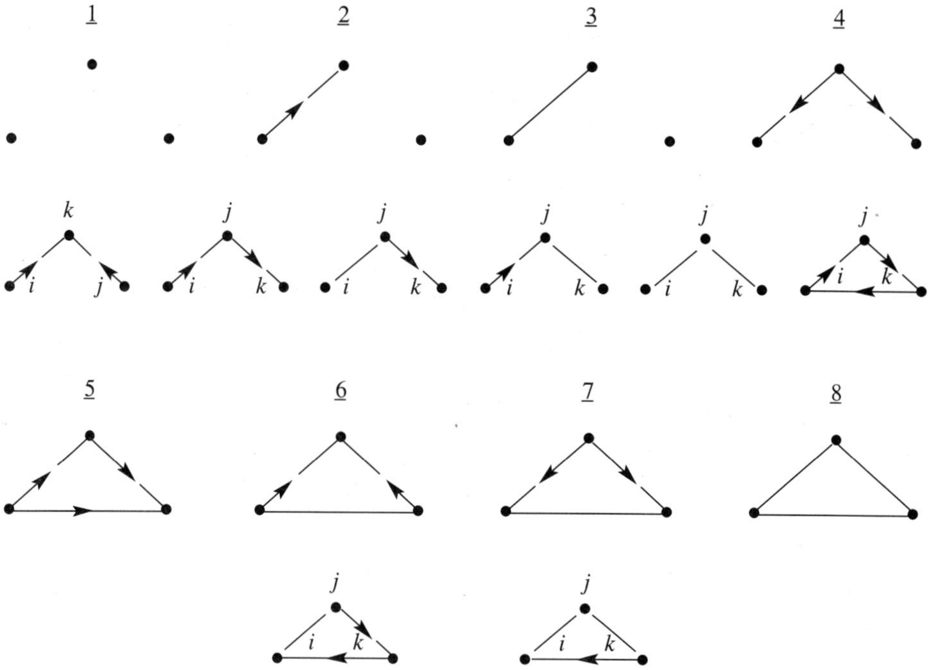

FIGURE 1

call a graph G locally "good" if every fragment of it with three vertices is of the form $\underline{1}$–$\underline{8}$. To prove the theorem we need only

LEMMA 6. *Every locally "good" graph G having at least two vertices is either reducible or complete.*

PROOF. Use induction on the number of vertices. The lemma is obviously true when their number is two. First we assume that G has no unoriented edges. If there exists a vertex e_i of the graph G without any edge having e_i as its source, then G is reducible (V_1 consists of e_i). Otherwise we can find a cycle: vertices $v_1, \dots, v_t, v_1$ that are sequentially connected by oriented edges. Suppose that $t \geq 2$ is as small as possible. If $t \geq 3$, then we have an edge from v_1 to v_3 (see all eight types of graph fragments) and we are able to decrease the cycle, otherwise $t = 2$; i.e., we have an unoriented edge. In the case of an unoriented edge in the graph we will contract this edge. Choose such an edge and denote it by E. It is shown in Figure 2 how to make a new graph G' with a new vertex v replacing the edge E in all four possible cases.

To verify the "local goodness" of the new graph G' we note that every "new" fragment of the graph G' having three vertices comes from a corresponding fragment with four vertices. By the "local goodness" of G each such four vertex fragment has to be one of those depicted in Figure 3.

The resulting three vertex fragments of the graph G' are of the forms $\underline{1}$, $\underline{2}$, $\underline{3}$, $\underline{2}$, $\underline{2}$, $\underline{3}$, $\underline{4}$, $\underline{4}$, $\underline{5}$, $\underline{7}$, $\underline{5}$, $\underline{6}$, $\underline{5}$, $\underline{6}$, $\underline{7}$, $\underline{8}$. The lemma and the theorem are proved. □

The second question of Professor E. B. Vinberg was:

— "What are the quasiderivations of a matrix algebra A over a field k?"

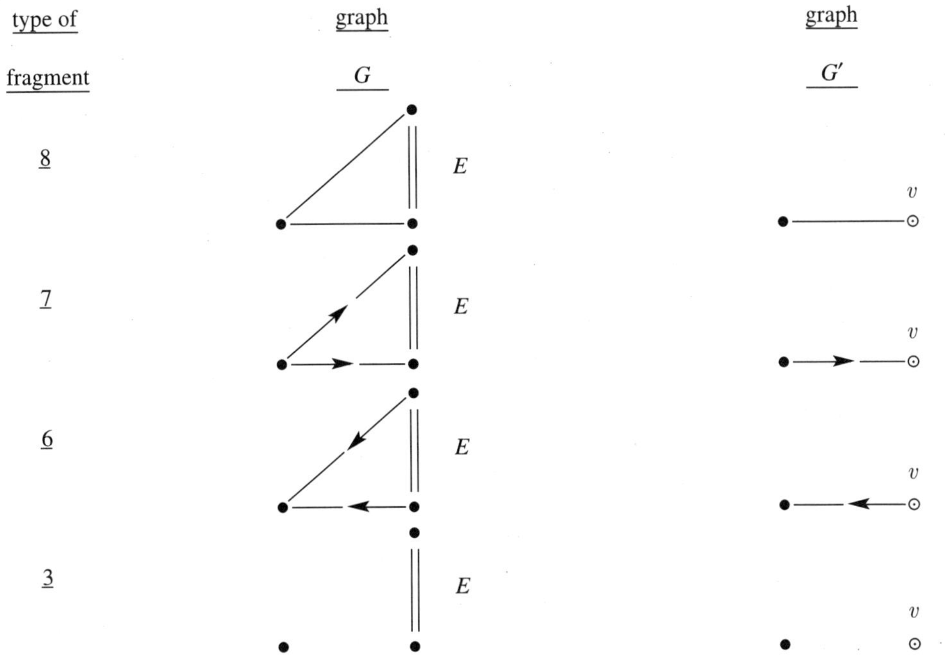

FIGURE 2

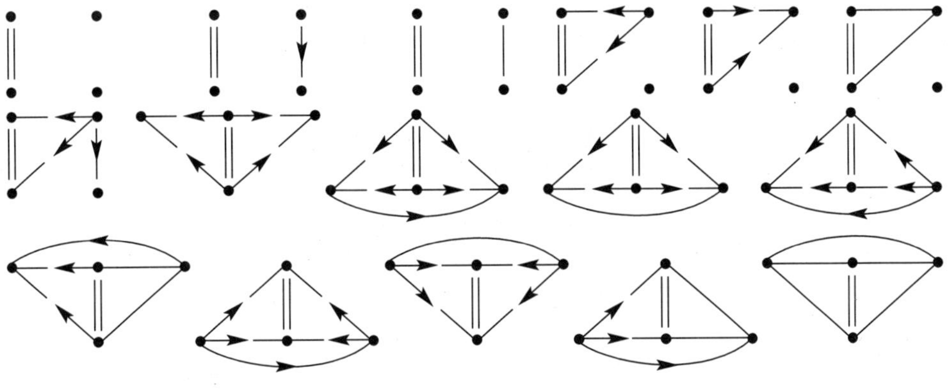

FIGURE 3

He pointed out some of them: $D\colon x \mapsto ax + xb$, where $a, b \in A$ satisfy $(a+b)^2 = ab - ba$.

The statement "Any quasiderivation D of a matrix algebra A satisfies the identity $D(xy) = (Dx)y + x(Dy) - x(D1)y$" is equivalent to the absence of quasiderivations different from those pointed out by E. B. Vinberg.

The author is grateful to Professor E. B. Vinberg for formulating the question and for showing interest in the solution.

Amer. Math. Soc. Transl.
(2) Vol. **163**, 1995

L^2 Atiyah-Bott-Lefschetz Theorem

A. Efremov

§0. Introduction

Let M be a C^∞-manifold vithout boundary, $\dim M = n$. Atiyah and Bott in **[1]** studied an elliptic complex

$$0 \to C^\infty(E_0) \xrightarrow{A_0} \dots \xrightarrow{A_{N-1}} C^\infty(E_N) \to 0 \tag{0.1}$$

of smooth sections of vector bundles E_j, $j = 1, \dots, N$, over M. The cohomology spaces $H^j = \operatorname{Ker} A_j / \operatorname{Im} A_{j-1}$ have finite dimension by the ellipticity of the complex. Then for any endomorphism T of (0.1) the Lefschetz number

$$L(T) = \sum_{j=0}^{N} (-1)^j \operatorname{tr} TH^j \tag{0.2}$$

is well defined. Here TH^j are the operators induced by T in cohomology. Atiyah and Bott considered the special case when T is a geometric endomorphism. Such an endomorphism is connected with a smooth map $f: M \to M$ that has only nondegenerate fixed points. The main result of **[1]** is a formula which expresses the Lefschetz number of such an endomorphism as a sum of contributions from the fixed points of f. This is an advanced generalization of the classical Lefschetz formula, which states that, for a continuous map of a finite polyhedron, the Lefschetz number is equal to the algebraic number of the fixed points of the map.

Many other fixed point theorems have appeared after **[1]** (see **[2–9]**). Most of them are devoted to the case of compact M and bundles E_j with fibers of finite dimension.

On the other hand, if M is not compact, the properties of an elliptic operator on M depend on the behavior of its coefficients at infinity and on the structure of M. There are different nonequivalent ways to define the cohomology, dimensions of solution spaces of elliptic equations may not be finite, etc. In many examples the Lefschetz number is infinite too. However there are some methods to define "nonstandard" numerical invariants that are finite, though their ordinary analogs are infinite. The first "nonstandard" or real-valued index theory was constructed in **[10, 11]** for almost periodic operators. Such theories were developed by many authors in different contexts **[12–17]**. Usually operator algebra techniques or similar ones are behind the "nonstandard" index. Such techniques are also useful in some

---

1991 *Mathematics Subject Classification*. Primary 57G05, 57G12.

other fields, e.g., in spectral theory **[18–21]** or Morse theory **[22]**. So, it is natural to use it in the Lefschetz number problem.

The aim of this paper is to prove a "nonstandard" L^2 Lefschetz fixed point theorem in the spirit of **[12, 13]**. This theorem was announced in **[23]**. Another paper will be devoted to an almost periodic Lefschetz fixed point theorem (see **[24]**). Before discussing the main result, we should point out that some variants of L^2 Lefschetz theorems are known for the Bergmann kernel in a strictly pseudoconvex domain **[9]**, manifolds with cylindrical ends **[25]**, homogeneous spaces **[26]**. A similar result was obtained by E. Troitskiĭ **[27]**.

Let $\mathfrak{A}$ be a finite von Neumann algebra in a Hilbert space F (see **[28]**) with a finite faithful normal trace $\operatorname{tr}_{\mathfrak{A}}$.

DEFINITION 0.1. Let us consider a locally trivial bundle $\mathscr{F}$ over M with fibers of the form $F \otimes G$, $\dim G = k < \infty$. Let $U(k, \mathfrak{A})$ be a group of unitary matrices with elements from $\mathfrak{A}$, supplied with the uniform topology. Suppose that transition functions of $\mathscr{F}$ are smooth ones with values in the $U(k, \mathfrak{A})$. Then it is called a Hilbert $\mathfrak{A}$-bundle.

We suppose that there is a measure $d\mu$ on M defined by a smooth density. Then we can define a Hermitian inner product on the sections of $\mathscr{F}$

$$(u, v) = \int_M \big(u(x), v(x)\big)_x \, d\mu. \tag{0.3}$$

So, one may define the space $L^2(\mathscr{F})$ of square-summable sections.

A natural example of such a bundle over M is a local system associated with a representation of $\pi_1(M)$ in a Hilbert space. The most important case is the left regular representation in $l_2(\Gamma)$, where Γ is a quotient group of $\pi_1(M)$. It arises when one considers a Galois covering $p\colon \widetilde{M} \to M$ with the group Γ. The space of L^2-sections of this local system is isomorphic to $L^2(\widetilde{M})$.

We shall consider an elliptic complex

$$0 \to C^\infty(\mathscr{F}_0) \xrightarrow{A_0} \dots \xrightarrow{A_{N-1}} C^\infty(\mathscr{F}_N) \to 0 \tag{0.4}$$

where the $\mathscr{F}_j$ are Hilbert $\mathfrak{A}$-bundles and the A_j are pseudodifferential operators of the same order m with coefficients in $\mathfrak{A}$ (similarly as in **[13, 17]**). We need to introduce the reduced cohomology

$$H^j = \operatorname{Ker} A_j / \overline{\operatorname{Im} A_{j-1}},$$

where the closure is taken in the natural $C^\infty(\mathscr{F})$-topology. These groups are isomorphic to the spaces of L^2 harmonic sections, i.e., spaces of L^2-solutions of the elliptic equations $\Delta_j u = 0$. Here the Δ_j are Laplacians of (0.4). The usual dimensions of such spaces are infinite, but not the von Neumann ones. So, having an appropriate endomorphism T, a finite Lefschetz number may be defined by (0.2) with tr replaced by $\operatorname{tr}_{\mathfrak{A}}$.

Now we shall state our main theorem. Let $f: M \to M$ be a smooth map with only nondegenerate fixed points. We shall use the concept of geometric endomorphism as in **[1]** with obvious modifications required by the difference between (0.1) and (0.4).

THEOREM 0.1 (L^2 Atiyah-Bott-Lefschetz). *Let T be a geometric endomorphism of* (0.4) *corresponding to f. Then*

$$L(T) = \sum_{f(x)=x} \nu(x), \tag{0.5}$$

where the expressions for $\nu(x)$ are the same as in **[1]** *with* tr *changed to* $\mathrm{tr}_{\mathfrak{A}}$.

See also §1 below for $\nu(x)$.

For Galois coverings Theorem 0.1 leads to an analytic fixed point formula (see Theorem 3.1 below). This formula is closely connected with Nielsen's fixed point theory **[29]**. It provides an analytic expression for the Nielsen fixed point class index of f if the elliptic complex is the usual de Rham complex on the universal covering manifold.

Our proof of Theorem 0.1 follows **[1]**. First we shall prove the alternating sum formula for smooth endomorphisms of the complex (0.4); then we shall approximate the geometric endomorphism by smooth ones. In the framework of operator algebras, which we use, it is natural to deal with complexes of Hilbert spaces. So we shall associate such a complex to the complex (0.4) in §5. The corresponding Hilbert spaces are the spaces of L^2-sections for Hilbert $\mathfrak{A}$-bundles.

The remainder of the paper is organized as follows. In §1 we briefly recall the Atiyah-Bott theorem and fix the notation. In §2 we introduce the necessary notions to explain the formulation of Theorem 0.1. In §3 we discuss some consequences in the case of local systems. The proof of the main theorem is given in §§4–7.

The author thanks M. Shubin, who suggested the problem and paid a lot of attention while work was in progress, and also A. Brenner, D. Efremov, and Yu. Kordukov for useful discussions.

§1. Atiyah-Bott fixed point theorem

Let us consider a C^∞-map $f: M \to M$, which has only simple or nondegenerate fixed points. This means that its graph is transversal to the diagonal in $M \times M$ or, equivalently, that $\det(1 - df(x)) \neq 0$ for every fixed point x. Here $df(x): T_xM \to T_xM$ is the induced morphism of the tangent spaces. Let

$$0 \to C^\infty(E_0) \xrightarrow{A_0} \dots \xrightarrow{A_{N-1}} C^\infty(E_N) \to 0 \tag{1.1}$$

be an elliptic complex of the C^∞-sections of the finite-dimensional vector bundles $E_j \to M$, $j = 0, \dots, N$. A family $T = \{T_j: C^\infty(E_j) \to C^\infty(E_j)\}_{j=0}^N$ of linear operators is called an endomorphism of the complex (1.1) if $T_{j+1}A_j = A_jT_j$. Any endomorphism T induces a family of maps TH^j in the cohomology $H^j = \operatorname{Ker} A_j / \operatorname{Im} A_{j-1}$ of (1.1), which are finite-dimensional due to the ellipticity. So the Lefschetz number of T is defined:

$$L(T) = \sum_{j=0}^N (-1)^j \operatorname{tr} TH^j.$$

Actually, the Atiyah-Bott theorem deals with a special class of endomorphisms, namely geometric ones. Given a smooth map $f\colon M \to M$ an endomorphism is called geometric, if there exist morphisms $\varphi_j\colon f^*E_j \to E_j$ such that

$$T_j s(x) = \varphi_j\big(f(x)\big)s\big(f(x)\big), \tag{1.2}$$

where $s \in C^\infty(E_j)$ and $\varphi_j\big(f(x)\big)\colon E_{j,f(x)} \to E_{j,x}$ are the corresponding maps of the fibers. If $x = f(x)$ is a fixed point, then $\varphi_j(x)$ is a linear operator in $E_{j,x}$ and $\operatorname{tr}\varphi_j(x)$ is well defined.

Now we formulate the Atiyah-Bott-Lefschetz theorem.

THEOREM 1.1 ([**1**]). *If T is a geometric endomorphism of the complex* (1.1), *corresponding to f, then*

$$L(T) = \sum_{f(x)=x} \nu(x),$$

where the summation is over the set of fixed points of f, and $\nu(x)$ is given by

$$\nu(x) = \frac{\sum_{j=0}^{N}(-1)^j \operatorname{tr}\varphi_j(x)}{|\det(1 - df(x))|}. \tag{1.3}$$

§2. Background for the main theorem

Our aim in this section is to collect the results necessary to formulate Theorem 0.1. The proofs are given mainly in the following sections. First we describe some properties of $\mathfrak{A}$-pseudodifferential operators ($\mathfrak{A}$-ΨDO) on sections of Hilbert $\mathfrak{A}$-bundles. Later we define a geometric endomorphism of an elliptic complex (0.4) and its Lefschetz number.

Let $\{U_j\}_{j=1}^k$ be a finite covering of M by coordinate patches and

$$\{\varphi_j \in C^\infty(M), \operatorname{supp}\varphi_j \subset U_j\}_{j=1}^k$$

a corresponding partition of unity. We also need the family

$$\{\psi_j \in C^\infty(M),\ \operatorname{supp}\psi_j \subset U_j,\ \psi_j|_{\operatorname{supp}\varphi_j=1}\}_{j=1}^k.$$

The definition of an $\mathfrak{A}$-ΨDO is almost the same, as in the ordinary case. It must be locally of the form

$$Bu(x) = (2\pi)^{-n}\int e^{i(x-y)\xi}\, b(x,\xi)u(y)\,dy\,d\xi + \int K(x,y)u(y)\,dy \tag{2.1}$$

where $b(x,\xi)$ is an $\mathfrak{A}$-matrix for any $(x,\xi) \in T^*M$ and $K(\cdot,\cdot)$ is a C^∞ function with values in the $\mathfrak{A}$-matrices with uniform topology. To be precise we shall give the following series of definitions.

DEFINITION 2.1. If U is a domain in $\mathbb{R}^n$, then the class $\mathfrak{A}S^m(U)$ of symbols of order $m \in \mathbb{R}$ includes all the smooth functions $b(\cdot,\cdot)$ on $U \times \mathbb{R}^n$ with values in the $\mathfrak{A}$-matrices, such that the following estimates hold for any compact K in U:

$$\|\partial_\xi^\alpha \partial_x^\beta b(x,\xi)\| \le C(1+|\xi|)^{m-|\alpha|}, \quad x \in K, \tag{2.2}$$

where the derivatives are taken with respect to the topology defined by the norm $\|\cdot\|$ in $\mathfrak{A}$.

Let $\mathscr{E}$, $\mathscr{F}$ be Hilbert $\mathfrak{A}$-bundles over M.

DEFINITION 2.2. Let $P\colon C^\infty(\mathscr{E}) \to C^\infty(\mathscr{F})$ be a linear operator and let there exist a smooth section $P(\cdot,\cdot)$ of the Banach bundle $\operatorname{Hom}_{\mathfrak{A}}\big(\pi_1^*(\mathscr{F}), \pi_2^*(\mathscr{E})\big)$ over $M \times M$, where $\pi_j\colon M \times M \to M$ is the projection onto the jth component, such that for any section $u \in C^\infty(\mathscr{E})$

$$Pu(\cdot) = \int_M (\pi_1)_* P(\cdot, y)(\pi_2)_* u(y)\, dy. \tag{2.3}$$

Then P is called a smoothing operator.

The class of the smoothing operators will be denoted by $\mathfrak{A}\Psi^{-\infty}(\mathscr{E},\mathscr{F})$.

DEFINITION 2.3. A linear operator $B\colon C^\infty(\mathscr{E}) \to C^\infty(\mathscr{F})$ is an $\mathfrak{A}$-ΨDO of order $m \in \mathbb{R}$, if

(1) $B - \sum_{j=1}^k \psi_j B \varphi_j \in \mathfrak{A}\Psi^{-\infty}(\mathscr{E},\mathscr{F})$;

(2) $\psi_j B \varphi_j$ are of the form (2.1) with $b_j(\cdot,\cdot) \in \mathfrak{A}S^m(U_j)$.

The class of $\mathfrak{A}$-ΨDO of order m will be denoted $\mathfrak{A}\Psi^m(\mathscr{E},\mathscr{F})$. Let $\mathfrak{A}\Psi^\infty(\mathscr{E},\mathscr{F}) = \bigcup_{m\in\mathbb{R}} \mathfrak{A}\Psi^m(\mathscr{E},\mathscr{F})$. For any $A \in \mathfrak{A}\Psi^\infty(\mathscr{E},\mathscr{F})$ one may define a formally adjoint operator A^+ by means of the inner product (0.3). We shall use the notation $\mathfrak{A}\Psi^*(\mathscr{F})$ instead of $\mathfrak{A}\Psi^*(\mathscr{F},\mathscr{F})$. We also need to introduce the class $\mathfrak{A}E\Psi^m(\mathscr{F})$ of elliptic operators. This class consists of the $B \in \mathfrak{A}\Psi^m(\mathscr{F},\mathscr{F})$ such that for any $\{U_j\}$, $\{\varphi_j\}$, $\{\psi_j\}$ as above, all the $b_j(x,\xi)$ from Definition 2.3 are invertable, if $x \in \operatorname{supp}\psi_j$ and $|\xi|$ is sufficiently large, and

$$\|b_j^{-1}(x,\xi)\| \le C(1+|\xi|)^{-m}.$$

The properties of $\mathfrak{A}$-ΨDO are summarized in the following statement which may be proved by standard arguments.

PROPOSITION 2.1.

a) *The class* $\mathfrak{A}\Psi^m(\mathscr{E},\mathscr{F})$ *is independent of the choice of the arbitrary elements in Definition* 2.3;

b) $\mathfrak{A}\Psi^{-\infty}(\mathscr{E},\mathscr{F}) = \bigcap_{m\in\mathbb{R}} \mathfrak{A}\Psi^m(\mathscr{E},\mathscr{F})$;

c) *if* $B \in \mathfrak{A}\Psi^m(\mathscr{E},\mathscr{F})$ *and* $A \in \mathfrak{A}\Psi^n(\mathscr{F},\mathscr{G})$, *then* $A \circ B \in \mathfrak{A}\Psi^{m+n}(\mathscr{E},\mathscr{G})$, *and the standard composition formula holds*;

d) *if* $A \in \mathfrak{A}\Psi^m(\mathscr{E},\mathscr{F})$, *then* $A^+ \in \mathfrak{A}\Psi^m(\mathscr{F},\mathscr{E})$;

e) *if* $A \in \mathfrak{A}\Psi^m(\mathscr{F})$, *then there exists a parametrix* $B \in \mathfrak{A}E\Psi^m(\mathscr{F})$, *i.e.*

$$I - A\circ B \in \mathfrak{A}\Psi^{-\infty}(\mathscr{F}), \qquad I - B\circ A \in \mathfrak{A}\Psi^{-\infty}(\mathscr{F}).$$

Now let us consider the complex (0.4), where all the A_j are $\mathfrak{A}$-ΨDOs of the same order m. Let $\{A_j^+\colon C^\infty(\mathscr{F}_{j+1}) \to C^\infty(\mathscr{F}_j)\}_{j=0}^{N-1}$ be the family of formally adjoint operators. Then one may define the *Laplacians*

$$\Delta_j = A_{j-1}\circ A_{j-1}^+ + A_j^+ \circ A_j.$$

By Proposition 2.1, we have $\Delta_j \in \mathfrak{A}\Psi^{2m}(\mathscr{F}_j)$.

DEFINITION 2.4. The complex (0.4) is called elliptic if all the Δ_j are elliptic operators.

For finite-dimensional bundles, the ellipticity leads to finite dimensions of the cohomology. This is not true for Hilbert bundles. Even the reduced cohomology spaces

$$H^j = \operatorname{Ker} A_j / \overline{\operatorname{Im} A_{j-1}}, \tag{2.4}$$

where the closure is taken in the C^∞-topology of sections, are often infinite-dimensional. However, one may use von Neumann techniques to define the dimension of (2.4).

Given a $\mathfrak{A}$-bundle on the $\mathfrak{A}$-bundle $\mathscr{F}$, we shall construct a semifinite von Neumann algebra $\mathfrak{A}_{\mathscr{F}}$, acting in L^2-sections of $\mathscr{F}$.

PROPOSITION 2.2. *We have*

$$L^2(\mathscr{F}) \cong (F \otimes G)\overline{\otimes} L^2(M), \tag{2.5}$$

where $\overline{\otimes}$ denotes the completion of the algebraic tensor product in the natural Hilbert topology.

PROOF. Using a smooth triangulation of M, we obtain $L^2(\mathscr{F}) \cong \oplus_j L^2(\mathscr{F}|_{V_j})$, where the sum runs over all simplexes of maximal dimension. Since $\mathscr{F}|_{V_j}$ is trivial, the proposition follows if we use the isomorphisms $L^2(\mathscr{F}|_{V_j}) \cong (F \otimes G)\overline{\otimes} L^2(V_j)$ and $L^2(M) \cong \oplus_j L^2(V_j)$. □

It can be easily seen from the proof of Proposition 2.2 that up to unitary equivalence the isomorphism (2.5) does not depend on the choice of a triangulation and trivializations over the simplexes.

Now Proposition 2.2 allows us to define $\mathfrak{A}_{\mathscr{F}}$ as the von Neumann algebra in $L^2(\mathscr{F})$ spanned by the algebraic tensor product of $\mathfrak{A}$ and $\mathscr{L}(G \otimes L^2(M))$, where $\mathscr{L}(\cdot)$ is the algebra of all bounded operators. Briefly, we have

$$\mathfrak{A}_{\mathscr{F}} \cong \mathfrak{A}\overline{\otimes}\mathscr{L}(G \otimes L^2(M)). \tag{2.6}$$

Then the commutant $\mathfrak{A}'_{\mathscr{F}}$ of $\mathfrak{A}_{\mathscr{F}}$ is spatially isomorphic to the algebra $\mathfrak{A}' \otimes I$ of constant functions on M with values in $\mathfrak{A}'$. Due to the isomorphism (2.5), the algebra $\mathfrak{A}'_{\mathscr{F}} \cong \mathfrak{A}'$ acts in $L^2(\mathscr{F})$ along the fibers; i.e., for $s \in L^2(\mathscr{F})$

$$As(x) = A(s(x)), \tag{2.7}$$

where $x \in M$, $A \in \mathfrak{A}'$.

The isomorphism (2.6) depends on the isomorphism (2.5) but actually the algebra $\mathfrak{A}_{\mathscr{F}}$ does not. It follows from the finiteness of $\mathfrak{A}$ that $\mathfrak{A}_{\mathscr{F}}$ is semifinite (see **[28,** Theorem V.2.30]). Moreover, (2.6) allows us to define the semifinite faithful normal trace $\operatorname{tr}_{\mathfrak{A}_{\mathscr{F}}} = \operatorname{tr}_{\mathfrak{A}} \otimes \operatorname{Tr}$ on $\mathfrak{A}_{\mathscr{F}}$. Here Tr is the usual operator trace. The so-defined trace also does not depend on the isomorphism (2.5). For simplicity we shall write $\operatorname{tr}_{\mathfrak{A}}$ instead of $\operatorname{tr}_{\mathfrak{A}_{\mathscr{F}}}$. Let $L \subset L^2(\mathscr{F})$ be a closed subspace affiliated to $\mathfrak{A}$; i.e., the orthogonal projection P_L onto L belongs to $\mathfrak{A}$. Then the $\mathfrak{A}$-dimension of L is defined by $\dim_{\mathfrak{A}} L = \operatorname{tr}_{\mathfrak{A}} P_L$. More generally, a linear operator B in $L^2(\mathscr{F})$ is affiliated to $\mathfrak{A}_{\mathscr{F}}$ if $CB \subset BC$ for every $C \in \mathfrak{A}'$. In this case we shall write $B\eta\mathfrak{A}_{\mathscr{F}}$. The main properties of such operators may be found, for instance, in **[18,** Proposition 7.1].

Returning now to the complex (0.4) we have a family of semifinite von Neumann algebras $\mathfrak{A}_j = \mathfrak{A}_{\mathscr{F}_j} \subset \mathscr{L}(L^2(\mathscr{F}_j))$. It is convenient to introduce the von Neumann algebra $\widetilde{\mathfrak{A}} = \mathfrak{A} \otimes \mathscr{L}(L^2(M))$, so that $\mathfrak{A}_j \cong \widetilde{\mathfrak{A}} \otimes \operatorname{End} G_j$.

Due to the local structure of $\mathfrak{A}$-ΨDO (see (2.1) and (2.3)) and (2.7) we have $B\eta\mathfrak{A}_j$ if $B \in \mathfrak{A}\Psi^*(\mathcal{F}_j)$.

The cohomology spaces (2.4) are connected with Laplacians in the usual way:

PROPOSITION 2.3. *For every* $j = 0, \dots, N$ *the following statements are true:*
a) *the subspace* $\operatorname{Ker}\Delta_j \subset C^\infty(\mathcal{F}_j)$ *is affiliated to* $\mathfrak{A}_j$;
b) $H^j \cong \operatorname{Ker}\Delta_j$;
c) $\dim_{\mathfrak{A}} \operatorname{Ker}\Delta_j < \infty$.

Now we shall define the Lefschetz number of a geometric endomorphism defined as in (1.2), where $\varphi_j \in C^\infty(\operatorname{Hom}_{\mathfrak{A}}(f^*E_j, E_j))$. Locally $\varphi_j(\cdot)$ are $\mathfrak{A}$–matrices at every point. By using Proposition 2.3, one can identify the operator TH^j, induced by T in cohomology with P_jTP_j, where P_j is the orthogonal projection onto $\operatorname{Ker}\Delta_j$, $j = 0, \dots, N$. It follows from (1.2) and (2.7) that $T_j \in \mathfrak{A}_j$. Further T_j is evidently continuous in the C^∞ topology. Moreover we have

PROPOSITION 2.4. $T_jP_j \in \mathcal{L}(L^2(\mathcal{F}_j))$.

Then $P_jT_jP_j \in \mathfrak{A}_j$, so by Proposition 2.3, c) *the Lefschetz number*

$$L(T) = \sum_{j=0}^{N}(-1)^j \operatorname{tr}_{\mathfrak{A}} P_jT_jP_j \tag{2.8}$$

is finite. To complete this section we recall that $\nu(x)$ in (0.5) is defined by (1.3) with tr replaced by $\operatorname{tr}_{\mathfrak{A}}$.

§3. Local systems and Nielsen fixed point classes

Let $\rho \colon \Gamma \to \mathcal{L}(\mathcal{F})$ be a unitary representation of the fundamental group $\Gamma = \pi_1(M)$. Then one may construct a local system $\mathcal{E}(\mathcal{F}, \rho)$ corresponding to this representation.

Let us also suppose that Γ acts by elements of a finite von Neumann algebra $\mathfrak{A}$ equipped with a finite faithful normal trace $\operatorname{tr}_{\mathfrak{A}}$. So it is natural to regard $\mathcal{E}(\mathcal{F}, \rho)$ as a flat $\mathfrak{A}$-Hilbert bundle, i.e., a bundle with locally constant transition functions.

Given an elliptic complex (0.1), the operators A_j may be extended to yield a complex of $\mathfrak{A}$-ΨDO. The complex obtained is

$$0 \to C^\infty(E_0 \otimes \mathcal{E}(\mathcal{F}, \rho)) \xrightarrow{\widetilde{A}_0} \dots \xrightarrow{\widetilde{A}_N} C^\infty(E_N \otimes \mathcal{E}(\mathcal{F}, \rho)) \to 0, \tag{3.1}$$

where $\widetilde{A}_j = A_j \otimes I$.

Let T be as in (1.2). We shall discuss the case when T can be extended up to a geometric endomorphism $\widetilde{T}$ of the complex (3.1). By $f^* \colon \Gamma \to \Gamma$ we mean the induced homomorphism of the fundamental group. It is defined uniquely up to an inner automorphism. Such a class of homomorphisms will be denoted as $[f^*]$. If $[f^*] \ni \mathrm{id}$, then the required extension is done as follows. The first step is to describe a family $\{\gamma_x, x \in M\}$ of paths, such that γ_x connects x with $f(x)$. After an identification of Γ with $\pi_1(M, x_0)$, $x_0 \in M$, we have $f^* \colon \gamma \to \alpha^{-1}f(\gamma)\alpha$, where α is a path connecting x_0 and $f(x_0)$. We suppose that $\alpha^{-1}f(\gamma)\alpha = \gamma$ in $\pi_1(M, x_0)$. Now let $\gamma_x = f(\delta^{-1})\alpha\delta$ for $x \in M$, where δ is a path, connecting x and x_0. It is important that the homotopy class of γ_x does not depend on the choice of δ. Indeed let ν be another such path, and $\tau_x = f(\nu^{-1})\alpha\nu$. Then $\nu \sim \rho\delta$ for some $\rho \in \pi_1(M, x_0)$, where by $\sim$ we mean homotopy with fixed ends. Hence

$$\tau_x \sim f(\delta^{-1})f(\rho^{-1})\alpha\rho\delta = f(\delta^{-1})\alpha\rho^{-1}\alpha^{-1}\alpha\rho\delta = \gamma_x .$$

The constructed family $\{\gamma_x\}$ induces a parallel translation map $f^*_{\mathscr{E}} : \mathscr{E}(\mathscr{F}, \rho) \to \mathscr{E}(\mathscr{F}, \rho)$. Namely, $f^*_{\mathscr{E}}$ is a family of fiber maps, $f^*_{\mathscr{E},x} : F_{f(x)} \to F_x$ being the parallel translation along γ_x^{-1}. The map $f^*_{\mathscr{E},x}$ depends only on the homotopy class of γ_x, so it does not depend on the choice of δ. The same is true for the induced map $T_{\mathscr{E}}$ on the sections of $\mathscr{E}(F, p)$:

$$T_{\mathscr{E}} s(x) = f^*_{\mathscr{E},x}\left(s\left(f(x)\right)\right).$$

Since we may always locally choose γ_x to be smooth with respect to x, the following proposition is true.

PROPOSITION 3.1. *$f^*_{\mathscr{E}}$ is a C^∞-map. Hence, $T_{\mathscr{E}}$ is also C^∞. In particular, it maps the C^∞-sections to C^∞ ones.*

Now we can put $\widetilde{T} = T \otimes T_{\mathscr{E}}$ to obtain a geometric endomorphism of the complex (3.1).

Then we have the following corollary of Theorem 0.1. If x is a fixed point, then there exists $\gamma_x \in \Gamma$ such that $f^*_{\mathscr{E},x} = \rho(\gamma_x)$. Then

$$L(\widetilde{T}) = \sum_{f(x)=x} \mathrm{tr}_{\mathfrak{A}}\left(\rho(\gamma_x)\right) \sum_{j=0}^{N} (-1)^j \frac{\mathrm{tr}\,\varphi_j(x)}{\left|\det(1 - df(x))\right|}. \tag{3.2}$$

An interesting class of local systems arises from a Galois covering $p : \widetilde{M} \to M$. The fiber of such a covering is the group $\Gamma = \pi_1(M)/\pi_1(\widetilde{M})$, provided $\pi_1(\widetilde{M})$ is a normal subgroup in $\pi_1(M)$. Let ρ be the left regular representation of $\pi_1(M)$ in $l_2(\Gamma) = \{ f : \Gamma \to \mathbb{C} \mid \sum_{\gamma \in \Gamma} |f(\gamma)|^2 < \infty \}$. If $[\gamma]$ is the right class of $\gamma \in \pi_1(M)$, then

$$\rho(\gamma) f[\gamma] = f[\delta^{-1}\gamma], \qquad \delta \in \pi_1(M). \tag{3.3}$$

Since $\pi_1(\widetilde{M})$ is a normal subgroup, we have $\rho(\delta) = \mathrm{id}$ for $\delta \in \pi_1(\widetilde{M})$, so $\rho(\cdot)$ is actually the left regular representation in $l_2(\Gamma)$. The operators $\{\rho(\gamma) \mid \gamma \in \Gamma\}$ generate a finite von Neumann algebra $\mathfrak{A}$. A faithful normal finite trace on $\mathfrak{A}$ is defined by the identities (see **[28,** §V.7]):

$$\begin{aligned} \mathrm{Tr}_{\mathfrak{A}}\left(\rho(e)\right) &= 1, \\ \mathrm{Tr}_{\mathfrak{A}}\left(\rho(\gamma)\right) &= 0, \quad \gamma \neq e. \end{aligned} \tag{3.4}$$

So, we can slightly modify (3.2) to obtain

$$L(\widetilde{T}) = \sum_{\substack{f(x)=x \\ [\gamma_x]=e}} \nu_x, \tag{3.5}$$

where ν_x is defined as in (1.3). The last formula actually involves the *Nielsen fixed point classes* **[29]**. The set of fixed points of f naturally falls into equivalence classes. Two fixed points x and y are equivalent, if there exists a path γ connecting x and y, such that $f(\gamma) \sim \gamma$. By definition $\gamma_x = e$ if and only if $\gamma_y = e$. So the summation on the right-hand side of (3.5) runs over some Nielsen classes. This means that it is the sum of the contributions of fixed Nielsen classes (i.e., such that $\gamma_x = e$). If p is the universal covering, the sum in (3.5) involves just one fixed point class.

It is well known that Nielsen fixed point classes are closely connected with liftings of f to the universal covering. We may rewrite (3.5) for the maps of $\widetilde{M}$. It leads to an analytic formula which does not explicitly involve von Neumann algebras.

Since p is a Galois covering, Γ acts on $\widetilde{M}$ freely.

DEFINITION 3.1. The Hermitian vector bundle $\pi\colon \widetilde{E} \to \widetilde{X}$ is a Γ-bundle if Γ acts on $\widetilde{E}$ and the following properties are fulfilled:

a) $\pi \circ \gamma = \gamma \circ \pi$ for any $\gamma \in \Gamma$;

b) for every $x \in \widetilde{M}$ the linear map $\gamma^*\colon \widetilde{E}_x \to \widetilde{E}_{\gamma x}$ is unitary.

Given a Γ-invariant measure $d\widetilde{\mu}$ on $\widetilde{M}$ we can consider the space $L^2(\widetilde{E})$ of square summable sections. The action of Γ on $\widetilde{E}$ yields the unitary action in $L^2(E)$ as in the space $C^\infty(E)$ of smooth sections or in the space $C_0^\infty(E)$ of compactly supported C^∞-sections by

$$L_\gamma s(x) = \gamma^* s(\gamma^{-1} x).$$

The orbit space $E = \widetilde{E}/\Gamma$ has a natural structure of a Hermitian vector bundle over M, with the same fiber as $\widetilde{E}$. Let V be a fiber of both E and $\widetilde{E}$. There exists an isomorphism

$$\mathrm{i}_\rho\colon L^2(\widetilde{E}) \cong V \otimes L^2\Big(\mathscr{E}(l_2(\Gamma), \rho)\Big),$$

where ρ is the same as in (3.3). If $u \in L^2(\widetilde{E})$, then $\mathrm{i}_\rho u(x)$ is just a vector function (defined up to the left shift (3.3)) $\gamma \mapsto \gamma^* u(\gamma^{-1}\widetilde{x})$ with $\widetilde{x} \in p^{-1}x$. The existence of i_ρ is essentially a consequence of the Fubini theorem. The image of $C_0^\infty(E)$ is the space $C^\infty_{\mathrm{comp}}\left(E \otimes \mathscr{E}(l_2(\Gamma), \rho)\right)$ of sections that are compactly supported over each point $x \in M$. This space is evidently dense in $L^2\left(E \otimes \mathscr{E}(l_2(\Gamma), \rho)\right)$.

Now let us consider an elliptic complex of Γ-invariant differential operators in sections of Γ-bundles

$$0 \to C_0^\infty(\widetilde{E}_0) \xrightarrow{\widetilde{A}_0} \cdots \xrightarrow{\widetilde{A}_{N-1}} C_0^\infty(\widetilde{E}_N) \to 0. \tag{3.6}$$

The operators $\widetilde{A}_j$ may be reduced to the operators A_j, acting in sections of quotient bundles E_j. Then $A_j \otimes I$, which is defined on $C^\infty_{\mathrm{comp}}\left(E \otimes \mathscr{E}(l_2(\Gamma), \rho)\right)$, corresponds to $\widetilde{A}_j$, defined on $C_0^\infty(\widetilde{E}_j)$ via i_ρ. So there is no danger to mix $\widetilde{A}_j$ in (3.6) with $\widetilde{A}_j$ in (3.1).

REMARK. We restrict our attention to the case of differential operators only for simplicity. Similar arguments hold for pseudodifferential operators with some conditions on the decrease of the Schwartz kernel. Then the $\widetilde{A}_j$ in (3.1) are not of the form $\widetilde{A}_j \otimes I$, but they are certainly $\mathfrak{A}$-ΨDOs.

A choice of x_0 and of a family $\{\gamma_x\}$ as above provides a lifting of f to a smooth map $\widetilde{f}\colon \widetilde{M} \to \widetilde{M}$. Let $x \in \widetilde{M}$ be the end of a path $\widetilde{\gamma}$ that covers γ and connects x_0 and $p(x)$. Then $\widetilde{f}(x)$ is the end of the lifting of the path $\gamma_x\gamma$. The identity $f^* = 1$ means that the map constructed is Γ-invariant. Further, let $\widetilde{T}$ be a Γ-invariant geometric endomorphism of (3.6). It corresponds to a geometric endomorphism of the complex (3.1) of the form $T \otimes I$ via i_ρ.

Now let $\widetilde{P}_j$ be the orthogonal projection to $\operatorname{Ker}\widetilde{\Delta}_j$ in $L^2(\widetilde{E}_j)$, where $\widetilde{\Delta}_j = \widetilde{A}_{j-1}\widetilde{A}_{j-1}^{+} + \widetilde{A}_j^{+}\widetilde{A}_j$ are the *Laplacians* of (3.6). Each $\widetilde{\Delta}_j$ belongs to the von Neumann algebra $\mathfrak{A}_{j,\Gamma}$ of all bounded Γ-invariant operators in $L^2(\widetilde{E}_j)$. These algebras are spatially isomorphic via i_p to the algebras $\mathfrak{A}_j$ constructed for (3.1) as in §2. The trace on $\mathfrak{A}_{j,\Gamma}$, arising from the trace (3.4), is just the Γ-trace of Atiyah **[12]**. As proved in **[12]**, the operators $\widetilde{P}_j$, hence $\widetilde{T}_j\widetilde{P}_j$, $\widetilde{P}_j\widetilde{T}_j\widetilde{P}_j$, have C^∞ L. Schwartz kernels. Their traces may be computed by integration of the matrix trace of the L. Schwartz kernels, restricted to the diagonal, over a fundamental domain (recall that a fundamental domain is an open set $\mathscr{U} \subset \widetilde{M}$ such that $\overline{\bigcup_{\gamma\in\Gamma}\gamma\mathscr{U}} = \widetilde{M}$ and $\mathscr{U} \cap \gamma\mathscr{U} = \varnothing$ if $\gamma \neq e$). Taking into account that $\operatorname{Tr}_{\mathfrak{A}} \widetilde{T}_j\widetilde{P}_j = \operatorname{Tr}_{\mathfrak{A}} \widetilde{P}_j\widetilde{T}_j\widetilde{P}_j$ we obtain an analytic expression for $L(\widetilde{T})$ in Theorem 3.1 below. Clearly, we can identify the fixed points of f with fixed points in $\mathscr{U}$ of the lifted map $\widetilde{f}$. It is always possible to choose $\mathscr{U}$ such that the boundary $\partial\mathscr{U}$ is fixed point free.

Summarizing, we obtain the following

THEOREM 3.1. *Let $\widetilde{f}\colon \widetilde{M} \to \widetilde{M}$ be a Γ-invariant smooth map such that $\widetilde{f}$ and the induced map $f\colon M \to M$ have only simple fixed points. Let $\widetilde{T}$ be a corresponding geometric endomorphism of the complex* (3.6). *Then*

$$L(\widetilde{T}) = \sum_{j=0}^{N}(-1)^j \int_{\mathscr{U}} \operatorname{tr}\widetilde{\varphi}_j(\widetilde{f}(x))\widetilde{P}_j(\widetilde{f}(x),x)\,d\widetilde{\mu}$$
$$= \sum_{j=0}^{N}(-1)^j \sum_{\substack{f(x)=x\\ x\in\mathscr{U}}} \frac{\operatorname{tr}\varphi_j(x)}{|\det(1-df(x))|},$$

where $\widetilde{P}_j(\cdot,\cdot)$ is the Schwartz kernel of $\widetilde{P}_j$.

To complete this section let us consider the case when (3.6) is a de Rham complex and p the universal covering. Then the right-hand side is just the sum of fixed point indices over a Nielsen class of f. The cohomological interpretation of such indices was suggested in **[30, 31]**. Theorem 3.1 provides both L^2-cohomological and analytic interpretations of them.

§4. Sobolev spaces

In this section we collect some technical results. The main ones are Theorem 4.1 about essential selfadjointness and Theorem 4.2 about topological invertibility. Almost all the proofs may be done by repeating standard ones, so we omit them (see **[32]** or **[33]**).

We shall start with the definition of the Sobolev space on $\mathbb{R}^n$ with values in a Hilbert space F. Let $S(\mathbb{R}^n, F)$ be a Schwartz space; i.e.,

$$\{u \in C^\infty(\mathbb{R}^n, F) \mid \forall\alpha,\beta\ |x|^\alpha\|\partial^\beta u(x)\| \to 0 \text{ as } x \to \infty\},$$

where the derivatives are taken in the natural topology of F. There is a locally convex topology in $S(\mathbb{R}^n, F)$ defined by seminorms as in the usual L. Schwartz space, and let $S'(\mathbb{R}^n, F)$ be the dual space. The Fourier transform $u \mapsto \hat{u}$ is well defined in $S'(\mathbb{R}^n, F)$, so we may introduce the Sobolev space for $s \in \mathbb{R}$ as follows:

$$H^s(\mathbb{R}^n, F) = \{ u \in S'(\mathbb{R}^n, F),\ \hat{u}(\cdot) \in L^2_{\mathrm{loc}}(\mathbb{R}^n, F),\ \|u\|_s = (u, u)'_s < \infty \},$$

where

$$(4.1) \qquad (u, u)'_s = \frac{1}{(2\pi)^n} \int_{\mathbb{R}^n} (\hat{u}(\xi), \hat{v}(\xi))(1 + |\xi|^2)^s \, d\xi < \infty.$$

It is a Hilbert space with inner product given by (4.1). Using convolution with a standard δ-sequence, one may prove that each $u \in H^s(\mathbb{R}^n, F)$ with compact support can be approximated by smooth functions in the $H^s(\mathbb{R}^n, F)$ topology. This allows us to define the spaces $H^s(\cdot)$ on manifolds. Let $\{U_j\}$ be a covering of M by coordinate patches, let $\{\varphi_j\}$ be a corresponding partition of unity, and let $\{\varkappa_j\}$ be coordinate diffeomorphisms.

Definition 4.1. The Sobolev sections space $H^s(\mathscr{F})$ is the completion of $C^\infty(\mathscr{F})$ in the norm defined by the inner product

$$(4.2) \qquad (u, v)_s = \sum_j (\varkappa_j^* \varphi_j u, \varkappa_j^* \varphi_j v)'_s.$$

As usual, we have the continuous embedding $H^s(\mathscr{F}) \subset H^t(\mathscr{F})$ if $s > t$ and $H^s(\mathscr{F}) \subset D'(\mathscr{F})$, where the last space is dual to $C^\infty(\mathscr{F})$. By the closed graph theorem, the norm (4.2) does not depend on the choice of $\{\varphi_j\}$ and $\{\varkappa_j\}$ up to equivalence.

Proposition 4.1 (Sobolev lemma). *There is a continuous embedding* $H^s(\mathscr{F}) \subset C^k(\mathscr{F})$ *if* $s > k + n/2$.

Hence, $H^\infty(\mathscr{F}) = \bigcap_{s \in \mathbb{R}} H^s(\mathscr{F}) \subset C^\infty(\mathscr{F})$.
In the next statement we collect the properties of $\mathfrak{A}$-ΨDO on Sobolev spaces.

Proposition 4.2. *If* $A \in \mathfrak{A}\Psi^m(\mathscr{F})$, *then*
(i) *A maps* $H^s(\mathscr{F})$ *to* $H^{s-m}(\mathscr{F})$ *continuously;*
(ii) *if A is elliptic and* $Au \in H^s(\mathscr{F})$, *then* $u \in H^{s+m}(\mathscr{F})$;
(iii) *there is a nondegenerate coupling* $H^s(\mathscr{F}) \times H^{-s}(\mathscr{F}) \to \mathbb{C}$ *that defines an isomorphism of* $H^{-s}(\mathscr{F})$ *and the dual space of* $H^s(\mathscr{F})$.

This proposition provides all the tools necessary for studying selfadjoint extensions of elliptic symmetric operators.

Theorem 4.1. *Let* $A \in \mathfrak{A}E\Psi^m(\mathscr{F})$ *be a formally selfadjoint operator. Then it is essentially selfadjoint in* $L^2(\mathscr{F})$ *on the domain* $C^\infty(\mathscr{F})$. *Its closure has* $H^s(\mathscr{F})$ *as its domain and coincides with A on it.*

Theorem 4.2. *Let* $A \in \mathfrak{A}E\Psi^m(\mathscr{F})$ *and* $A \geq cI > 0$. *Then A provides a topological isomorphism* $A\colon H^s(\mathscr{F}) \xrightarrow{\sim} H^{s-m}(\mathscr{F})$ *for every* $s \in \mathbb{R}$.

Proof. The operator A is invertible on $L^2(\mathscr{F})$ and A^{-1} maps $L^2(\mathscr{F})$ onto $H^m(\mathscr{F})$ by Theorem 4.1. Then by (i) of Proposition 4.2 and the closed graph theorem, A^{-1} maps $H^s(\mathscr{F})$ onto $H^{s+m}(\mathscr{F})$ continuously if $s > 0$. The case $s < 0$ is dealt with by duality (Proposition 4.2, (iii)). □

Now we are in a position to prove Proposition 2.4. It follows from the coincidence of the C^∞- and L^2-topologies on Ker Δ. Indeed, let $\{u_k\}_{k=1}^\infty$ be a sequence of elements in $\operatorname{Ker}\Delta$. If u_k converges in the C^∞-topology, then it evidently does in L^2 one. Further, by Theorem 4.2 the Sobolev norm $\|u_k\|_s$ is equivalent to

$$\|(1+\Delta)^{s/2m}u_k\| = \|u_k\|.$$

Then if u_k converges in the L^2-topology, it also does in the H^s-topology for any $s>0$ and, in view of Proposition 4.1, in the C^∞-topology as well.

§5. The associated Hilbert space complex

In this section we begin the proof of the main theorem.

Let us consider the A_j as unbounded operators in $L^2(\mathscr{F}_j)$ defined on the C^∞-sections and let the $\overline{A}_j$ be their closures. It is easily seen that $\overline{A}_j\overline{A}_{j-1}=0$ on the domain of $\overline{A}_{j-1}$. So, we have the following complex of Hilbert spaces:

$$0 \to L^2(\mathscr{F}_0) \xrightarrow{\overline{A}_0} \dots \xrightarrow{\overline{A}_{N-1}} L^2(\mathscr{F}_N) \to 0. \tag{5.1}$$

The following Hodge decomposition theorem allows us to study the cohomology of (5.1) instead of the cohomology of (0.4).

THEOREM 5.1. *There exists an orthogonal decomposition*

$$\operatorname{Ker}\overline{A}_j = \overline{\operatorname{Im}A_{j-1}} \oplus \operatorname{Ker}\Delta_j, \tag{5.2}$$

where A_j, A_{j-1} and the Laplacians Δ_j are unbounded operators defined on C^∞-sections.

PROOF. Let us consider the operators $\Lambda_j = \overline{A}_{j-1}(\overline{A}_{j-1})^* + (\overline{A}_j)^*\overline{A}_j$ with natural domains

$$D(\Lambda_j) = \left\{ u \in D(\overline{A}_j)\cap D(\overline{A}_{j-1})^* \mid \overline{A}_j u \in D(\overline{A}_j^*),\ (\overline{A}_{j-1})^* u \in D(\overline{A}_j) \right\}.$$

Evidently, $\operatorname{Ker}\Lambda_j = \operatorname{Ker}\overline{A}_j \cap \operatorname{Ker}(\overline{A}_{j-1})^*$. Taking into account that

$$\operatorname{Ker}\overline{A}_j \supset \overline{\operatorname{Im}A_{j-1}} = (\operatorname{Ker}\overline{A}_{j-1}^*)^\perp,$$

we obtain the orthogonal decomposition

$$\operatorname{Ker}\overline{A}_j = \overline{\operatorname{Im}A_{j-1}} \oplus \operatorname{Ker}\Lambda_j. \tag{5.3}$$

It remains to prove that $\operatorname{Ker}\Lambda_j = \operatorname{Ker}\Delta_j$.

The identity $\operatorname{Ker}\overline{A}_{j-1} = (\operatorname{Im}\overline{A}_j^*)^\perp$ and (5.3) show that Λ_j is the direct sum of the selfadjoint operators

$$\overline{A}_{j-1}\overline{A}_{j-1}^*|_{\overline{\operatorname{Im}A_{j-1}}}, \qquad 0|_{\operatorname{Ker}\Lambda_j}, \qquad \overline{A}_j^*\overline{A}_j|_{\overline{\operatorname{Im}\bar{A}_j^*}},$$

so that it is selfadjoint on $D(\Lambda_j)$. Moreover $D(\Lambda_j)\supset D(\Delta_j)$.

The rest of the proof will be divided into two steps. First we shall prove that $\Delta_j \subset \Lambda_j$, so that $\overline{\Delta}_j = \Lambda_j$ by Theorem 4.1. After this, we shall show that $\operatorname{Ker}\Delta_j = \operatorname{Ker}\overline{\Delta}_j$.

For the formally adjoint operator $A_j^+ \colon L^2(\mathscr{F}_{j+1}) \to L^2(\mathscr{F}_j)$ with domain $C^\infty(\mathscr{F}_{j+1})$ we have

$$(A_j^+u, v) = \lim_{k\to\infty}(A_j^+u, v_k) = \lim_{k\to\infty}(u, A_j v_k) = (u, \overline{A}_j v),$$

if $u\in C^\infty(\mathscr{F}_{j+1})$, $v\in D(\overline{A}_j)$, and $(v_k, A_j v_k)\to(v, A_j v)$ as $k\to\infty$. So, $A_j^+\subset\overline{A}_j^*$ and

$\Delta_j \subset \Lambda_j$. Now let $u \in \operatorname{Ker} \overline{\Delta}_j$. Then there exists a sequence $\{u_k\} \subset C^\infty(\mathscr{F}_j)$ such that $u_k \to u$, $\Delta_j u_k \to u$ in $L^2(\mathscr{F}_j)$. Due to the ellipticity, we have (see Proposition 2.1)

$$u_k = BAu_k + Ru_k,$$

where $R \in \mathfrak{A}\Psi^{-\infty}(\mathscr{F}_j)$. Then $u = Ru$ and by §4, $u \in C^\infty(\mathscr{F}_j) = D(\Delta_j)$. □

Recall that $\operatorname{Ker} \Delta_j$ in $D'(\mathscr{F}_j)$ is actually embedded in $C^\infty(\mathscr{F}_j)$ because of the ellipticity of Δ_j. Besides, $P_j \in \mathfrak{A}_j$, since $\Delta_j \in \mathfrak{A}_j$. This proves assertion a) of Proposition 2.3.

The decomposition (5.2) lets us identify the L^2-cohomology $\operatorname{Ker} A_j / \overline{\operatorname{Im} A_{j-1}}$ with $\operatorname{Ker} \Delta_j$. On the other hand, by Proposition 2.3, which will be proved below, the same is true for the cohomology of (0.4) defined in the C^∞-topology. So, by Theorem 5.1, we can replace the study of C^∞-objects by the study of L^2 ones. This is useful when we deal with operator algebras.

Proof of Proposition 2.3. The assertion a) is already proved.

To prove b) it is sufficient to state the orthogonal (in $L^2(\mathscr{F}_j)$) decomposition, involving subspaces of $C^\infty(\mathscr{F}_j)$:

$$\operatorname{Ker} A_j = \overline{\operatorname{Im} A_{j-1}} \oplus \operatorname{Ker} \Delta_j .$$

Here the closure is taken in the C^∞-topology. The C^∞-closure is embedded in the L^2-closure so the parts of the direct sum are orthogonal by Theorem 5.1. If $u \in \operatorname{Ker} A_j$, then $u = h + v$, where $h \in \operatorname{Ker} \Delta_j$, both h and v are from $C^\infty(\mathscr{F}_j)$, and $h \perp v$. We have to prove that $v \in \overline{\operatorname{Im} A_{j-1}}$. Our arguments here are as in **[34]**. Let us fix $p \in \mathbb{Z}$ and consider $v^{(p)} = (1+\Delta_j)^p v \in C^\infty(\mathscr{F}_j)$. Then $v^{(p)} \in \operatorname{Ker} \Delta_j$. Indeed

$$0 = (h, v) = \left(h, (1+\Delta_j)^{-p} v^{(p)}\right) = \left((1+\Delta_j)^{-p} h, v^{(p)}\right) = (h, v^{(p)}).$$

Since $v^{(p)} \in \operatorname{Ker} A_j$, there exists a sequence $\{v_k^{(p)} \in \operatorname{Im} A_{j-1}\}$ such that $v_k^{(p)} \to v^{(p)}$ in $L^2(\mathscr{F}_j)$. Hence, $(1+\Delta_j)^{-p} v_k^{(p)} \to v$ in $H^{\mathrm{mp}}(\mathscr{F}_j)$. Since $A_j \Delta_j = \Delta_{j+1} A_j$, the same is true for functions of Δ_j. So, $(1+\Delta_j)^{-p} v_k^{(p)} \in \operatorname{Im} A_{j-1}$. Then v may be approximated by elements of $\operatorname{Im} A_{j-1}$ in all the Sobolev spaces, hence in $C^\infty(\mathscr{F}_j)$, by Proposition 4.1.

The proof of c) is based on a generalized Fredholm alternative stated by Breuer **[35]**. It follows from Proposition 2.1 e) that $\operatorname{Ker} \Delta_j \subset \operatorname{Ker}(I - R)$, where $R \in \mathfrak{A}\Psi^{-\infty}(\mathscr{F}_j)$. So we have only to prove that $\dim_{\mathfrak{A}} \operatorname{Ker}(I - R) < \infty$. The generalized Fredholm alternative states that this is true if R is $\mathfrak{A}$-compact; i.e., it can be approximated by operators from $\mathfrak{A}_j$ with $\mathfrak{A}$-finite-dimensional ranges. Now we shall explain a construction of such an approximation for a smoothing operator. Let $R(\cdot,\cdot)$ be the smooth Schwartz kernel of R (see (2.3)). Having a fixed covering of M by coordinate simplexes and trivializations over them one can approximate $R(\cdot,\cdot)$ by a sequence $\{R_k(\cdot,\cdot)\}_{k=1}^\infty$ of locally constant functions on $M \times M$ with values in $\mathfrak{A}$. Moreover, we can assume that each of them takes only a finite number of values and that the integral operators R_k in $L^2(\mathscr{F}_j)$, defined by $R_k(\cdot,\cdot)$ (see (2.3)), converge to R uniformly as $k \to \infty$. All ranges of the R_k are finite dimensional by the finiteness of $\mathfrak{A}$.

§6. Alternating trace formula

In this section we shall study the complex (5.1). Actually we shall consider a more abstract situation. The main object is a complex of Hilbert spaces

$$(6.1) \qquad 0 \to \mathscr{H}_0 \xrightarrow{d_0} \dots \xrightarrow{d_{N-1}} \mathscr{H}_N \to 0.$$

We shall suppose that the following properties hold.

a) $\mathscr{H}_j = \mathscr{H} \otimes M_j$, where $\mathscr{H}$ is a complex Hilbert space and $\dim M_j = n_j < \infty$.

b) There is a semifinite von Neumann algebra A in $\mathscr{L}(\mathscr{H})$ equipped with a faithful normal semifinite trace $\mathrm{tr}_{\mathfrak{A}}$.

c) The d_j are closed linear operators with dense domains $D(d_j)$, such that the compositions $d_j d_{j-1}$ are densely defined and equal to 0 on their domains. Briefly, $d_j d_{j-1} = 0$.

d) The d_j are affiliated to $\mathfrak{A}_j$ in the following sense. There is a family of semifinite von Neumann algebras $\{\mathfrak{A}_j = \mathfrak{A} \otimes \operatorname{End} M_j \subset \mathscr{L}(\mathscr{H}_j)\}_{j=0}^N$. Then all the commutants $\{\mathfrak{A}'_j\}_{j=0}^N$ are naturally isomorphic to the commutant $\mathfrak{A}'$ of $\mathfrak{A}$. What we suppose exactly is that

$$(6.2) \qquad V^* d_j V = d_j$$

for every unitary $V \in \mathfrak{A}'$. It follows from (6.2) and the uniqueness of the polar decomposition that

$$(6.3) \qquad V^* d_j^* V = d_j^*.$$

We note that bounded operators acting from $\mathscr{H}_j$ to $\mathscr{H}_{j+1}$ are represented by matrices with elements from $\mathscr{L}(\mathscr{H})$. Then (6.2) means that the elements are actually from $\mathfrak{A}$. But we cannot use such an interpretation when the operators are not bounded.

e) The cohomology spaces $H^j = \operatorname{Ker} d_j / \overline{\operatorname{Im} d_{j-1}}$ are finite dimensional in the von Neumann sense.

We have to explain e) in more detail. Consider the Laplacians $\Delta_j = d_{j-1} d_{j-1}^* + d_j^* d_j$ of (6.1). Each Δ_j is a selfadjoint nonnegative operator, affiliated to $\mathfrak{A}_j$. One can prove the first property as in the proof of Theorem 5.1. The second follows from (6.2) and (6.3).

Moreover, by using arguments from the proof of Theorem 5.1 one can take $\operatorname{Ker} d_j = \overline{\operatorname{Im} d_{j-1}} \oplus \operatorname{Ker} \Delta_j$ which identifies H^j and $\operatorname{Ker} \Delta_j$. So e) means that $\dim_{\mathfrak{A}} \operatorname{Ker} \Delta_j < \infty$.

Because of our previous considerations the complex (5.1) satisfies the conditions a)–e) with $d_j = \overline{A}_j$, $\mathscr{H}_j = L^2(\mathscr{F}_j)$, $\mathfrak{A} = \widetilde{\mathfrak{A}}$.

DEFINITION 6.1. A family $T = \{T_j\}_{j=0}^N$, where $T_j \in \mathfrak{A}_j$, is called an endomorphism of (6.1) if $T_j d_{j-1} \subset d_{j-1} T_{j-1}$.

An endomorphism T induces operators HT^j in cohomology. They may be identified with the operators $P_j T_j P_j$, where P_j is the orthogonal projection onto $\operatorname{Ker} \Delta_j$. By e) we have $\mathrm{tr}_{\mathfrak{A}} P_j T_j P_j < \infty$. So, the *Lefschetz number* of T can be defined by

$$L(T) = (-1)^j \,\mathrm{tr}_{\mathfrak{A}} P_j T_j P_j.$$

If $\mathfrak{A}$ is a semifinite von Neumann algebra, let $(\mathfrak{A})_+$ be the cone of nonnegative elements of $\mathfrak{A}$, let $S_2(\mathfrak{A}) = \{A \in \mathfrak{A},\ \mathrm{tr}_{\mathfrak{A}} A^* A < \infty\}$ be the Hilbert-Schmidt ideal, and let $S_1(\mathfrak{A}) = (S_2(\mathfrak{A}))^2 = \{A \in \mathfrak{A}, \operatorname{tr}_{\mathfrak{A}} |A| < \infty\}$ be the ideal of operators with finite trace.

Returning now to the complex (6.1), we see that each $\mathfrak{A}_j$ is an algebra of square matrices $A = \|A_{ik}\|$ with $A_{ik} \in \mathfrak{A}$. The trace $\operatorname{tr}_{\mathfrak{A}}$ can be extended to $\mathfrak{A}_j$ by

$$\operatorname{tr}_{\mathfrak{A}} A = \sum_i \operatorname{tr}_{\mathfrak{A}} A_{ii}$$

for $A \in (\mathfrak{A})_+$.

LEMMA 6.1. *$A \in S_1(\mathfrak{A}_j)$ (resp. $S_2(\mathfrak{A}_j)$) if and only if $A_{kl} \in S_1(\mathfrak{A}_j)$ (resp. $S_2(\mathfrak{A}_j)$) for all k, l.*

PROOF. Since

$$\operatorname{tr}_{\mathfrak{A}} A^*A = \sum_{k,l} \operatorname{tr}_{\mathfrak{A}} A^*_{kl} A_{kl},$$

$\operatorname{tr}_{\mathfrak{A}} A^*A < \infty$ if and only if $\operatorname{tr}_{\mathfrak{A}} A^*_{kl}A_{kl} < \infty$ for every k, l. The inclusion $A \in S_1(\mathfrak{A})$ is equivalent to the existence of a factorization $A = B_1B_2$, where $B_{1,2} \in S_2(\mathfrak{A})$. Hence, $\|(B_j)_{kl}\| \in S_2(\mathfrak{A})$ as above, $j = 1, 2$. So the lemma follows from

$$A_{kl} = (B_1)_{kr}(B_2)_{rl}.$$

□

The main result of this section is the following

THEOREM 6.1. *Let T be an endomorphism of (6.1) such that $T_j \in S_1(\mathfrak{A}_j)$, $j = 0, \dots, N$. Then*

$$L(T) = (-1)^j \operatorname{tr}_{\mathfrak{A}} T_j.$$

PROOF. Let $\{R_j\}_{j=0}^N$, $\{Q_j\}_{j=0}^N$ be the orthogonal projections on $\overline{\operatorname{Im} d_{j-1}}$ and $\overline{\operatorname{Im} d_j^*}$ respectively, $R_0 = Q_N = 0$. By (6.2), (6.3) we have R_j, $Q_j \in \mathfrak{A}_j$. Furthermore $I_j = R_j + P_j + Q_j$, where I_j is the identity operator in $\mathscr{H}_j$. By the mutual orthogonality of R_j, P_j, Q_j, we have

$$\operatorname{tr}_{\mathfrak{A}} T_j = \operatorname{tr}_{\mathfrak{A}}(R_jT_jR_j) + \operatorname{tr}_{\mathfrak{A}}(P_jT_jP_j) + \operatorname{tr}_{\mathfrak{A}}(Q_jT_jQ_j).$$

So it is sufficient to prove that

$$\operatorname{tr}_{\mathfrak{A}}(Q_jT_jQ_j) = \operatorname{tr}_{\mathfrak{A}}(R_{j+1}T_{j+1}R_{j+1})$$

for all $j = 0, \dots, N$. Let $d_j = U_jS_j$ be the polar decomposition, $S_j^* = S_j$. The uniqueness of the polar decomposition and (6.2) imply that $S_j \in A_j$ and $V^*U_jV = U_j$ for every $V \in \mathfrak{A}'$. Then U_j can be represented as a matrix $\|(U_j)_{kl}\|$, $k = 1, \dots, n_{j+1}$, $l = 1, \dots, n_j$, with elements from $\mathfrak{A}$. Since $R_{j+1} = U_jU_j$, $Q_j = U_jU_j$ we have

$$R_{j+1}T_{j+1}R_{j+1} = U_jU_j^*T_{j+1}U_jU_j^*, \tag{6.4}$$

$$Q_jT_jQ_j = U_j^*U_jT_jU_j^*U_j. \tag{6.5}$$

Our first step is to prove that

$$\operatorname{tr}_{\mathfrak{A}} R_{j+1}T_{j+1}R_{j+1} = \operatorname{tr}_{\mathfrak{A}} U_j^*T_{j+1}U_j. \tag{6.6}$$

This immediately follows from (6.4) and the following lemma.

LEMMA 6.2. *Let* $B \in \mathfrak{A}_{j+1}$ *and* $U\colon \mathscr{H}_j \to \mathscr{H}_{j+1}$ *a matrix with elements from* A. *Then*

a) $U^*BU \in \mathfrak{A}_j$;

b) *if* $B \in S_1(\mathfrak{A}_{j+1})$, *then* $U^*BU \in S_1(\mathfrak{A}_j)$;

c) *if, moreover,* U *is a partial isometry and if* $\operatorname{Im} B \subset \operatorname{Im} U$, *then*

$$\operatorname{tr}_{\mathfrak{A}} U^*BU = \operatorname{tr}_{\mathfrak{A}} B.$$

PROOF. Let $B = \|B_{kl}\|_{k,l=1}^{n_{j+1}}$; then $U^*BU = \|\widetilde{B}_{rs}\|_{r,s=1}^{n_j}$ where

$$\widetilde{B}_{rs} = (U^*)_{rk} B_{kl} U_{ls}.$$

So $\widetilde{B}_{rs} \in \mathfrak{A}$, which proves a). The assertion b) follows from a) and Lemma 6.1. To prove c) we use the chain of equalities

$$\begin{aligned}\operatorname{tr}_{\mathfrak{A}} U^*BU &= \sum_{r=1}^{n_j} \operatorname{tr}_{\mathfrak{A}} \widetilde{B}_{rr} = \sum_{r=1}^{n_j} \sum_{k,l=1}^{n_{j+1}} \operatorname{tr}_{\mathfrak{A}} \big((U^*)_{rk} B_{kl} U_{lr}\big) \\ &= \sum_{l=1}^{n_{j+1}} \sum_{k=1}^{n_{j+1}} \sum_{r=1}^{n_j} \operatorname{tr}_{\mathfrak{A}} \big(U_{lr} (U^*)_{rk} B_{kl}\big) = \sum_{l=1}^{n_{j+1}} \operatorname{tr}_{\mathfrak{A}} (UU^*B)_{ll} = \operatorname{tr}_{\mathfrak{A}} B.\end{aligned}$$

The last equality holds because $UU^*B = B$. By Lemma 6.1, all the traces in the sums are finite. □

The next step is to show that

$$S_j Q_j T_j Q_j \supset U_j^* T_{j+1} U_j S_j. \tag{6.7}$$

Since $\operatorname{Im} S_j = \operatorname{Im} d_j^*$ and S_j is a selfadjoint operator, we have

$$S_j = Q_j S_j = S_j Q_j.$$

Then, by (6.5),

$$S_j Q_j T_j Q_j = U_j^* d_j T_j Q_j \supset U_j^* T_{j+1} d_j Q_j = U_j^* T_{j+1} U_j S_j.$$

Now the final step is to prove that (6.7) leads to the equality

$$\operatorname{tr}_{\mathfrak{A}} Q_j T_j Q_j = \operatorname{tr}_{\mathfrak{A}} U_j^* T_{j+1} U_j. \tag{6.8}$$

Passing now from $\mathscr{H}_j$ to $Q_j \mathscr{H}_j$, we see that this is a corollary of the following lemma.

LEMMA 6.3. *Let* G *be a Hilbert space and let* $B \in \mathscr{L}(G)$ *be a von Neumann algebra equipped with a normal semifinite trace* tr. *If* $A, B \in S_1(\mathfrak{B})$ *and* $SA \supset BS$ *for a selfadjoint nonnegative operator* $S \in \mathfrak{B}$ *such that* $\operatorname{Ker} S = 0$, *then* $\operatorname{tr} A = \operatorname{tr} B$.

REMARK. This was proved in **[36]** for selfadjoint A and B.

PROOF. For $n \in \mathbb{N}$ let $F_n = I - E_{1/n}$, where E_λ is the spectral projection of S corresponding to $(-\infty, \lambda]$. The following properties hold:

$$\left(I + \tfrac{1}{n}S\right)^{-1} \in \mathfrak{B}, \qquad F_n \in \mathfrak{B}, \qquad S^{-1}\eta\mathfrak{B}, \qquad S^{-1}F_n \in \mathfrak{B}.$$

Multiplying the inclusion $SA \supset BS$ by $S^{-1}F$ from the left and by $(I + \frac{1}{n}S)^{-1}$ from the right, we have

$$F_nA\left(I + \tfrac{1}{n}S\right)^{-1} \supset S^{-1}F_nBS\left(I + \tfrac{1}{n}S\right)^{-1}.$$

Both the left and the right side operators are bounded, so they coincide. Then since $S^{-1}F_n \in \mathfrak{B}$

$$\text{(6.9)} \qquad \operatorname{tr} F_nA\left(I + \tfrac{1}{n}S\right)^{-1} = \operatorname{tr} BS\left(I + \tfrac{1}{n}S\right)^{-1}S^{-1}F_n = \operatorname{tr} F_nB\left(I + \tfrac{1}{n}S\right)^{-1}.$$

It is easy to see that $F_n \uparrow I$, $\left(I + \frac{1}{n}S\right)^{-1} \uparrow I$ as $n \to \infty$. Let $D_n = F_n\left(I + \frac{1}{n}S\right)^{-1}$. Then $D_n \geq 0$, $D_n \uparrow I$ as $n \to \infty$, since F_n, $\left(I + \frac{1}{n}S\right)^{-1}$ are commuting operators. Now, by normality, for any $C \in \mathfrak{B}$, $C \geq 0$ we have

$$\operatorname{tr} D_nC = \operatorname{tr} C^{1/2}D_nC^{1/2} \to \operatorname{tr} C \quad \text{as } n \to \infty.$$

So, one can represent B as a linear combination of four nonnegative operators from $S_1(\mathfrak{B})$ to obtain

$$\operatorname{tr} B = \lim_{n\to\infty} \operatorname{tr} D_nB = \lim_{n\to\infty} \operatorname{tr} F_nB\left(I + \tfrac{1}{n}S\right)^{-1}.$$

The same holds for A. So, the lemma follows by passing to the limit in (6.9). □

The theorem follows now from (6.6) and (6.9). □

§7. Fixed point contributions

This section completes the proof of Theorem 0.1. We need to introduce a parametrix $\{B_j\}_{j=0}^{N-1}$ of the complex (0.4). If all the A_j are of the same order m, then the $B_j \in \mathfrak{A}\Psi^{-m}(\mathcal{F}_j)$ have to satisfy

$$\text{(7.1)} \qquad K_j = I - A_{j-1}B_{j-1} - B_jA_j \in \mathfrak{A}\Psi^{-\infty}(\mathcal{F}_j)$$

for all $j = 0, \ldots, N$. Here we put $B_{-1} = B_N = 0$. The standard construction of $\{B_j\}$ [**1**] is to put $B_j = A_j^+Q_{j+1}$, where Q_j is a parametrix of Δ_j (see Proposition 2.1). It is easily seen that $\{K_j\}$ is an endomorphism of (0.4). Moreover $\{K_j\}$ is an endomorphism of (5.1) in the sense of Definition 6.1. Indeed, if $u \in D(\overline{A}_j)$, then there exists a sequence $\{u_k \in C^\infty(\mathcal{F}_j)\}_{k=1}^\infty$ such that $u_k \to u$, $A_ju_k \to \overline{A}_ju$ as $k \to \infty$. Then $C^\infty(\mathcal{F}_j) \ni K_ju_k \to K_ju$ and $A_jK_ju_k = K_{j+1}A_ju_k \to K_{j+1}\overline{A}_ju$. So, $K_ju \in D(\overline{A}_j)$ and $\overline{A}_jK_ju = K_{j+1}\overline{A}_ju$.

Let T be the geometric endomorphism from §2. Then it is not hard to see that the family $\{T_jK_j\}_{j=0}^\infty$ is an endomorphism of both complexes (0.4) and (5.1) and that $T_jK_j \in \mathfrak{A}\Psi^{-\infty}(\mathcal{F}_j)$. Furthermore, we have $K_jP_j = P_j - A_{j-1}B_{j-1}P_j$, where P_j is the orthogonal projection onto $\operatorname{Ker}\Delta_j$ in $L^2(\mathcal{F}_j)$. So,

$$\begin{aligned} P_jT_jK_jP_j &= P_jT_jP_j - P_jT_jA_{j-1}B_{j-1}P_j \\ &= P_jT_jP_j - P_jA_{j-1}T_{j-1}B_jP_j = P_jT_jP_j. \end{aligned}$$

Thus $L(T) = L(KT)$ for any K from (7.1). The idea of the rest of the proof is to approximate the endomorphism T by endomorphisms KT, then to use Theorem 6.1

and to compute the limit value of the right-hand side. In order to do this we need the following

PROPOSITION 7.1. *If $\mathscr{F}$ is a Hilbert $\mathfrak{A}$-bundle and if $R \in \mathfrak{A}\Psi^{-\infty}(\mathscr{F})$, then $R \in S_1(\mathfrak{A}_{\mathscr{F}})$ and*

$$\operatorname{tr}_{\mathfrak{A}} R = \int_M \operatorname{tr}_{\mathfrak{A}} R(x,x)\, d\mu, \tag{7.2}$$

where $R(\cdot,\cdot)$ is the Schwartz kernel from (2.3).

REMARK. The operator $R(x,x) \in \mathfrak{A} \otimes \operatorname{End} G$ is well defined up to the similarity for any $x \in M$. So, $\operatorname{tr}_{\mathfrak{A}} R(x,x)$ is uniquely defined.

PROOF. Let $A \in \mathfrak{A}E\Psi^m(\mathscr{F})$, where m is a sufficiently large positive number. Then the parametrix B of A belongs to the class $\operatorname{Int}(\mathscr{F})$ of integral operators whose Schwartz kernels are continuous sections of $\operatorname{Hom}_{\mathfrak{A}}(\pi_1^*(\mathscr{F}), \pi_2^*(\mathscr{F}))$. So, R can be represented as the following expression involving operators from $\operatorname{Int}(\mathscr{F})$:

$$R = (RA)B + RR_1.$$

Here $R_1 = I - AB \in \mathfrak{A}\Psi^{-\infty}(\mathscr{F})$. Now we have to prove that if $R = R_1R_2$ with $R_j \in \operatorname{Int}(\mathscr{F})$, then $R \in S_1(\mathfrak{A}_{\mathscr{F}})$ and (7.2) holds. This is a consequence of the following

LEMMA 7.1. *If $R \in \operatorname{Int}(\mathscr{F})$, then $R \in S_2(\mathfrak{A}_{\mathscr{F}})$ and*

$$\operatorname{tr}_{\mathfrak{A}} R^*R = \int_M \operatorname{tr}_{\mathfrak{A}} (R(x,y))^* R(x,y)\, d\mu.$$

PROOF. We can suppose without loss of generality that $\dim G = 1$ (see Definition 0.1). Let us fix an isomorphism $\Lambda^2(\mathscr{F}) \cong L^2(M) \otimes F$ as in §2. Then R will become an integral operator in $L^2(M, F)$ with a piecewise smooth kernel $R(\cdot,\cdot)$. To be more precise, we call an $\mathfrak{A}$-valued function piecewise smooth, if it is smooth outside a set of measure 0, consisting of a finite number of manifolds of codimension 1, and has finite limit values (in the uniform topology) on this set. If $R(\cdot,\cdot)$ satisfies these conditions, it belongs to the Hilbert space $L^2(M \times M, \mathfrak{A}^{(2)})$. Here $\mathfrak{A}^{(2)}$ is the completion of $\mathfrak{A}$ in the topology defined by the inner product $\operatorname{tr}_{\mathfrak{A}}(A^*B)$ for A, $B \in \mathfrak{A}$. Let $\{e_j\}_{j=1}^{\infty}$ be a basis of smooth functions in $L^2(M)$. Given this basis, one can represent R as a matrix $\|R_{ij}\|_{i,j=1}^{\infty}$, where

$$R_{ij} = \int_{M\times M} R(x,y)\overline{e_i(x)}e_j(y)\, dx\, dy.$$

By standard arguments **[28,** §IV.1], $R \in S_2(\mathfrak{A}_{\mathscr{F}})$ if and only if

$$\operatorname{tr}_{\mathfrak{A}} R^*R = \sum_{i,j=1}^{\infty} \operatorname{tr}_{\mathfrak{A}}(R_{ij}^*R_{ij}) < \infty.$$

So the required equality is simply the Parseval identity in $L^2(M \times M, \mathfrak{A}^{(2)})$. Indeed $\{e_i \otimes \overline{e_j}\}_{i,j=1}^{\infty}$ is a basis in $L^2(M \times M)$ and standard arguments show that there is an orthogonal decomposition

$$R(\cdot,\cdot) = \sum_{i,j=1}^{\infty} R_{ij} \otimes e_i(\cdot) \otimes \overline{e_j(\cdot)}.$$

□

Proposition 7.1 is proved. □

Now we are in a position to compute the contributions of the fixed points. We need to introduce a bounded family of operators from $\mathfrak{A}\Psi^k(\mathscr{F})$. A family $\{D_j\}$ is by definition bounded if the constants in (2.2) can be chosen independently of the elements of the family. Let $\chi \in C_0^\infty(\mathbb{R}^n)$, $\int \chi = 1$, $\chi(x) = 0$ if $|x| \geq 1$. Let R_ε be the convolution operator corresponding to $\chi_\varepsilon(\cdot) = \varepsilon^{-n}\chi(\varepsilon^{-1})$. We fix a coordinate covering $\{U_j\}$ of M and a subordinated partition of unity $\{\varphi_j\}$. Additionally we suppose that either each U_j contains no fixed points of f or the only fixed point in U_j is the origin in local coordinates. Moreover, the measure $d\mu$ is supposed to coincide in the local coordinates with Lebesgue measure. Thus, we can put $R_\varepsilon = \sum_i r_i R_\varepsilon \varphi_i$, where r_i is extended by 0 outside U_j. Then R_ε is well defined if ε is sufficiently small. Now let $\{B_\varepsilon = (I - R_\varepsilon)B_\varepsilon\}$ be a family of parametrices and $\{T_{j,\varepsilon} = T_j K_{j,\varepsilon}\}$ the corresponding family of endomorphisms of the complex (5.1). Then we have

$$
\begin{aligned}
T_{j,\varepsilon} &= T_j K_j + T_j R_\varepsilon B_j A_j + T_j A_{j-1} R_\varepsilon B_{j-1} \\
&= T_j K_j + T_j R_\varepsilon (I - K_j - A_{j-1}B_{j-1}) + T_j A_{j-1} R_\varepsilon B_{j-1} \qquad (7.3)\\
&= T_j R_\varepsilon + T_j (I - R_\varepsilon) K_j + T_j (A_{j-1} R_\varepsilon - R_\varepsilon A_{j-1}) B_{j-1}.
\end{aligned}
$$

We have to prove that

$$\operatorname{tr}_{\mathfrak{A}} T_{j,\varepsilon} \to \sum_{x=f(x)} \frac{\operatorname{tr}_{\mathfrak{A}} \varphi_j(x)}{|\det(1 - df(x))|}$$

as $\varepsilon \to 0$. This is a consequence of the following statements.

PROPOSITION 7.2. *Let $S_{j,\varepsilon}$ be the sum of the two last members in (7.3). Then $S_{j,\varepsilon} \in \mathfrak{A}\Psi^{-\infty}(\mathscr{F}_j)$ and the family $\{S_{j,\varepsilon}\}$ is bounded in $\mathfrak{A}\Psi^0(\mathscr{F}_j)$. Moreover, $S_{j,\varepsilon} \to 0$ as $\varepsilon \to 0$ in the strong operator topology in $\mathscr{L}(C^\infty(\mathscr{F}_j), C^\infty(\mathscr{F}_j))$.*

PROOF. This is obvious, because $\{R_\varepsilon\}$ satisfies these properties. □

If $P \in \mathfrak{A}\Psi^{-\infty}(\mathscr{F})$, then one can define a regular distribution $\Delta^* P$ by

$$\langle \Delta^* P, u\rangle = \int_M u(x) \operatorname{tr}_{\mathfrak{A}} P(x,x)\, d\mu_x$$

for $u \in C^\infty(M)$. The next proposition can be proved by using the same arguments as in **[1]** (see (4.4), (4.9), (4.11), and below).

PROPOSITION 7.3. *If the family $\{P_\varepsilon \in \mathfrak{A}\Psi^{-\infty}(\mathscr{F})\}$ is bounded in $\mathfrak{A}\Psi^k(\mathscr{F})$ and if $P_\varepsilon \to 0$ as $\varepsilon \to 0$ in the strong operator topology in $\mathscr{L}(C^\infty(\mathscr{F}), C^\infty(\mathscr{F}))$, then $\Delta^* P_\varepsilon \to 0$ in the weak topology of distributions.*

So, $\operatorname{tr}_{\mathfrak{A}} S_{j,\varepsilon} = \langle D^* S_{j,\varepsilon}, 1\rangle \to 0$ as $\varepsilon \to 0$.

PROPOSITION 7.4.

$$\Delta^* T_j R_\varepsilon \to \sum_{x=f(x)} \frac{\operatorname{tr}_{\mathfrak{A}} \varphi_j(x)}{|\det(1 - df(x))|} \delta_x(\cdot)$$

as $\varepsilon \to 0$.

PROOF. Let $u \in C^\infty(M)$. Then

$$\langle \Delta^* T_j R_\varepsilon, u\rangle = \sum_i \int_{U_j} d_{i,j,\varepsilon}(x)\,dx,$$

where

$$d_{i,j,\varepsilon}(x) = \chi_\varepsilon(f(x) - x)\varphi_i(x)u(x)\,\mathrm{tr}_{\mathfrak{A}}\,\varphi_j(f(x)).$$

Clearly, the $d_{i,j,\varepsilon}(\cdot)$ are well defined when ε is sufficiently small, because $f(x) \in \mathfrak{A}_j$ if $x \in \operatorname{supp} d_{i,j,\varepsilon}$. If U_j is fixed point free, then $d_{i,j,\varepsilon} \equiv 0$ when ε is sufficiently small. In the other case, after the substitution $y = f(x) - x$ we have

$$\int_{U_j} d_{i,j,\varepsilon}(x)\,dx = \int_{\mathbb{R}^n} \chi_\varepsilon(y) \frac{\mathrm{tr}_{\mathfrak{A}}\,\varphi_j(f(x(y)))\,u(x(y))\varphi_i(x(y))}{|\det(1 - df(x(y)))|}\,dy$$

$$\to \frac{\mathrm{tr}_{\mathfrak{A}}\,\varphi_j(f(x(0)))\,u(x(0))\varphi_i(x(0))}{|\det(1 - df(x(0)))|} = \frac{\mathrm{tr}_{\mathfrak{A}}\,\varphi_j(v)}{|\det(1 - df(v))|}u(v),$$

where v is the fixed point. □

References

1. M. F. Atiyah and R. Bott, *A Lefschetz fixed point formula for elliptic complexes.* I, II, Ann. of Math. (2) **86** (1967), 374–407; **88** (1968), 451–491.
2. A. V. Brenner and M. A. Shubin, *The Atiyah-Bott-Lefschetz fixed point theorem for manifolds with boundary*, Funktsional. Anal. i Prilozhen. **15** (1981), 67–68; English transl. in Functional Anal. Appl. **15** (1981).
3. A. V. Brenner, Uspekhi Mat. Nauk **43** (1988), no. 4, 167. (Russian)
4. J. M. Bismut, *The Atiyah-Singer theorems: A probabilistic approach.* II. *The Lefschetz fixed point formulas*, J. Funct. Anal. **57** (1984), 329–348.
5. N. Berline and M. Vergne, *A computation of the equivariant index of the Dirac operator*, Bull Soc. Math. France **113** (1985), 305–345.
6. P. Gilkey, *Lefschetz fixed point formulas and the heat equation*, Lecture Notes in Pure and Appl. Math., vol. 48, Springer-Verlag, Berlin and New York, 1979, pp. 91–147.
7. H. Donnelly, *Spectrum and the fixed point sets of isometries.* I, Math. Ann. **224** (1976), 161–170.
8. H. Donnelly and V. K. Patodi, *Spectrum and the fixed point sets of isometries.* II, Topology **16** (1977), 1–11.
9. H. Donnelly and C. Fefferman, *Fixed point formula for the Bergman kernel*, Amer. J. Math. **108** (1986), 1241–1258.
10. L. A. Coburn, R. G. Douglas, D. G. Schaeffer, and I. M. Singer, *C^*-algebras of operators on a half space.* II. *Index theory*, Inst. Hautes Études Sci. Publ. Math. **1971**, no. 40, 69–79.
11. L. A. Coburn, R. D. Moyer, and I. M. Singer, *C^*-algebras of almost periodic pseudo-differential operators*, Acta. Math. **130** (1973), 279–307.
12. M. F. Atiyah, *Elliptic operators, discrete groups and von Neumann algebras*, Astérisque, no. 32–33, Soc. Math. France, Paris, 1976, pp. 43–72.
13. I. M. Singer, *Some remarks on operator theory and index theory*, Lecture Notes in Math., vol. 575, Springer-Verlag, Berlin and New York, 1977, pp. 128–138.
14. B. V. Fedosov and M. A. Shubin, *Index of random elliptic operators.* I, II, Mat. Sb. **106** (1978), no. 1, 108–140; **106**, no. 3, 455–483; English transl. in Math. USSR-Sb. **34** (1978); **35** (1979).
15. A. Connes, *Sur la théorie non commutative de l'integration*, Lecture Notes in Math., vol. 725, Springer-Verlag, Berlin and New York, 1979, pp. 19–143.
16. A. Connes and H. Moscovici, *The L^2-index theorem for homogeneous spaces of Lie groups*, Ann. of Math. (2) **115** (1982), 291–330.
17. A. S. Mishchenko and A. T. Fomenko, *Index of elliptic operators over C^*-algebras*, Izv. Akad. Nauk SSSR Ser. Mat. **43** (1979), no. 4, 831–859; English transl. in Math. USSR-Izv. **15** (1980).
18. M. A. Shubin, *Pseudodifferential almost periodic operators and von Neumann algebras*, Trans. Moscow Math. Soc. **1979**, no. 1, 103–166.

19. D. V. Efremov and M. A. Shubin, *Spectrum distribution function and variational principle for automorphic operators on hyperbolic space*, Séminare sur les Équations aux Dérivées Partielles, 1988-1989, Exp. No. VIII, Ecole Polytech., Palaiseau, 1989.
20. A. V. Efremov and D. V. Efremov, *The spectrum asymptotics of an elliptic operator invariant with respect to discrete group of diffeomorphisms*, Vestnik Moskov. Univ. Ser. I Mat. Mekh. **1986**, no. 1, 57–59; English transl. in Moscow Univ. Math. Bull. **41** (1986).
21. Yu. Kordyukov, *A theorem on the identity of spectra for tangentially elliptic operators on foliated manifolds*, Funktsional. Anal. i Prilozhen. **19** (1985), no. 4, 90–91; English transl. in Functional Anal. Appl. **19** (1985).
22. S. P. Novikov and M. A. Shubin, *Morse inequalities and von Neumann* II_1*-factors*, Dokl. Akad. Nauk SSSR **289** (1986), no. 2, 289–292; English transl. in Soviet Math. Dokl. **34** (1987).
23. A. V. Efremov, *The Atiyah-Bott-Lefschetz formula for Hilbert bundles*, Vestnik Moskov. Univ. Ser. I Mat. Mekh. **1988**, no. 4, 92–95; English transl. in Moscow Univ. Math. Bull. **43** (1988).
24. ———, *Lefschetz-type theorems for elliptic complexes on noncompact manifolds*, Thesis, Moskov. Gos. Univ., Moscow, 1989. (Russian)
25. S. Seifarth and M. A. Shubin, *A Lefschetz fixed point formula for manifolds with cylindrical ends*, C. R. Acad. Sci. Paris Sér. I Math. **310** (1990), 849–853.
26. J. Arthur, *Characters, harmonic analysis, and an* L^2*-Lefschetz formula*, Proc. Sympos. Pure Math., vol. 48, Amer. Math. Soc., Providence, RI, 1988, pp. 167–179.
27. E. V. Troitskiĭ, *An equivariant index of* C^**-elliptic operators*, Izv. Akad. Nauk SSSR Ser. Mat. **50** (1986), no. 4, 849–865; English transl. in Math. USSR-Izv. **29** (1987).
28. M. Takesaki, *Theory of operator algebras*. I, Springer-Verlag, Berlin and New York, 1979.
29. Bo-Ju Jiang, *Lectures on Nielsen fixed point theory*, Contemp. Math., vol. 14, Amer. Math. Soc., Providence, RI, 1983.
30. E. Fadell and S. Husseini, *Fixed point theory for non-simply-connected manifolds*, Topology **20** (1981), 53–92.
31. S. Husseini, *Generalized Lefschetz numbers*, Trans. Amer. Math. Soc. **272** (1982), 247–274.
32. M. A. Shubin, *Pseudodifferential operators and spectral theory*, Springer-Verlag, Berlin and New York, 1987.
33. M. E. Taylor, *Pseudodifferential operators*, Princeton Univ. Press, Princeton, NJ, 1981.
34. J. Dodziuk, *de Rham-Hodge theory for* L^2*-cohomology of infinite coverings*, Topology **16** (1977), 157–165.
35. M. Breuer, *Fredholm theories in von Neumann algebras*. II, Math. Ann. **180** (1969), 313–325.
36. M. A. Shubin, *Spectral theory and the index of elliptic operators with almost periodic coefficients*, Uspekhi Mat. Nauk **34** (1979), no. 2, 95–135; English transl. in Russian Math. Surveys **34** (1979).

Council for Computer Science, 117333 Vavilova st. 40, Moscow, USSR

Amer. Math. Soc. Transl.
(2) Vol. **163**, 1995

Topological Structure of k-saddle Surfaces

V. V. Glazyrin

§0. Introduction

A complete n-dimensional surface F^n in E^m is called a k-saddle surface if

$$H_k(F^n, F^n \cap E^r) = 0$$

for any r-dimensional plane E^r, $2 \le r < m$, where H_k is the k-dimensional Vietoris homology group with coefficients in the group of integers.

In the sequel by a surface we shall understand a proper submanifold of E^m.

The definition of k-saddle surface is due to Sefel′.

We shall denote by U_k^n the class of n-dimensional k-saddle surfaces of class C^3, by T_xF^n (N_xF^n) the tangent (normal) space of F^n at a point x, and by $k(e_1, e_2)$ the Riemannian curvature of F^n in the two-dimensional direction $e_1\ e_2$ ($e_1, e_2 \in T_xF^n$).

It was shown by the author in **[G]** that the following assertions are equivalent (Theorem 0.1):

α) $F^n \in U_k^n$;

β) for every point $x \in F^n$ and $v \in N_xF^n$ the second quadratic form of v relative to F^n has at most $k-1$ eigenvalues of the same sign;

$\gamma 1$) every k-dimensional plane in T_xF^n contains a two-dimensional plane such that the Riemannian curvature of F^n in the direction of this plane is nonpositive;

$\gamma 2$) property $\gamma 1$) is preserved under nondegenerate affine transformations of E^m.

The purpose of this paper is to investigate the topological structure of k-saddle surfaces. We shall prove

THEOREM 2.1. *Let $F^n \in U_k^n$ ($n \ge 6$, $k \le n-2$). Then the following assertions are equivalent*:

- α) *there exists a locally tame compact $(k-1)$-dimensional polyhedron that is a strong deformation retract of F^n*;
- β) *F^n is homeomorphic to the interior of some compact manifold V^n with boundary ∂V^n.*

The next theorem is influenced by a paper of Browder, Levin, and Livesay **[B-L-L]**.

THEOREM 2.2. *Let $F^n \in U_k^n$ ($n \ge 6$, $k \le n-2$) be a simply connected surface. If the homology groups of F^n are finitely generated, then there exists a compact locally tame $(k-1)$-dimensional polyhedron that is a strong deformation retract of F^n.*

1991 *Mathematics Subject Classification*. Primary 53C40.

§1. Behavior of k-saddle surfaces at infinity

DEFINITION. Let V^n be a topological manifold. We say that V^n has $\geq n$ ends if there exists a compact set $K \subseteq V^n$ that has no less than n components with non-compact closure. A manifold V^n has n ends if it has $\geq n$ ends, but not $\geq n+1$ ends.

LEMMA 1.1. *If $F^n \in U^n_{n-1}$, then F^n has exactly one end.*

PROOF. It was shown in [G] that F^n is an unbounded surface and, consequently, has no less than one end. Suppose that there exists a compact set $K \subseteq F^n$ for which $F^n \backslash K$ has no less than two unbounded components.

Let $x_0 \in E^m$ and $L_{x_0}(x) = -|x - x_0|$. By standard arguments of differential topology [M] and Theorem 0.1 we may assume that the restriction of the function L_{x_0} on F has only nondegenerate critical points with index no less than $n - (n-1) + 1 = 2$. Hence, there exists a regular value a of the function L_{x_0} such that $K \subseteq \{x \in F^n : L_{x_0} > a\}$. Let

$$G = \{x \in F^n : L_{x_0} \leq a\}.$$

By the fundamental theorem of Morse theory [M] the pair (F^n, G) is homotopy equivalent to some relative CW-complex (X, G) that has no cells of dimension 1 or 0. Hence,

$$\pi_1(F^n, G) = \pi_1(X, G) = 0.$$

This, together with connectivity of F^n, proves the connectivity of G. But G has no less than two unbounded components. This contradiction completes the proof of Lemma 1.1. □

Note that $F^n \in U^n_{n-1}$ is a necessary condition: there is the example of the cylinder with one-dimensional element which is a surface in U^n_n with two ends.

Let V^n be a topological manifold with one end, ∞.

We say that $\pi_j(V^n, \infty) = 0$ for $j \leq m$, if there exists a system of neighborhoods of ∞: $\{U_i\}_{i=1}^{\infty}$, $\cap U_i = \varnothing$ for which the following holds:

$$\pi_j(V^n, U_i) = 0 \qquad \text{for } j \leq m.$$

LEMMA 1.2. *Let $F^n \in U^n_k$. Then $\pi_j(F^n, \infty) = 0$.*

The proof is similar to that of Lemma 1.1.

We say that a manifold V^m is simply connected at infinity if for any compact set $K \subseteq V^m$ there is a compact set $K_1 \supseteq K$ with simply connected complement.

LEMMA 1.3. *Let $F^n \in U^n_{n-2}$. The surface F^n is simply connected if and only if it is simply connected at infinity.*

PROOF. Let $x_0 \in E^m$. We may assume that the restriction of $-L_{x_0}$ on F^n has only nondegenerate critical points. Take an increasing sequence of regular values of $-L_{x_0}$: $\{a_n\}_{n=1}^{\infty}$, $a_n \to \infty$, and consider the sets $U_i = \{x \in F^n,\ -L_{x_0}(x) \geq a_i\}$. A similar argument as in Lemma 1.1 yields

$$\begin{aligned} &\pi_j(F^m, U_i) = 0 \qquad \text{for } j \leq 2,\ i \geq 1, \\ &\pi_j(U_s, U_i) = 0 \qquad \text{for } 1 \leq s < i, \quad j \leq 2. \end{aligned} \tag{1.1}$$

Let F^n be simply connected, K any compact subset of F^n. There exists $i_0 \geq 1$ such that

$$K \subseteq \{x \in F^n : L_{x_0}(x) < a_{i_0}\}.$$

Consider the exact homotopy sequence of the pair (F^n, U_{i_0}):

$$\pi_2(F^n, U_{i_0}) \longrightarrow \pi_1(U_{i_0}) \longrightarrow \pi_1(F^n) \longrightarrow \pi_1(F^n, U_{i_0}).$$

Then it follows from (1.1) that $\pi_1(U_i) \simeq \pi_1(F^m) = 0$. Hence, F^n is simply connected at infinity.

Conversely, let F^n be simply connected at infinity. Hence there is a compact set

$$K \supseteq \{x \in F^n : L_{x_0}(x) \leq a_{i_0}\}$$

with simply connected complement K^c. We may assume that $K^c \supseteq U_2$. By the exact homotopy sequence of the pair (U_1, U_2) and (1.1) we obtain that $\pi_1(U_1)$ is isomorphic to $\pi_1(U_2)$. Using the commutativity of the diagram

$$\begin{array}{ccc} \pi_1(U_1) & \longrightarrow & \pi_1(U_2) \\ & \searrow \quad \nearrow & \\ & \pi_1(K^c) = 0 & \end{array}$$

we obtain that

$$\pi_1(U_1) = \pi_1(U_2) = \pi_1(F^m) = 0.$$

This completes the proof of the lemma. □

An inverse sequence of groups $A_1 \longleftarrow A_2 \longleftarrow \cdots$ is called stable if there exists a subsequence $\{B_1, B_2, \dots\} \subseteq \{A_1, A_2, \dots\}$ for which the homomorphisms $B_1 \xleftarrow{f_1} B_2 \xleftarrow{f_2} B_3 \longleftarrow \cdots$ induce homomorphisms of images $\operatorname{im}(f_1) \simeq \operatorname{im}(f_2) \simeq \operatorname{im}(f_3) \simeq \cdots$ Obviously, $\operatorname{im}(f_1) \simeq \varprojlim A_j$.

Let V^m be a topological manifold with one end. We say that π_1 is stabilized at infinity if there exists a sequence of neighborhoods of ∞: $U_1 \subseteq U_2 \subseteq \dots$ for which the corresponding sequence of fundamental groups

$$\pi_1(U_1) \underset{i_2^*}{\longleftarrow} \pi_1(U_2) \underset{i_3^*}{\longleftarrow} \dots$$

is stable.

It was shown by Siebenmann [S] that this is well defined.

Let $\pi_1(\infty) = \varprojlim_j \pi_1(U_j)$. The inclusion $U_i \to V^n$ induces a homomorphism $\pi_1(\infty) \to \pi_1(V^n)$ for large i, defined up to conjugacy.

LEMMA 1.4. *If $F^n \in U^n_{n-2}$, then π_1 is stabilized at infinity and $\pi_1(\infty) \to \pi_1(F^m)$ is an isomorphism.*

PROOF. Let U_i be the same as in the proof of Lemma 1.3. Then the lemma follows from the sequence of isomorphisms $\pi_1(U_1) \simeq \pi_1(U_2) \simeq \cdots$. □

§2. The skeleton of k-saddle surfaces

Suppose that the polyhedron P is topologically imbedded in the topological manifold V^n. We say that P is locally tame if there exists a covering of it by open sets $U \subseteq V^n$ with piecewise linear structure and also the imbedding $P \cap U \to U$ is piecewise linear.

THEOREM 2.1. *Let $F^n \in U_k^n$ $(n \geq 6,\ k \leq n-2)$. Then the following assertions are equivalent*:

α) *there exists a locally tame compact $(k-1)$-dimensional polyhedron P that is a strong deformation retract of F^n*;

β) *F^n is homeomorphic to the interior of some compact manifold V^n with boundary ∂V^n.*

PROOF. Suppose β) holds. Consider the manifold V_1^n obtained from V^n by attaching a collar $\partial V^n \times [0,1)$. By Connelly **[C]** we have that V_1^n is homeomorphic to the interior of V^n and the manifold V^n is a strong deformation retract of V_1^n. Hence V_1^n is homeomorphic to F^n and by Lemma 1.2 we see that

$$\pi_j(V^n, \partial V^n) = 0 \quad \text{for} \quad j \leq m-k.$$

It was shown by Pedersen **[P]** that for any triad $(M^n, \partial_- M^n, \partial_+ M^n)$ $(n \geq 6)$ with the condition $\pi_j(M^n, \partial_+ M^n) = 0$ for $j < m-r$, $r \leq m-3$ there exists some r-dimensional locally tame compact polyhedron P for which $\partial M^n \cup K$ is strong deformation retract of M^n. In our case $r = k-1$, $\partial M = \varnothing$. Hence α) follows from β).

Suppose α) holds. F^n is a differentiable manifold. Hence, F^n can be triangulated and we consider F^n as a piecewise linear manifold. Since $n - \dim P \geq 3$, there exists a homeomorphism $h : F^n \to F^n$ that is close to the identity map and the restriction of h on P is piecewise linear **[K-S]**.

Let $Q = h(P)$, let $i : P \to F^n$, $i_1 : Q \to F^n$ denote the imbeddings, and let r be the retraction of F on P. Hence, the diagram

$$\begin{array}{ccc} P & \underset{r}{\overset{i}{\leftrightarrows}} & F^n \\ {\scriptstyle h}\downarrow & & \downarrow{\scriptstyle h} \\ Q & \xrightarrow{i_1} & F^n \end{array}$$

is commutative. Let $r_1 = h \circ r \circ h^{-1}$. Then

$$i_1 \circ r_1 = i_1 \circ h \circ r \circ h^{-1} = h \circ i \circ r \circ h^{-1} \simeq \mathrm{Id}_{F^n},$$

$$r_1 \circ i_1 = h \circ r \circ h^{-1} \circ i = h \circ r \circ i \circ h^{-1} \simeq \mathrm{Id}_Q .$$

Hence, i_1 is a homotopy equivalence and a piecewise linear map. Since Q is a compact set, Q is a subpolyhedron of F^n **[R-S]**. □

Let $\mathfrak{R}$ be a regular neighborhood of Q in F^n. For example, it may be a closed star in the second barycentric refinement of the triangulation of F^n. $\mathfrak{R}$ is a compact piecewise linear manifold with boundary $\partial\mathfrak{R}$. Note that, by Lemma 1.4, F^n has one end, π_1 is stabilized at infinity, and $\pi_1(\infty)$ is isomorphic to $\pi_1(F^n)$. Hence, as was shown by Siebenmann **[S]**, F^n is piecewise linear homeomorphic to the interior of the compact piecewise linear manifold $\mathfrak{R}$. This completes the proof of Theorem 2.1.

There is the result of Browder et al. **[B-L-L]** which states that a piecewise linear manifold V^n $(n \geq 6)$ is piecewise linear homeomorphic to the interior of a compact manifold with simply connected boundary if and only if $H_*(V^n)$ is finitely generated and V^n is simply connected at infinity.

Hence, by using Theorem 2.1 and Lemma 1.3 we obtain the following assertion:

THEOREM 2.2. *Let $F^n \in U_k^n$ be a simply connected surface. If the homology groups of F^n are finitely generated, then there exists a compact locally tame $(k-1)$-dimensional polyhedron that is a strong deformation retract of F^n.*

References

[C] R. Connelly, *A new proof of Brown's collaring theorem*, Proc. Amer. Math. Soc. **27** (1971), 180–182.

[B-L-L] W. Browder, J. Levine, and G. Livesay, *Finding a boundary for an open manifold*, Amer. J. Math. **87** (1965), 1017–1028.

[G] V. V. Glazyrin, *Topological and metric structure of k-saddle surfaces*, Sibirsk. Mat. Zh. **19** (1978), no. 3, 555–563; English transl. in Siberian Math. J. **19** (1978).

[K-S] R. C. Kirby and L. C. Siebenmann, *Foundational essays on topological manifolds, smoothings, and triangulations*, Princeton Univ. Press, Princeton, NJ, 1977.

[M] J. Milnor, *Morse theory*, Princeton Univ. Press, Princeton, NJ, 1963.

[P] E. K. Pedersen, *Spines of topological manifolds*, Comment. Math. Helv. **50** (1975), 41–44.

[S] L. C. Siebenmann, *On detecting open collars*, Trans. Amer. Math. Soc. **142** (1969), 201–227.

[R-S] C. P. Rourke and B. J. Sanderson, *Introduction to piecewise linear topology*, Springer-Verlag, Berlin and New York, 1972.

ENERGETIC DEPARTMENT, CHITINSKII POLYTECHNICAL INSTITUTE, CHITA, RUSSIA

Amer. Math. Soc. Transl.
(2) Vol. **163**, 1995

On Uniqueness of Reconstruction of the Form of Convex and Visible Bodies from Their Projections

V. P. Golubyatnikov

Inverse problems of reconstructing the shape of convex bodies from their projection has been considered in many publications; they are closely connected with the Minkowski and Christoffel problems **[1]**. It was shown in **[2]** that if the projections of two closed analytic surfaces in $\mathbb{R}^3$ on any plane have equal areas and perimeters, then these surfaces are equal. S. Campy **[3]** showed the necessity of the analyticity conditions: he constructed in $\mathbb{R}^3$ pairs of convex C^∞-surfaces of revolution that are not pairwise equal, but whose projections on any plane have pairwise equal areas and perimeters. All these surfaces have a constant width.

The main results of the present paper concern a classical question **[4–6]**: if two convex bodies in $\mathbb{R}^3$ have congruent projections on any plane, how different can they be? The same question can be posed for $\mathbb{R}^k$, and in the sequel we shall suppose that $k > 2$ and that all the projections are orthogonal. We shall call two compact sets V_1 and V_2 in Euclidean space parallel if one of them can be obtained from the other by a parallel translation. We shall call them centrally symmetric if one of them can be obtained from the other by a central symmetry.

W. Süss proved **[7]** (see also **[4]**) that if the projections of two convex compact bodies in $\mathbb{R}^k$ on any hyperplane are parallel, then these convex compact sets are parallel in $\mathbb{R}^k$ themselves. We shall generalize this statement in three different directions.

1. Sometimes it happens that one does not have full information necessary to determine a function by its integral transformation, or to reconstruct the shape of a (convex) body from its projections **[8]**. Let us relax the condition in Süss's lemma that all the projections be parallel. We shall suppose that for a certain subset $L \subset S^{k-1}$ of the unit sphere in $\mathbb{R}^k$ the projections of two convex compact bodies $V_1, V_2 \subset \mathbb{R}^k$ in the directions of the vectors from L are parallel. Let $P(n)$ be an oriented hyperplane in $\mathbb{R}^k$ that contains the origin of $\mathbb{R}^k$ and has a unit normal vector n. The intersection $E(n) = P(n) \cap S^{k-1}$ will be called a big $(k-2)$-dimensional sphere. We shall denote by $V_i(n)$ the projection of the convex compact V_i on $P(n)$ and by (n, m) the scalar product of the vectors in $\mathbb{R}^k$.

LEMMA 1. *Let $L \subset S^{k-1}$ be a set of unit vectors that has nonempty intersection with any big $(k-2)$-dimensional sphere. If the projections of convex compact sets*

1991 *Mathematics Subject Classification*. Primary 52A20; Secondary 53C65.

$V_1, V_2 \subset \mathbb{R}^k$ on any hyperplane $P(n)$, $n \in L$, coincide, $V_1(n) = V_2(n)$, then the convex compact sets V_1, V_2 coincide.

The proof is almost obvious.

LEMMA 2. *Let $n_1, n_2, n_3 \subset S^{k-1}$ be noncoplanar unit vectors, let the projections of convex compact sets V_1, V_2 on $P(n_1), P(n_2)$ coincide, and let the projections $V_1(n_3)$, $V_2(n_3)$ be parallel; then $V_1(n_3) = V_2(n_3)$.*

PROOF. Let $Q_{i,j} = P(n_i) \cap P(n_j)$, $i, j = 1, 2, 3$, be the intersections of the hyperplanes. The projections of the compact sets V_1, V_2 on $(n-2)$-dimensional planes $Q_{1,3}, Q_{2,3} \subset P(n_3)$ coincide because the projection on $Q_{1,3}$ can be done in two steps: first on $P(n_1)$, where the projections of V_1 and V_2 coincide and then on $Q_{1,3}$, so these projections also coincide. Similarly for $Q_{2,3}$.

If $V_1(n_3)$ can be obtained from $V_2(n_3)$ by a parallel translation on a vector $m \neq 0$, $m \perp n_3$, then the projections of this m on $Q_{1,3}$ and $Q_{2,3}$ are zero because of the coincidence of the projections of V_1 and V_2 on these $(k-2)$-dimensional planes. But $Q_{1,3}$ and $Q_{2,3}$ generate all of the hyperplane $P(n_3)$, so this vector m equals zero and $V_1(n_3) = V_2(n_3)$. □

THEOREM 1. *If $L \subset S^{k-1}$ contains three noncoplanar vectors and intersects each big $(k-2)$-dimensional sphere and the projections of two convex compact sets $V_1, V_2 \subset \mathbb{R}^k$ on any hyperplane $P(n)$, $n \in L$, are parallel, then these V_1, V_2 are parallel in $\mathbb{R}^k$ themselves.*

PROOF. Consider three noncoplanar vectors $n_1, n_2, n_3 \in L$ and let us translate V_1 in order to make the projections on $P(n_1)$ of the thus obtained V_1' and of V_2 coincide. The projection of V_2 on $P(n_2)$ can be obtained from the corresponding projection of V_1' by a parallel translation by a vector m orthogonal to $P(n_1) \cap P(n_2)$ as was shown in Lemma 2. Let us translate V_1' in the direction of n_1 in order to make the projections $V_2(n_2)$ and $V_1''(n_2)$ coincide. Here V_1'' is the result of the last parallel translation. The length of this translation is equal to $|m|(\sin\alpha)^{-1}$, where α is the angle between n_1 and n_2. As was shown above, the projections of V_1'' and V_2 on $P(n_3)$ also coincide. Let n be any vector in L. One of the triples $\{n, n_2, n_3\}$, $\{n, n_3, n_1\}$, $\{n, n_1, n_2\}$ is linearly independent. It follows from Lemma 2 that the projections of V_1'' and V_2 on $P(n)$ also coincide, and our theorem follows from Lemma 1. □

It is easy to see that this theorem holds for those $L \subset S^{k-1}$ whose closure intersects any big $(k-2)$-dimensional sphere.

The condition of linear independence of some three vectors in L is essential: let $L \subset S^2$ be an equator. Consider in the plane of this equator all the figures of a constant width w and let us construct cylinders of height h over all these figures. The projections of all these cylinders on the vertical planes are the rectangles with base w and height h, so they are parallel but one cannot make all the projections of such a priori noncongruent cylinders on all the vertical planes coincide at once.

2. In this section we shall generalize Süss's lemma in the following way: let us call two convex plane figures congruent if one of them can be obtained from the other by an orientation-preserving motion of the plane (we shall omit the nonorientable case). We shall say that a plane compact figure has no rotation symmetries if it can be made to coincide with itself only by the identity motion of the plane. One of the main results of this section is

THEOREM 2. *If $V_1, V_2 \subset \mathbb{R}^k$ are convex compact sets and their projections on any two-dimensional plane are congruent and have no rotation symmetries, then V_1 and V_2 are either parallel or central symmetric.*

Consider first the case $k = 3$. Let us denote by $\varphi(n)$ the least angle φ in absolute value such that the projection $V_1(n)$ is obtained from $V_2(n)$ by rotation through the angle φ with a suitable center of rotation. If these projections coincide after a parallel translation, we set $\varphi(n) = 0$. Note that $\varphi(\pi) = \varphi(-\pi)$ and $\varphi(-n) = -\varphi(n)$, where the sign is determined by a unit normal n, $n \perp P(n)$, $-n \perp P(-n)$.

LEMMA 3. *If all the projections $V_1(n)$, $V_2(n)$ have no rotation symmetries, then $\varphi(n)$ is a continuous function of n.*

PROOF. If $\lim(n_i) = n_0$ and $\lim(\varphi(n_i)) \neq \varphi(n_0)$ we can choose a subsequence $\{m_j\} \subseteq \{n_i\}$ whose image $\varphi(\{m_j\}) \subset S^1$ has a limit $\varphi_1 \neq \varphi(n_0)$, because S^1 is compact.

We can make the projections $V_1(n_0)$ and $V_2(n_0)$ coincide by rotation through the angles $\varphi(n)_0$ and φ_1, but this is impossible because of the absence of rotation symmetries of the projections considered. □

If $\varphi(n) \equiv 0$ for all $n \in S^2$ our theorem follows from Süss's lemma. If $\varphi(n) \equiv \pi$ for all $n \in S^2$, then consider V_1', obtained by a certain central symmetry of the compact body V_1. The bodies V_1', V_2 have parallel projections on any plane, so they satisfy the conditions of Süss's lemma and we can make the bodies V_1, V_2 coincide by a central symmetry.

Let us suppose that $\pi > \varphi(n_1) > 0$ for some $n_1 \in S^2$. Consider all meridians $m(t)$ on S^2 which join the points $+n_1$ and $-n_1$, $0 \leq t \leq 2\pi$, t parametrizes the points of the circle $E(n_1)$. For a continuous $\varphi : S^2 \to S^1$ denote by $\varphi^{-1}(0)$ and $\varphi^{-1}(\pi)$, respectively, the preimages of 0 and π and by $[\varphi^{-1}(0)]$ and $[\varphi^{-1}(\pi)]$ the sets of nonisolated points of these preimages. For $\beta \in S^1$ the set $\varphi^{-1}(\beta)$ can be defined similarly.

LEMMA 4. *One of the sets $\varphi^{-1}(0), \varphi^{-1}(\pi)$ intersects all the meridians $m(\alpha)$.*

PROOF. The meridian $m(0)$ determines a continuous map of triples

$$g_0\colon ([-\pi/2;\pi/2];\ -\pi/2,\ \pi/2) \to (S^1;\ -\varphi(n_1),\ \varphi(n_1)).$$

The induced homomorphism of integer homology groups

$$\mathbb{Z} \approx H_1\Big(([-\pi/2,\pi/2]; -\pi/2, \pi/2)\Big) \to H_1(S^1; -\varphi(n_1), \varphi(n_1)) \approx \mathbb{Z} \oplus \mathbb{Z}$$

maps the generators of $\mathbb{Z}$ into the element $(l_1, l_2) \in \mathbb{Z} \oplus \mathbb{Z}$ so that $l_1 + l_2$ is an odd number because $\varphi(-n_1)$ and $\varphi(n_1)$ lie in the different half-planes with respect to the horizontal line containing 0 and π in $S^1 \subset \mathbb{R}^2$ and the intersection index of $g_0(m_0)$ and this line is equal to 1. l_1 corresponds to the left arc $(-\varphi(n_1), \varphi(n_1))$ of the circle and l_2 to the right one. If l_1 is odd, the left arc $(-\varphi(n_1), \psi(n_1))$, which contains π, is covered by the map φ of all the meridians $m(t)$; if l_2 is odd, then the right arc $(-\varphi(n_1), \varphi(n_1))$ is covered by this φ. □

COROLLARY 1. *One of the sets $[\varphi^{-1}(0)], [\varphi^{-1}(\pi)]$ intersects all meridians $m(t)$.*

PROOF. If all the meridians $m(t)$ are intersected by $\varphi^{-1}(\pi)$, we change as above the body V_1 to V_1', which is obtained by a central symmetry; in this act the sets

$\varphi^{-1}(0), \varphi^{-1}(\pi)$ interchange their roles and for the convex bodies V_1', V_2 all the meridians $m(t)$ are intersected by $\varphi^{-1}(0)$; so we shall suppose in the sequel that such a change is executed if necessary.

Let C be the set of all the unit vectors n such that the projections $V_i(n)$, $i = 1, 2$, have constant width. The value of this width does not depend on $n \in C$ because any two big circles on S^2 have nonempty intersection. □

LEMMA 5. *If $[\varphi^{-1}(0)]$ is not a big circle on S^2 and intersects all the meridians $m(t)$, there are two nonparallel vectors $j_1, j_2 \in [\varphi^{-1}(0)]$ such that for a dense set in $E(n_1)$ of parameters t the corresponding meridians $m(t)$ intersect the set $[\varphi^{-1}(0)]$ at points not coplanar with j_1, j_2.*

PROOF. If there are two different points of $[\varphi^{-1}(0)]$ on a certain meridian $m(t_1)$, we take these two points for the required j_1, j_2. If on every $m(t)$ there is strictly one point of $[\varphi^{-1}(0)]$, we shall denote it as $n(t)$ and note that $[\varphi^{-1}(0)]$ is homeomorphic to a circle. For every $j_1 \in [\varphi^{-1}(0)]$ consider the family of all big circles $E(j_1, t)$ containing j_1 and $n(t) \neq j_1$. It is easy to see that the intersection of at least one of these circles $E(j_1, t_1)$ with $[\varphi^{-1}(0)]$ has a dense complement in $[\varphi^{-1}(0)]$. We put $j_2 = n_2$ and the lemma is proved. □

For the j_1, j_2 just obtained perform a parallel translation of V_1 (or V_1', see above) to make the projections on $P(j_1), P(j_2)$ of the V_1'' and V_2 obtained coincide.

Take any t that belongs to the dense subset of $E(n_1)$ from Lemma 5 and $n(t) \in [\varphi^{-1}(0)]$ not coplanar to j_1, j_2. It follows from Lemma 2 that the projection $V_1''(n(t))$ coincides with $V_2(n(t))$ for any t. The set $[\varphi^{-1}(0)]$ intersects all the big circles on S^2 and contains a noncoplanar triple of vectors, so that Lemma 1 yields Theorem 2 for $k = 3$ and for $\varphi^{-1}(0)$ not contained in any big circle of S^2.

If the projections of convex bodies $V_1, V_2 \subset \mathbb{R}^3$ on any plane $P(n)$ – $V_1(n)$, $V_2(n)$ are congruent, their perimeters $L_1(n), L_2(n)$ are equal and can be expressed in the terms of the widths $H_1(s)$, $H_2(s)$ of the bodies V_1, V_2 as follows:

$$L_1(n) = L_2(n) = 1/2 \int_{E(n)} H_1(s)\, ds = 1/2 \int_{E(n)} H_2(s)\, ds \tag{1}$$

(see **[9]**). As was shown in **[10]**, this integral equation has a unique solution. Thus, convex bodies V_1, V_2 under consideration have the same width for any $n \in S^2$; i.e., $H_1(n) = H_2(n)$.

LEMMA 6. *If the projections of the convex bodies $V_1, V_2 \subset \mathbb{R}^3$ on any plane are congruent, then $S^2 = \varphi^{-1}(0) \cup \varphi^{-1}(\pi) \cup C$.*

PROOF. Here we do not suppose that the projections considered have no rotation symmetries. We shall denote by $\{\varphi(n)\}$ the set of all the angles $\varphi(n) \in S^1$ such that one can make the projection $V_1(n)$ coincide with $V_2(n)$ by a rotation through the angle $\varphi(n)$.

If the open set $F = S^2 \backslash (\varphi_{-1}(0) \cup \varphi_{-1}(\pi) \cup C)$ is not empty, let $n_1 \in F$. If the set $\{\varphi(n_1)\}$ contains the angle πd for an irrational d , the angles πdi (where i is an integer) form a dense set in the circle $E(n_1)$, so the widths of V_1 and V_2 are constant on this circle and $n_1 \in C$, which contradicts the choice of n_1.

So if F is not empty it is the countable union of the sets F_r such that $r = p/q$ are irreducible fractions and for $n \in F_r$, $\pi r \in \{\varphi(n)\}$.

It follows from the Baire category theorem that at least one of these F_r is dense in an open $D \subset S^2$. The rest of the proof of Lemma 6 is based on

LEMMA 7. *If $V_1, V_2 \subset \mathbb{R}^3$ and $D \subset S^2$ are as above, p, q are positive integers $p < q$, and for any $n \in D$, $r = \pi p/q \in \{\varphi(n)\}$, then for these $n \in D$ the projections $V_1(n), V_2(n)$ have constant width.*

PROOF OF LEMMA 7. If the statement of this lemma is false, there exists an $n_2 \in D$ such that for a certain $m_1 \in E(n_2)$ the width of V_1 $H_1(m_1) = M$ is a maximum on $E(n_2)$ and is not constant in any neighborhood of m in $E(n_2)$. Let m_2 be the unit vector in $E(n_1)$ which is obtained by the rotation of m_1 through the angle $\varphi(n_2)$ with the direction of this rotation determined by the normal n_2. Also let $S(m_1, r) \subset S^2$ be the circle with center m_1 and radius $r = \pi p/q$ in the metric of the sphere. Since D is open, there is an open arc l_1 on this circle such that $m_2 \in l_1$ and the widths of V_1 and V_2 in the directions of the vectors generating l_1 are equal to $M = H_1(m_2) = H_2(m_2)$.

Similarly, on the circle $S(m_2, r) \subset S^2$ with the same radius and center at m_2 there is an open arc l_2 containing m_1 and such that the widths of V_1, V_2 in the directions of the vectors generating l_2 are also equal to M.

Take any point $m_4 \in l_1$, $m_4 \neq m_2$, and consider an arc l_4 on the circle $S(m_4, r)$ such that $m_2 \in l_4$ and the widths of V_1, V_2 in the directions of the vectors belonging to l_4 are equal to M. In the same manner we construct an arc $l_3 \subset S(m_3, r)$ with center $m_3 \in l_2$, $m_3 \neq m_2$. Let m_c be a vector from a small neighborhood of m_1 in $E(n_2)$ such that $H_1(m_c) = H_2(m_c) = c \neq M$. Constructing for these m_c the figures "X", $X = l_1 \cup l_3$ or $X = l_2 \cup l_4$, we see that on different X's corresponding to different values of c the widths of V_1, V_2 are constant, so that the different figures "X" are mutually disjoint. The cardinality of these c is that of the continuum but it is well known that the cardinality of a set of mutually disjoint figures "X" on the plane is strictly less than that of the continuum. Therefore, in some neighborhood of m_1 on $E(n_2)$ H_1 must be constant, which contradicts the choice of m_1. Hence, the widths H_1, H_2 are constant on the big circle $E(n_2)$ and Lemma 7 is proved. □

Since $\varphi^{-1}(\pi d), \varphi^{-1}(\pi p/q) \subset C$ for an irrational d and for natural $q > p$, Lemma 6 is also proved and it is easy to see that if for convex compact bodies $V_1, V_2 \subset \mathbb{R}^3$ as in Theorem 2 $\varphi(n)$ is not a constant, $\varphi^{-1}(\pi)$ does not intersect all the big circles on S^2, and $\varphi^{-1}(0)$ is a big circle on S^2, then the compact sets V_1, V_2 have a constant width. If $\varphi^{-1}(\pi)$ intersects all the big circles and is not a big circle itself, we can obtain Theorem 2 by interchanging the roles of $\varphi^{-1}(0)$ and $\varphi^{-1}(\pi)$ as above. If $\varphi^{-1}(\pi) \cup \varphi^{-1}(0)$ is contained in a big circle, it follows from Lemma 6 that $S^2 = C$.

Let $\varphi^{-1}(0)$ be contained in a big circle and let P be the plane containing this circle. Consider P_1, P_2, the support planes to V_2 parallel to P. Since this convex body has a constant width, the common points N_1, N_2 of V_2 and P_1, P_2, respectively, belong to the common perpendicular to the planes P_1, P_2. Hence, the compact set V_1 can be moved by a parallel translation so that the obtained V_1' is tangent to P_1, P_2 at the same points N_1, N_2. The projections of V_2, V_1' in the directions of the vectors from $\varphi^{-1}(0)$ coincide; $\varphi^{-1}(0)$ intersects all the big circles on the unit sphere. Therefore, for the plane $\varphi^{-1}(0)$ Theorem 2 follows from Lemmas 1 and 5. This completes the proof of our theorem in the most difficult case $k = 3$, the first induction step.

For $k > 3$ this theorem can be proved by induction.

In $\mathbb{R}^k$ the projections of the V_1, V_2 on any two-dimensional plane P^2 can be done in two steps: first on a three-dimensional plane $P^3 \supset P^2$ and then on P^2.

It follows from the previous arguments that the projections of V_1 and V_2 on any three-dimensional plane are either centrally symmetric or parallel. The type of this transformation is the same for all three-dimensional planes in $\mathbb{R}^k$, since the projections of V_1, V_2 on two-dimensional planes have no rotation symmetries. The induction step is based on Süss's lemma.

Let us call two plane figures similar if one of them can be obtained from the other with the help of a composition of a motion and a homothety.

THEOREM 3. *If $V_1, V_2 \subset \mathbb{R}^k$ are convex compact bodies and their projections on any two-dimensional plane are similar and have no rotation symmetries (the ratio of the similitude is not supposed to be constant, independent of the plane), then the compact sets V_1, V_2 are either homotethic or parallel in $\mathbb{R}^k$.*

We shall consider the case $k = 3$. The function $\varphi(n)$, defined as above, is continuous and it is easy to see that for all $n \in \varphi^{-1}(0) \cup \varphi^{-1}(\pi)$ the ratios of similitude of the projections V_1, V_2 do not depend on n; so we can assume that they are equal to 1. If there exists $n_1 \notin \varphi^{-1}(0) \cup \varphi^{-1}(\pi)$, then for any vector $n \in E(n_1)$ one can find a vector from $\varphi^{-1}(0)$ orthogonal to n (see Corollary 1). Performing a parallel translation (or a central symmetry) of V_1, as in Theorem 2, we can make the support functions of the projections of the thus obtained V_1' and V_2, $V_1'(m)$ and $V_2(m)$, coincide for all m that are perpendicular to some vector in $\varphi^{-1}(0)$. Therefore, the projections $V_2(n_1)$, $V_2(n_2)$ also coincide, which contradicts the existence of n_1 and our theorem holds for $k = 3$. The case $k > 3$ can be treated by induction as in Theorem 2.

3. It is obvious that the restrictions on rotation symmetries of the projections $V_i(n)$ in the previous section are caused by the method of proof. Here we shall remove some of these restrictions. In the next section Theorems 2 and 3 will be generalized to objects more complicated than convex bodies and in that case these restrictions will be essential. Note that the set of all convex compact bodies in $\mathbb{R}^k$ whose projections on any two-dimensional plane have no rotation symmetries form a dense set with respect to the Hausdorff metric.

THEOREM 4. *Let $V_1, V_2 \subset \mathbb{R}^k$ be convex bodies. If the restrictions of their support functions on any big circle of the unit sphere S^{k-1} are of class C^2, their projections on any two-dimensional plane are congruent, and each such projection that has nontrivial rotation symmetries and a constant width is a circle, then V_1 and V_2 are either parallel or centrally symmetric.*

PROOF. As above, we shall give the proof in the case $k = 3$. It is easy to see that a convex figure of constant width that has rotational symmetries of an even order has a center of symmetry and, hence, is a circle. It follows from the conditions of our theorem and from Lemmas 6 and 7 that on each connected component of the open set $F = (S^2 \backslash (\varphi^{-1}(0) \cup \varphi^{-1}(\pi))) \subset C$ the function $\varphi(n)$ is continuous because the projections of V_1, V_2 corresponding to the vectors of C have no rotation symmetries. If F is empty, Theorem 4 follows directly from the Süss lemma. Let $n_1 \in F$. There are two cases to study:

a) $\varphi(n_1) = \pi d$, where d is irrational. Consider the radii of curvature of the boundaries $\partial V_1(n_1)$, $\partial V_2(n_1)$ of the projections $V_1(n_1), V_2(n_1)$. Since these projections have constant width and since the angles $\pi d i$ for integral i form a dense set in the unit circle, one can show that the boundaries of these projections have constant curvature. Therefore, $V_1(n_1)$ and $V_2(n_1)$ are circles and $n_1 \in \varphi^{-1}(0) \cap \varphi^{-1}(\pi)$ which contradicts the choice of n_1. The author needs about seven pages for a careful proof of this fact.

b) For any $n \in F$, the number $\varphi(n)/\pi$ is rational, so that $\varphi(n)$ is constant on any connected component of F. Let $F_1 \subset F$ be such a component and $\varphi(n) = \pi p/q < \pi$. For any boundary point $n_2 \in F_1$ the projection $V_1(n_2)$ can be obtained from $V_2(n_2)$ by rotation through the angle $\pi p/q$ and by a subsequent parallel translation or by a central symmetry, because $n_2 \in \varphi^{-1}(0) \cup \varphi^{-1}(\pi)$. But $n_2 \in C$; so, the projections $V_i(n_2)$ have constant width and rotation symmetry; hence, they are circles of the same radius and $n_2 \in \varphi^{-1}(0)$. So for $n_1 \in F_1$ every meridian of the unit sphere with ends at $\pm n_1$ intersects the set $\varphi^{-1}(0)$. Theorem 4 in the case $k = 3$ follows from Lemma 5 and the last part of the proof of Theorem 2. The higher-dimensional cases can be obtained by induction, as above. □

It can be shown that Theorems 2 and 3 cannot be generalized in the direction of Theorem 1; i.e., it is impossible to essentially decrease the set of projections under consideration. Let $\varepsilon > 0$ and let $D \subset S^2$ be the set of unit vectors $x \in S^2$ such that the angles between x and the vectors $\pm e_3$ are larger than ε. Let e_3 be the vertical vector $(0; 0; 1)$. On the equator $E(e_3)$ consider the points $N = (-0.6; 0.8; 0.0)$ and sequences $M_i^1 = (-\cos(a_i); -\sin(a_i); 0)$ $M_i^0 = (\cos(a_i); \sin(a_i); 0)$ where $i \geq 0$, $a_i \in (\pi/6; \pi/4)$, $a_i > a_{i+1}$. We denote by SM_i^0, SM_i^1 the shell segments which are symmetric pairwise and are obtained by intersection of the unit ball $B \subset \mathbb{R}^3$ by planes orthogonal to the vectors OM_i^0, OM_i^1, where $O = (0; 0; 0)$ is the zero point in $\mathbb{R}^3$. Let SN be an analogous segment with summit N. The sizes of all these segments must be so small that every circle $E(n)$ for all $n \in D$ intersects no more than one segment $SN, SM_0^1, \dots, SM_i^1, \dots$. All these segments are mutually disjoint and their volumes tend to zero as $i \to \infty$. It is obvious that if the circle $E(n)$ intersects SM_i^1 it intersects the segment SM_1^0 as well because of the pairwise symmetry of these segments.

Let g be a countable sequence of 0's and 1's indexed by the positive integers. It is well known that the cardinality of the set $G = \{g\}$ of all such sequences is continuum. Let us construct a convex compact set V_g as follows: cut off the segments SN, SM_0^1 and the segments SM_i^m, $i > 1$, from the unit ball B, where m equals 0 or 1 according to the value of the ith index of g. It is easy to see that all the compact sets $\{V_g\}$, $g \in G$, are mutually noncongruent, but their projections in the directions of the vectors belonging to D are mutually congruent: if for $n \in D$ the circle $E(n)$ intersects the segments SN, SM_0^1 or does not intersect any of the segments above, the projections $V_g(n)$ of all the V_g coincide. If $E(n)$ intersects the segments SM_j^0, SM_j^1 for some $j > 0$, the projections of all the $V_g(n)$ either coincide or are mutually centrally symmetric, depending on the jth index. Naturally the projections of these V_g in the directions of the vectors belonging to $S^2 \backslash D$ need not be congruent, but the measure of $S^2 \backslash D$ depends on ε and can be arbitrarily small. Also, it is easy to see that for all $n \in S^2$ the areas and the perimeters $S(V_g(n))$ and $L(V_g(n))$ of the projections $V_g(n)$ depend only on n and do not depend on g.

The Süss lemma and some of its generalizations obtained above can be extended to the infinite-dimensional case.

LEMMA 8. *Let V_1, V_2 be convex compact bodies in Hilbert space and suppose that the orthogonal projections of V_1, V_2 on any finite-dimensional plane are parallel. Then V_1, V_2 are parallel in the Hilbert space.*

The proof is based on the consideration of a sequence of finite ε, $\varepsilon/2$, $\varepsilon/4, \dots$ nets of compact sets V_i. Each of these finite nets lies in a corresponding finite-dimensional subspace of the Hilbert space.

Such an analogy of the Süss lemma allows us to generalize Theorems 2, 3, and 4 to Hilbert space. In all these generalizations one considers the projections of the compact convex bodies V_1, V_2 on all two-dimensional planes. The congruence or the similarity of such projections allows us to conclude that the initial convex bodies in the Hilbert space are parallel or homothetic.

4. The next direction of generalization of the Süss lemma is related to a class of compact sets more general than convex compact sets.

We say that a compactum $W \subset \mathbb{R}^k$ is q-convex if for any point $x \in \mathbb{R}^k \setminus W$ there exists a q-dimensional plane containing x and mutually disjoint with W **[11]**. An analogous definition can be given in the complex and the projective cases. In complex analysis the complement to a q-convex compactum is said to be $(k-q-1)$ linearly concave **[12]**.

We say that a compactum $W \subset \mathbb{R}^k$ is q-visible if any $(q-1)$-dimensional plane in $\mathbb{R}^k \setminus W$ is contained in some q-dimensional plane which is also contained in $\mathbb{R}^k \setminus W$. Here $0 < 1 \leq q < k$.

It is obvious that a connected $(k-1)$-visible compactum in $\mathbb{R}^k$ is convex, that the projection of a q-visible compactum on any hyperplane is $(q-1)$-visible, and the projections of $(k-2)$-visible compact set on any 3-dimensional plane in $\mathbb{R}^k$ are 1-visible, or, equivalently, 1-convex. The intersection of 1-visible compact sets is a 1-visible compact.

We shall describe some examples in $\mathbb{R}^3$.

1. Let $W = I^3 \setminus V$, where I^3 is the cube $-1 \leq x, y, z \leq 1$ and V is the vertical cylinder $x^2 + y^2 < 0.01$. W is homeomorphic to a torus. It is obvious that any point not contained in W lies on some straight line disjoint with W.

The standard smooth torus of revolution in $\mathbb{R}^3$ is not a 1-visible compactum because in its "hole" there are some points which cannot be seen outside of any projection of this torus. A. V. Kuz′minykh has constructed a smooth embedding of the torus in $\mathbb{R}^3$ which gives a 1-visible compact: consider the cube $ABCDA'B'C'D'$ and the closed polygonal path $ABB'C'D'DA$; let us smooth the angles of this path and consider a small tubular neighborhood of the smooth path obtained. This neighborhood is 1-visible.

One can construct in $\mathbb{R}^5$ a 2-convex compactum that is not 2-visible.

If two 1-visible compact sets in $\mathbb{R}^3$ have homeomorphic projections on any plane, it is not true that they are homeomorphic themselves. If V is as in the first example and V_1, V_2 are the cylinders $x^2 + z^2 < 0.01$ and $(x - 0.5)^2 + z^2 < 0.01$, then the 1-visible compact sets $W_1 = I^3 \setminus (V \cup V_1)$, $W_2 = I^3 \setminus (V \cup V_2)$ are not homeomorphic, because $V \cap V_2$ is empty and $V \cap V_1$ is not. However the projections of W_1 and W_2 are pairwise homeomorphic.

In contrast to the convex case the 1-visible compact sets do not satisfy the stability theorem. Indeed let $B \subset \mathbb{R}^3$ be a unit ball and let $W \subset B$ be obtained as follows: draw on the boundary ∂B a sufficiently dense spiral going from one pole to other, cut off from B a small neighborhood of this spiral using a "rectilinear knife" (as one does when peeling potatoes). If this neighborhood is sufficiently small, it is a 1-visible body and its distance from B in the Hausdorff metric is almost 1, although any of its projections is close to the corresponding projection of B.

THEOREM 5. *If $W_1, W_2 \subset \mathbb{R}^k$ are $(k-2)$-convex compact sets and their projections on any two-dimensional plane $\mathbb{R}^k$ are similar and the convex hulls of these projections have no rotation symmetries, then W_1 and W_2 are either homothetic or parallel in $\mathbb{R}^k$.*

PROOF. Since the convex hulls of W_1, W_2, $\operatorname{conv}(W_1)$, $\operatorname{conv}(W_2)$ have similar projections on any two-dimensional plane, it follows from Theorem 3 that one of them can be obtained from the other by a homothety or by a parallel translation. Let us make these convex hulls coincide. If the obtained W_1' does not coincide with W_2 we can find a point x in W_1' and not in W_2 (or vice versa). Consider a $(k-2)$-dimensional plane P containing x and disjoint with W_2. Let Q be an orthogonal complement of P in $\mathbb{R}^k$. The projections of the $(k-2)$-convex compact sets W_1', W_2 on Q are congruent and the convex hulls of these projections coincide. Therefore, one cannot obtain the projection $W_1'(Q)$ of W_1' from the corresponding projection of W_2 by a nonzero translation. Moreover, one cannot obtain $W_1'(Q)$ from $W_2(Q)$ by a rotation, because their convex hulls have no rotation symmetries. This contradicts the existence of the point x. □

THEOREM 6. *If $W_1, W_2 \subset \mathbb{R}^k$ are $(k-2)$-visible, simply connected compact sets and their projections on any two-dimensional plane are congruent and have no rotation symmetries then W_1, W_2 are either parallel or centrally symmetric in $\mathbb{R}^k$.*

PROOF. As above, we consider the case $k = 3$. As in Theorem 2, one can define a continuous function $\varphi(n)$ because the projections of W_1, W_2 have no rotation symmetries. It follows from the uniqueness of the solution of the integral equation (1) that the width of the convex hull $\operatorname{conv}(W_1)$ coincides with the corresponding width of $\operatorname{conv}(W_2)$ in any direction.

If $0 \neq \varphi(n_1) \neq \pi$ for some $n_1 \in S^2$, in any neighborhood of n_1 there are points where the values of the function $\varphi()/\pi$ are irrational. So the projections $\operatorname{conv}(W_1(n_1))$, $\operatorname{conv}(W_2(n_1))$ have constant width. For all n close to n_1 the projections $\operatorname{conv}(W_i(n))$ have also constant width, so for the projection in the direction n_1 the preimage of any point in the boundary $\partial(\operatorname{conv}(W_1(n_1)))$ or $\partial(\operatorname{conv}(W_2(n_1)))$ consists of a single point of the boundary $\partial(\operatorname{conv}(W_1))$ (resp. $\partial(\operatorname{conv}(W_2))$); if such a preimage contains an interval, then this interval can be seen on some projections $\operatorname{conv}(W_1(n))$, $\operatorname{conv}((W_2(n))$ for some n close to n_1 but these projections must have constant width. It follows from the uniqueness of these preimages that the preimages of the strictly convex boundaries $\partial(\operatorname{conv}(W_1(n_1)))$, $\partial(\operatorname{conv}(W_2(n_1)))$ are closed lines on the boundaries of the compact sets W_1, W_2. Since these compact sets are simply connected, the boundaries $\partial(\operatorname{conv}(W_i(n_1)))$, $i = 1, 2$, are contractible in the projections $W_i(n_1)$; hence the projections $W_i(n_1)$ are convex, have constant width, and have no rotation symmetries, $\operatorname{conv}(W_1(n_1)) = W_i(n_1)$. It follows from Theorem 2 that the convex hulls $\operatorname{conv}(W_1)$, $\operatorname{conv}(W_2)$ are either parallel or centrally symmetric, so the projection $W_1(n_1) = \operatorname{conv}(W_1(n_1))$ can be obtained from $W_2(n_1)$ by the same transformation, which contradicts the choice of n_1.

The inductive step of the proof for the higher-dimensional cases utilizes the $(q-1)$-visibility of the projection of a q-visible compactum on any hyperplane. □

The absence of rotation symmetries of the projections of $(k-2)$-convex or $(k-2)$-visible compact sets, which was supposed in Theorems 5 and 6, is essential:

Let $Q = I^3 \cup \Lambda$ be the union of the cube $I^3 \subset \mathbb{R}^3$ and the cone Λ with summit at $(-0.5; 0.5; 1.1)$ and base $z = 1$, $(x + 0.5)^2 + (y - 0.5)^2 = 0.01$. If

$$G_1 = \{(x, y, z) \mid [0.6 < x < 0.7;\ y > 0.8] \text{ or } [-0.7 < x < -0.6;\ z > 1.8 + y] \text{ or } [-0.7 < x < -0.6;\ z > -1.8 - y]\},$$

$$G_2 = \{(x, y, z) \mid [0.6 < z < 0.7;\ z > 1.8 - y] \text{ or } [0.6 < x < 0.7;\ z < y - 1.8] \text{ or } [-0.7 < x < -0.6;\ y < -0.8]\},$$

and $W_1 = Q \backslash G_1$, $W_2 = Q \backslash G_2$ (see Figure 1), then these W_1, W_2 are 1-visible and noncongruent, but their projections on which the cone Λ is seen and the sets $J = [0.6 < |x| < 0.7;\ |y| > 0.8]$ cannot be seen through coincide and for this case $\varphi(n) = 0$. Their projections where these J can be seen through are pairwise centrally symmetric, $\varphi(n) = \pi$. All their projections in the directions of the unit vectors (x, y, z) such that $|z| \in [0.2; 0.5]$ coincide and have central symmetry so for these vectors $0 \in \{\varphi(n)\}$, $\pi \in \{\varphi(n)\}$.

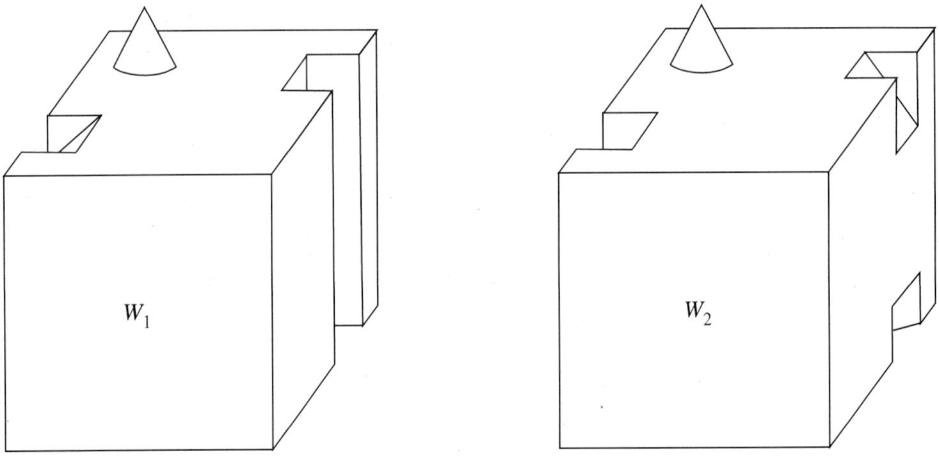

FIGURE 1

An analogous counterexample in the multiply connected case was constructed by A.V. Kuz′minykh.

Theorems 5 and 6 can be formulated in the infinite-dimensional case; let us call a compactum W in the Hilbert space $\mathscr{H}$ $(-k)$-convex if for any point $x \in \mathscr{H} \backslash W$ there exists a k-codimensional plane in $\mathscr{H}$ that contains x and does not intersect W.

Let us say that a compactum $W \subset \mathscr{H}$ is $(-k)$-visible if any plane in $\mathscr{H} \backslash W$ of codimension greater than k is contained in a plane of codimension k that does not intersect W.

THEOREM 7. *Let $W_1, W_2 \subset \mathscr{H}$ be (-2)-convex compact bodies and suppose that their projections on any two-dimensional plane are similar and that the convex hulls of these projections have no rotation symmetries. Then W_1, W_2 are either homothetic or parallel in $\mathscr{H}$.*

The theorem follows from Lemma 8 and Theorem 5 applied to the convex hulls $\operatorname{conv}(W_1)$ and $\operatorname{conv}(W_2)$.

An analogous formulation exists for the Hilbert space variant of Theorem 6.

References

1. A. V. Pogorelov, *Extrinsic geometry of convex surfaces*, Transl. Math. Monographs, vol. 35, Amer. Math. Soc., Providence, RI, 1973.
2. Yu. E. Anikonov, *A uniqueness theorem for convex surfaces*, Mat. Zametki **6** (1969), 115–117; English transl. in Math. Notes **6** (1969).
3. S. Campi, *Reconstructing a convex surface from certain measurements of its projections*, Boll. Un. Mat. Ital. B(6) **5** (1986), no. 3, 945–959.
4. A. D. Aleksandrov, *Zur Theorie der gemischten Volumina von konvexen Körpern*, Recuell Math. **2** (1937), no. 6, 1205–1235.
5. V. P. Golubyatnikov, *On the unique determination of visible bodies from their projections*, Siberian Math. J. **29** (1988), 761–764.
6. ———, *On reconstructing the shape of a body from its projections*, Soviet Math. Dokl. **25** (1982), 62–63.
7. W. Süss, *Zusammensetzung von Eikörpern und homotetische Eiflächen*, Tôhoku Math. J. **35** (1932), 47–50.
8. A. A. Kirillov, *A problem of I. M. Gel′fand*, Soviet Math. Dokl. **2** (1961), 268–269.
9. W. Blaschke, *Kreis und Kugel*, de Gruyter, Berlin, 1956.
10. H. Minkowski, *Über die Körper konstanter Breite*, Gesammelten Werke, Vol. 2, Teubner, Leipzig and Berlin, 1911, pp. 277–279.
11. Yu. B. Zelinskiĭ, *The generalized convexity in the complex analysis*, Abstracts, Baku International Topological Conference. Part 2. (Russian)
12. S. G. Gindikin and G. M. Henkin, *Transformation de Radon pour la d''-cohomologie des domaines q-linéairement concaves*, C. R. Acad. Sci. Paris Sér. A **287** (1978), 209–212.

Amer. Math. Soc. Transl.
(2) Vol. **163**, 1995

On Subalgebras of Maximal Rank of Semisimple Lie Algebras

P. Ya. Grushko and L. A. Osipenko

§ 1. Introduction

A. I. Mal′cev is the founder of an important branch in the theory of Lie algebras. The results of Mal′cev **[5]** on semisimple subalgebras have been further developed in papers of E. B. Dynkin, A. L. Onishchik, F. I. Karpelevich, and others.

Let L be a semisimple Lie algebra, $L_0 \subset L$ a subalgebra of maximal rank. Then the adjoint representation of L_0 on L induces the isotropy representation $\varepsilon\colon L_0 \longrightarrow \operatorname{End}(V')$ of L_0 on $V' = L/L_0$. We state the problem of describing all the cases when the image $\varepsilon(L_0)$ in $\operatorname{End}(V')$ may be extended to a Lie subalgebra G acting irreducibly on V'.

This problem is the infinitesimal version of a problem on a certain class of homogeneous G-structures and we use the name "G structure algebra" for our object (L, L_0, G, V').

The case when V' is the standard representation of a classical Lie algebra is well known **[6, 7, 8]**. There are rather weak restrictions on the pairs (L, L_0) in this case. In our paper we suppose that V' is not the standard representation of a classical Lie algebra, so there is the complete list of all possibilities. Some particular cases of this problem were studied in **[10, 11, 12, 13]**.

First we deal with algebras over the complex number field. There is a Cartan subalgebra $H \subset L_0 \subset L$ such that $L_0 = H + \sum_{\alpha\in\Sigma_0} \mathbb{C}\, e_\alpha$, where Σ_0 is a closed subsystem of the root system Σ of L. We can reformulate our problem as follows. Give an irreducible representation V of a structure Lie algebra G (which we call the structure representation), a linear isomorphism $l\colon V' \to V$, and a Lie algebra homomorphism $k\colon L_0 \to G$ such that

$$lgv = kg \cdot lv, \qquad g \in L_0, \quad v \in V'. \tag{1}$$

In our case either the center Z of G consists of scalar operators or $Z = 0$. Obviously, if we can find all our objects when $Z = 0$, then we have the easy task to determine all cases when it is possible to remove the center. If $\operatorname{Tr} k(L_0) = 0$ (i.e., $\operatorname{Tr} A = 0$ for each $A \in k(L_0)$), we list only the semisimple groups G. The other algebras G must be constructed by adding a scalar center.

Further, if the semisimple component P of G is not simple we first study in §2 the special case of $G = \widetilde{G} = Z + sl(n_1, \mathbb{C}) + sl(n_2, \mathbb{C})$, $V = \mathbb{C}^{n_1} \otimes \mathbb{C}^{n_2}$. In fact,

1991 *Mathematics Subject Classification*. Primary 17B20.

if $G = Z + \sum_{i=1}^{r} P_i$, $V = \bigotimes_{i=1}^{r} W_i$, where W_i is an irreducible representation of the simple Lie algebra P_i and $k(L_0) \subset G$, then we put $\widetilde{W}_1 = \bigotimes_{i=1}^{r-1} W_i$; $\widetilde{W}_2 = W_r$, $\widetilde{G} = Z + sl(m_1, \mathbb{C}) + sl(m_2, \mathbb{C})$, where $m_i = \dim \widetilde{W}_i$. Then $k(L_0) \subset \widetilde{G}$. Once we have found all the pairs (L, L_0) corresponding to $\widetilde{G} = Z + sl(n_1, \mathbb{C}) + sl(n_2, \mathbb{C})$ it is sufficient to list all irreducible subalgebras G such that $k(L_0) \subset G \subset \widetilde{G}$.

Further, there is the Cartan subalgebra $HG \subset G$ such that $k(H) \subset HG$. So the restrictions to $k(H) \subset HG$ of the weights of the structure representation of G on V are weights of the representation of L_0 on V', i.e., made up of the roots of L in a subsystem $Q = \Sigma \backslash \Sigma_0$.

In particular, there are no multiple weights. So any linear dependence among the weights of V induces the same linear dependences on the roots of the subsystem Q. By this argument it is easy to prove that L is simple. In our paper we use the function $d(\lambda, \mu)$ which is the number of ways of writing a sum of two weights of the given representation and the analogous function $r(\lambda, \mu)$ for the difference of weights. We use the same function for the roots of L. For example, if $L = A_l$ then for any roots α, β we have

$\langle\alpha, \beta\rangle$	$d(\alpha, \beta)$	$r(\alpha, \beta)$
0	2	4
1	1	$2(l-1)$
-1	$l-1$	2
2	1	$(l+1)l/2$
-2	$(l+1)l/2$	1

where $\langle\ ,\ \rangle$ denotes the Cartan-Killing form. We denote by V_Λ the irreducible representation corresponding to the highest weight $\Lambda = \sum_{i=1}^{m} c_i w_i$, where the w_i are the fundamental weights. Using the notations of Bourbaki **[1]** for the basic vectors of the ε_i space associated to Cartan subalgebra $H \subset L$, we denote by δ_i the analogous vectors of the Cartan subalgebra $HG \subset G$. For example, $\{\varepsilon_i - \varepsilon_j\}$ are the roots of $L = A_l$ and $\frac{1}{2} \sum_{i=1}^{m} (\pm\delta_i)$ are the weights of the spin representation of $G = B_m$. Finally, $\mathscr{V} = \sum_{\beta \in Q} \mathbb{C} e_\beta$ is the canonical complement of L_0 in L. The main result of our paper is the complete list of $\{L, L_0, G, V\}$ over $\mathbb{C}$. Most of the pairs (L, L_0) are symmetric; i.e., $L_0 = \{x \in L : fx = x\}$ where f is an involutive automorphism of L. But the most interesting cases occur when L_0 is not a reductive subalgebra of L and $L_0 = G$. We consider our problem also in the category of real algebras provided the structure representation is absolutely irreducible.

§ 2. The case of tensor representations

First we consider a particular case of our problem.

Let $V = W_1 \otimes W_2$ and $G = sl(n_1, \mathbb{C}) + sl(n_2, \mathbb{C}) + Z$, where Z is the one-dimensional scalar center of G, $W_1 = \mathbb{C}^{n_1}$, $W_2 = \mathbb{C}^{n_2}$. We can suppose that $n_1 \geq n_2$.

If $\{\lambda_i\}$ are the weights of the $sl(n_1, \mathbb{C})$-module W_1 and $\{\mu_j\}$ are the weights of the $sl(n_2, \mathbb{C})$-module W_2, then $\{\lambda_i + \mu_j + z\}$ are the weights of the G-module

$W_1 \otimes W_2$. The restriction of the weights $\lambda_i + \mu_j + z$ to H are all different and their union forms the set of roots $Q \subset \Sigma$.

LEMMA 1.2. *The Lie algebra L has at most two components. If the number of the components equals two, then* $n_2 = 2$.

PROOF. Let $L = L_1 + L_2 + \cdots + L_N$, where the L_i are simple Lie algebras. Assume that $L_1 \cap \mathscr{V}$ has a minimal dimension among $L_i \cap \mathscr{V}$. Let

$$(\lambda_1 + \mu_1 + z)|_H \subset \Sigma_1 = \Sigma(L_1).$$

If $\dim(L_1 \cap \mathscr{V}) - 1 < (n_1 - 1)(n_2 - 1)$, there exists a weight $(\lambda_i + \mu_j + z)$, $i, j \neq 1$, such that $(\lambda_i + \mu_j + z)|_H \in \Sigma_k = \Sigma(L_k)$, $k \neq 1$, and $(\lambda_1 + \mu_1 + z)|_H + (\lambda_i + \mu_j + z)|_H = (\lambda_1 + \mu_j + z)|_H + (\lambda_i + \mu_1 + z)|_H$. This is impossible. Hence,

$$\dim(L_1 \cap \mathscr{V}) > (n_1 - 1)(n_2 - 1) + 1.$$

On the other hand,

$$\dim(L_1 \cap \mathscr{V}) \leq \frac{1}{N} \dim V = \frac{n_1 n_2}{N}.$$

This inequality is true only if $N = 2$ and $n_2 = 2$. □

LEMMA 2.2. *If* $n_2 = 2$ *and L is not simple, then* $L = 2A_1$, $L_0 = H$.

The proof is in **[9]**.

Since $\dim V = 4$ for $L = 2A_1$, for every algebra G this representation is standard. Thus, things can be confined to simple algebras L.

Now let L be a simple classical algebra. We denote by $\{\varepsilon_i\}$ the standard basis in $H^\#$, where $H^\# = H$ if $L \neq A_l$, and H is hyperplane in $H^\#$ if $L = A_l$. The sets

$$S_i = \{(\lambda_i + \mu_j + z)|_H,\ j = 1, \ldots, n_2\} \in \Sigma$$

will be called configurations.

Evidently, Q is the union of n_1 nonintersecting configurations.

The configurations $S = \{a + x_1, a + x_2, \ldots, a + x_m\}$ (the "star"), $S = \{a, a + x_1, a + x_2, \ldots, a + x_m\}$ (the "star with vertex"), $S = \{x_1, x_2, \ldots, x_m\}$ (the "constellation"), $S = \{2a, a + x_1, a + x_2, \ldots, a + x_m\}$ (the "star with double vertex"), $S = \{x_1 + x_2, x_1 + x_3, x_2 + x_3\}$ (the "triangle"), $S = \{x_1, x_2, x_1 + x_2\}$ (the "dumb-bell"), where $a \neq x_k$, $a, x_k \in \{\pm\varepsilon_i\}$ will be called regular; the other configurations will be called exceptional.

If S is the "star" a will be called the vertex and x_k will be called the extremity. Note that for every indices i and j we have $S_i = S_j + f$, where $f = (\lambda_i - \lambda_j)|_H$. Moreover $f = a + b + c + d$ or $f = a + b + c$, where $a, b, c, d \in \{\pm\varepsilon_i\}$.

LEMMA 3.2. *If* $n > 2$, *then all configurations are regular.*

PROOF. Let $L \neq B_l$. Then a root can be written as $x + y$, $x, y \in \{\pm\varepsilon_i\}$. Assume that $S = \{x_i + y_i\}$, $1 \leq i \leq n_2$, is exceptional and $n_2 \geq 4$. If $\overline{S}$ is another configuration, then $\overline{S} = \{x_i + y_i + f\}$, where $f = a + b + c + d$ or $f = a + b + c$.

Let f be four-term. Then it is easy to see that either $S = \{-a - b, -a - c, -a - d, -b - c, -b - d, -c - d\}$, $n_2 = 6$, or $S = \{-a - c, -a - d, -b - c, -b - d, -c - d\}$, $n_2 = 5$. The corresponding $\overline{S}$ are $\{c + d, b + d, b + c, a + d, a + c, a + b\}$ and $\{b + d, b + c, a + d, a + c, a + b\}$.

Moreover, is no other f such that $S + f$ is a set of roots. Hence, $n_1 \leq 2$. This contradicts the assumption that $n_1 \geq n_2$.

TABLE 1

L	L_0	n_1	n_2
D_{p+q}	$D_p + D_q$	$2p$	$2q$
C_{p+q}	$C_p + C_q$	$2p$	$2q$
B_{p+q}	$D_p + B_q$	$2p$	$2q+1$
D_{p+2}	$D_p + 2A_1$	$4p$	2
B_{p+2}	$B_p + 2A_1$	$4p+2$	2
A_{p+q+1}	$A_p + A_q + C + N$	$p+1$	$q+1$
C_{p+q+1}	$C_p + C_q + C + N$	$2p+1$	$2q+1$

If f consists of two terms, then also $n_1 \geq 2$.

Now let $n_2 \leq 4$. Then $n_2 = 3$ and there are four configurations (up to transpositions of a, b, c, d): $\{-a-b, -b-c, -c-d\}$; $\{c+d, a+d, a+b,\}$; $\{-a+d, d-c, b-c\}$; $\{c-b, a-b, a-d\}$. Their union forms the system Q. In addition, if $L = A_l$, then either $x+y$ or $x-y$ is not a root. Hence $n_1 = 2$, but $n_1 \geq n_2$.

If $L = D_l$ or $L = C_l$, then $a-c, -a+b \in \Sigma_0$ and $b-c \in Q$. Hence, Σ_0 is not closed.

Now let $L = B_l$. Assume that there exists $f = a+b+c$ such that $S + f \subset \Sigma$. If there exists a four-term $\widetilde{f}$ such that $S + \widetilde{f} \subset \Sigma$, then $S = \{-a-b, -a-c, -b-c\}$. So S is regular. If there exists a three-term $\widetilde{f}$, then S is the "dumb-bell". If there exists a two-term $\widetilde{f}$, then $n_1 = 2$, which is impossible.

All other cases are considered similarly. □

We denote by C the one-dimensional center and by N the nilradical of L_0. We have

THEOREM 1.2. *Let L be a classical Lie algebra. Then all possible L, L_0, n_1, n_2 are listed in Table* 1.

The structure of the nilradical is described by (2) *and* (3).

PROOF. Assume first that $n_2 > 2$.

Let $L = A_l$. By Lemma 3.2 all configurations are "stars". We have (up to isomorphism) $(\lambda_i + \mu_j + z)|_H = \varepsilon_i - \varepsilon_{n_1+j}$, $1 \leq i \leq n_1$, $1 \leq j \leq n_2$. Since Σ_0 is closed, $n_1 + n_2 = l+1$. Hence, $Q = \{\varepsilon_i - \varepsilon_k\}$, $\Sigma_0 = \{\pm(\varepsilon_i - \varepsilon_j);\ \pm(\varepsilon_r - \varepsilon_k);\ \varepsilon_k - \varepsilon_r\}$, $1 \leq i, j \leq n_1$, $n_1 + 1 \leq k,\ r \leq l+1$. Thus $L = A_p + A_q + C + N$, where $p = n_1 - 1$, $q = n_2 - 1$.

$$(2) \qquad N = \sum_{\alpha \in R} \mathbb{C} e_\alpha, \qquad R = \{\varepsilon_k - \varepsilon_i\}.$$

Evidently, $\ker k = \ker \varepsilon = N$.

Let $L = B_l$. Assume that there are "triangles" among the configurations. It is easy to see that all other configurations are "stars" or "constellations". If there are no "constellations" then Σ_0 contains all the short roots. This implies that $\Sigma_0 = \Sigma$. Hence there is a "constellation". We have $S_1 = \{\varepsilon_1 + \varepsilon_2; \varepsilon_1 + \varepsilon_3; \varepsilon_2 + \varepsilon_3\}$, $S_2 = \{-\varepsilon_1, -\varepsilon_2, -\varepsilon_3\}$, $S_t = \{x_t - \varepsilon_1; x_t - \varepsilon_2; x_t - \varepsilon_3\}$, where $x_t \in \{\pm\varepsilon_4, \pm\varepsilon_5, \dots, \pm\varepsilon_l\}$. Since $\varepsilon_1, \varepsilon_2 \in \Sigma_0$ and $\varepsilon_1 + \varepsilon_2 \in Q$, Σ_0 is not closed.

Thus there are no "triangles" among the configurations.

By the same reason there are no "dumb-bells".

Assume that all configurations are "stars with vertices". If $\mathfrak{M}$ is the set of the vertices and $\mathfrak{N}$ is the set of extremities, then we easily see that $\mathfrak{M} = \{\pm\varepsilon_1; \pm\varepsilon_2; \ldots; \pm\varepsilon_p\}$, $\mathfrak{N} = \{\pm\varepsilon_{p+1}, \ldots, \pm\varepsilon_l\}$. Thus, $L = D_p + B_q$, $n_1 = 2p$, $n_2 = 2q+1$, where $p+q=l$.

Assuming that there is a "constellation" among the configurations, we have $L_0 = B_p + D_q$, $n_1 = 2p+1$, $n_2 = 2q$, where $p+q=l$.

The case of $L = D_l$ is considered in a similar manner as $L = B_l$. We have $L = D_{p+q}$, $L_0 = D_p + D_q$, $n_1 = 2p$, $n_2 = 2q$, where $p+q=l$.

Now let $L = C_l$. If all configurations are "stars," then $\pm 2\varepsilon_i \in \Sigma_0$, $i = 1, \ldots, l$, and we have $L_0 = C_p + C_q$, $n_1 = 2p$, $n_2 = 2q$, where $p+q=l$.

If there is a "star with double vertex," then we easily see that all other configurations are "stars". This implies that $Q = \{2\varepsilon_1; \varepsilon_1 \pm \varepsilon_i; \pm\varepsilon_k \pm \varepsilon_j\}$, $2 \le i \le l$, $2 \le k \le \frac{n_2+1}{2}$; $\frac{n_2+3}{2} \le j \le l$. Hence $L_0 = C_p + C_q + C + N$, $n_1 = 2p+1$, $n_2 = 2q+1$, where

$$N = \sum_{\alpha \in R} \mathbb{C}\, e_\alpha, \qquad R = \{-2\varepsilon_1; -\varepsilon_1 \pm \varepsilon_i\}. \tag{3}$$

Evidently $\ker k = \ker \varepsilon = \mathbb{C}\, e_{-2\varepsilon_1}$.

In addition, $k(L_0)$ can be written as follows:

$$\begin{pmatrix} h & 0 & 0 \\ a & A & C_1 \\ c & C_4 & -S_1^T A S_1 \end{pmatrix} \otimes \begin{pmatrix} h & 0 & 0 \\ b & B & C_2 \\ d & C_3 & -S_2^T B S_2 \end{pmatrix},$$

where S_1 is the square matrix of order p with all elements equal to zero except the elements of the secondary diagonal which are equal to 1. The matrix S_2 has a similar structure but is of order q **[1]**. A, C_1, C_4 are square matrices of order p, and B, C_2, C_3 are square matrices of order q; moreover,

$$C_k = S_1^T C_k S_1, \quad C_r = S_2^T C_r S_2, \quad k = 1, 4, \ r = 2, 3$$
$$a = {}^T(a_1, a_2, \ldots, a_p), \quad d = {}^T(d_1, d_2, \ldots, d_p);$$
$$b = {}^T(b_1, b_2, \ldots, b_q), \quad c = {}^T(c_1, c_2, \ldots, c_q).$$

It is easy to prove that (1) is true.

Thus the theorem is proved for $n_2 > 2$.

Let $n_2 = 2$. Using the identity

$$(\lambda_1 + \mu_1 + z) - (\lambda_1 + \mu_2 + z) = \big[(\lambda_i + \mu_1 + z) - (\lambda_i + \mu_2 + z)\big], \ i = 2, \ldots, n_1,$$

we find the subsystem Q. Observing this, we obtain that (2) and (3) are also true for $n_2 = 2$.

Moreover, by virtue of $D_2 = 2A_1$, we obtain two more pairs: $L = D_{p+2}$, $L_0 = D_p + 2A_1$ and $L = B_{p+2}$, $L_0 = B_p + 2A_1$. □

Now let us consider the case of an exceptional Lie algebra L.

It is proved in **[13]** that there is no exceptional L such that L_0 is nonreductive.

Let L_0 be a reductive subalgebra of L.

LEMMA 4.2. *Let $L = E_i$, $i = 6, 7, 8$. The following inequalities are true*: $n_1 \ge 14$ *if* $L = E_8$; $n_1 \ge 10$ *if* $L = E_7$; $n_1 \ge 7$ *if* $L = E_6$.

TABLE 2

L	L_0	n_1	n_2
E_8	E_7+A_1	56	2
E_7	D_6+A_1	32	2
E_6	A_5+A_1	20	2
F_4	C_3+A_1	14	2
G_2	$2A_1$	4	2
G_2	A_1+C	5	2

PROOF. Let $L=E_8$. If $n_1\leq 11$ and $n_2\leq 10$, then $\dim V=121$ and $\dim L_0=127$. There is no algebra of rank 8 of such dimension. If $n_2\leq 10$, then $\dim L_0\leq 138$, and this is impossible. If $n_1=n_2=12$, then $\dim L_0=108$, and hence $L_0=2F_4$. But it is impossible to insert F_4 into $sl(12,\mathbb{C})$. If $n_2<12$, then we obtain inconsistent values of $\dim L_0$.

If $n_1=13$, then $L_0=D_7+C$, but it is impossible to insert D_7 into $sl(n,\mathbb{C})$ if $n\leq 13$. Thus $n_1\geq 14$.

If $L=E_6, E_7$ the proof proceeds in the same way as that for $L=E_8$. □

DEFINITION. The roots $\pm\varepsilon_i\pm\varepsilon_j$ will be called "edges"; the roots $\frac{1}{2}\sum\limits_{i=1}^{8}(\pm\varepsilon_i)$ will be called "tails". The set of roots Σ is called regular if:

a) two "edges" have a common vertex with the same signs and the remaining extremities belong to different vertices;
b) a "tail" and an "edge" have the same signs at common vertices;
c) two "tails" have two changes of signs.

LEMMA 5.2. *All configurations are regular. Each configuration contains either only "edges" or only "tails".*

PROOF. The proof proceeds by ordinary selection of configurations, using the function r.

Assume that all configurations consist of "tails". Then all the "edges" belong to Σ_0 and we obtain $L=E_7$, $L_0=D_6+A_1$, $n_1=32$, $n_2=2$.

If all configurations consist of "edges" then it is easy to see that there are two more pairs (L, L_0). They are given in Table 2.

The case $L=F_4$ is considered similarly. □

Thus, we can formulate

THEOREM 2.2. *Let L be an exceptional Lie algebra. Then all possible L, L_0, n_1, n_2 are listed in Table* 2.

The subalgebra A_1 is given by the long root. We have $k(A_1)\subset sl(5)$; moreover, the $k(A_1)$-module $\mathbb{C}^5$ is isomorphic to $\mathbb{C}^2+\mathbb{C}^{2*}+\mathbb{C}^1$, $k(\mathbb{C})\subset sl(2)$.

§3. The case of second alternation

In this section we consider the case $V=\Lambda^2\mathbb{C}^n$, $G=gl(n,\mathbb{C})$. Since the representation of $sl(4,\mathbb{C})$ in $V=\Lambda^2\mathbb{C}^4$ is isomorphic to the standard representation of $SO(6,\mathbb{C})$, we can suppose that $n\geq 5$.

If $\{\delta_i\}$ are the weights of the $sl(n_1,\mathbb{C})$-module V then $(\delta_i+\delta_j+z)$, $1\leq i<j\leq n$, are the weights of the G-module V.

LEMMA 1.3. *L is a simple Lie algebra.*

The proof proceeds in the same way as that of Lemma 1.2.

Throughout this section a set $S_i = \{(\delta_i + \delta_j + z)|_H;\ 1 \le j \le n;\ j \ne i\} \subset Q$ will be called a configuration. Note that in this case every two configurations have nonempty intersection.

Let $\alpha, \beta \in S_i$ for certain i. Then it is easy to see that the following lemma holds.

LEMMA 2.3. *If $n \ge 6$, then the following roots cannot belong to the same configuration: a) $\pm\varepsilon_i;\ \pm\varepsilon_j \pm \varepsilon_k$, $i \ne j \ne k$ (if $L = B_l$); b) $\pm 2\varepsilon_i;\ \pm\varepsilon_j \pm \varepsilon_k$, $i \ne j \ne k$ (if $L = C_l$); c) $\pm\varepsilon_i \pm \varepsilon_j;\ \pm\varepsilon_k \pm \varepsilon_r$, $i \ne j \ne k \ne r$.*

PROOF. Let $(\delta_1 + \delta_2 + z)|_H = \varepsilon_i$, $(\delta_1 + \delta_3 + z)|_H = \varepsilon_j + \varepsilon_k$, $i \ne j \ne k$. Since $r(\varepsilon_i; \varepsilon_j + \varepsilon_k) = 6$, we have ε_i, $\varepsilon_i + \varepsilon_j$ cannot belong to the same configuration if $n > 8$. If $n = 6, 7$ the proof proceed by ordinary selection of configurations.

The cases b) and c) are considered similarly. □

Let $L = A_l$. Applying Lemma 2.3 we see that the roots $\varepsilon_i - \varepsilon_j$, $\varepsilon_k - \varepsilon_r$, $i \ne j \ne k \ne r$ cannot belong to the same configuration if $n \ge 5$. Thus we have, up to isomorphism, $(\delta_1 + \delta_j + z)|_H = \varepsilon_1 - \varepsilon_j$, $j = 2, \ldots, n$. Therefore, $(\delta_i + \delta_j + z)|_H = -\varepsilon_i - \varepsilon_j$, $i, j = 2, \ldots, n$. But $-\varepsilon_i - \varepsilon_j$ is not a root.

Let $L = B_l$, $l \ge 2$. If all the short roots belong to Σ_0, then $\Sigma_0 = \Sigma$. If $Q = \{\pm\varepsilon_i\}$, $i = 1, \ldots, n$, then $\dim V = 2l$. This is is possible only if $n = 5$. Thus if $n \ge 6$, Lemma 1.3 implies that $(\delta_1 + \delta_i + z)|_H = \varepsilon_1 + \varepsilon_i$, $i = 2, \ldots, n-1$, $(\delta_1 + \delta_n + z)|_H = \varepsilon_1$. Hence $Q = \{\varepsilon_i; \varepsilon_j + \varepsilon_k\}$, $i = 1, \ldots, l$, $1 \le j < k \le l$. This implies that

(4) $$L_0 = A_{l-1} + C + N, \quad N = \sum_{\alpha \in R} \mathbb{C}\, e_\alpha; \quad R = \{-\varepsilon_i; -\varepsilon_j - \varepsilon_k\}$$

If $n = 5$ then a straightforward calculation shows that $L_0 = A_3 + C + N$ and $V = \Lambda^2\mathbb{C}^5$. This is a special case of (3).

The cases $L = C_l, D_l$ are considered similarly. We obtain $L = D_l$,

(5) $$L_0 = A_{l-1} + C + N;\ N = \sum_{\alpha \in R} \mathbb{C}\, e_\alpha;\ R = \{-\varepsilon_i - \varepsilon_j\}.$$

The cases of exceptional L are considered in a similar manner **[12]**. There are no pairs (L, L_0) in this case.

Thus, we can formulate

THEOREM 1.3. *If $G = gl(n, \mathbb{C})$, $V = \Lambda^2\mathbb{C}^n$, then either $L = B_l$, $L_0 = A_{p-1} + C + N$ or $L = D_l$, $L_0 = A_{l-1} + C + N$.*

The structure of the nilradical is described by (4) and (5).

§4. The general case

We adopt the restriction that if $\operatorname{Tr} k(L_0) = 0$, then we look for semisimple G only.

1. Let the semisimple component P of G be not simple. Then for each L, L_0, n_1, n_2 described in §2 it is sufficient to take all the irreducible subalgebras G such that $k(L_0) \subset G \subset Z + sl(n_1, \mathbb{C}) + sl(n_2, \mathbb{C})$. Obviously, if $L = sl(p+q, \mathbb{C})$, $L_0 = sl(p, \mathbb{C}) + sl(q, \mathbb{C}) + Z + \mathbb{C}^{p*} \otimes \mathbb{C}^{q*}$ then $G = k(L_0) = sl(p, \mathbb{C}) + sl(q, \mathbb{C}) + Z$.

If $L = G_2$, $L_0 = gl(2,\mathbb{C})$ then $\operatorname{Tr} K(L_0) = 0$ and we have two semisimple groups $G = sl(5,\mathbb{C}) + sl(2,\mathbb{C})$ and $S = so(5,\mathbb{C}) + sl(2,\mathbb{C})$.

If $L_0 = X + A_1$, where X is a simple subalgebra and $V = \mathbb{C}^m \otimes \mathbb{C}^2$ then we must take $G = so(m,\mathbb{C}) + sl(2,\mathbb{C})$ if the representation $\mathbb{C}^m$ of X is orthogonal and $G = sp(m/2,\mathbb{C}) + sl(2,\mathbb{C})$ if $\mathbb{C}^m$ is symplectic. If the isotropy representation is irreducible, then we must also take $G = L_0$, etc.

2. Suppose $G = Z + P$ where P is a simple algebra. There are no multiple weights. So it is sufficient to consider the fundamental representations of P and $V = S^k\mathbb{C}^n$ for $G = gl(n,\mathbb{C})$, $k \geq 2$, and some representations of exceptional algebras.

Further, since $L_0 \supset H$ the inequality $\dim(L/L_0) \leq \frac{15}{4}(\operatorname{rank} L)^2$ is true for any semisimple L. If every simple component of L is classical, then we have $\dim(L/L_0) \leq 2(\operatorname{rank} L)^2$. For any parabolic subalgebra L_0 we have $\dim(L/L_0) \leq \frac{15}{8}(\operatorname{rank} L)^2$ and $\dim(L/L_0) \leq (\operatorname{rank} L)^2$ for a classical algebra L. These inequalities will be used in our computations. Further, let α be a root of P and let $\Lambda, \Lambda - \alpha, \dots, \Lambda - k\alpha$ be weights. Then their restriction to $H \subset HG$ is a set of roots $\gamma_i = \beta - it$ where β is a root, $0 \leq i \leq k$. This is possible only if $k \leq 3$; moreover, $k = 3$ if and only if $L = G_2$. Using these ideas, let us consider every simple Lie algebra P.

LEMMA 1.4. *If $P = B_m$ and V is the spin representation, then $m = 4$, $L = F_4$, $L_0 = B_4$.*

PROOF. If $m = 2$, then V_{ω_m} is the standard representation of C_2, so $m \geq 3$. Suppose $G + B_m + Z$. Then the weights are $Z + \frac{1}{2}\sum_{i=1}^{m} C_i\delta_i$, $C_i = \pm 1$. Write $\Delta = \frac{1}{2}\sum_{i=1}^{m}\delta_i$. If $\lambda + \widetilde{\lambda} \neq 0$, then $r(\lambda, \widetilde{\lambda}) > 1$ or $d(\lambda, \widetilde{\lambda}) > 1$, so L is simple. Let L_0 be not reductive. Then there is a root $\alpha \in Q$ such that $-\alpha \overline{\in} Q$. Let $\alpha = (Z + \Delta)|_H$. Then $\beta = (Z - \Delta)|_H$ is a root such that $\beta \in Q$, $\beta \neq -\alpha$. Since $d(Z + \Delta, Z - \Delta) = 2^{m-1}$, we have

$$d(\alpha, \beta) \geq 2^{m-1}, \qquad l \leq m + 1. \tag{6}$$

If L is classical, (6) is true only if $m = 3$. But if $m = 3$, then $\Sigma\backslash Q$ is not a closed subsystem. Similarly, (6) is not true if $L = G_2, E_6, E_7, E_8$. If $L = F_4$ then $m \leq 5$. Since there is no subalgebra such that $\dim L_0 = 44$, we have $m > 3$. If $m = 4$ we get the reductive subalgebra $L_0 = B_4$. If $m = 5$, then condition (6) is not true. Thus L_0 is reductive, so $k(L_0) \subset B_m$ and we take $G = B_m$. Now $l \leq m$. If λ, μ are weights such that there is one change of sign, then $r(\lambda, \mu) = 2^{m-1}$. For classical L we have $r(\alpha, \beta) \leq 4(l - 2)$ if $\alpha \neq \beta$. So

$$2^{m-1} \leq 4(l - 2), \qquad l \leq m. \tag{7}$$

This inequality holds only if $m = 3$ or 4. If $m = 4$ then $l = 4$ and $L = B_4$ or C_4. But there are no reductive subalgebras such that $\dim L_0 = 20$. Similar arguments are used in other cases. The cases $L = E_7, E_8$ are removed by dimension arguments. If $L = E_6$ then $r(\alpha, \beta) \leq 20$, $\alpha \neq \beta$, $m \leq 6$ and (7) is not true. If $L = F_4$, then $m = 4$ or 5. If $m = 5$, then $\dim L_0 = 20$, but this is impossible. If $m = 4$, then we get the symmetric pair (F_4, B_4). Finally, suppose $L = G_2$. Then $m = 3$ and $L_0 = 2A_1$. Since the isotropy representation of $L_0 = 2A_1$ on L/L_0 is irreducible, the same would hold for the restriction of structure representation to $2A_1 \subset G$, but this contradicts a result of Dynkin [**4**]. □

LEMMA 2.4. *If* $P = D_m$, *V is spin representation, then* $\{L, L_0, G, V\}$ *are* $\{F_4, B_4, B_4, \mathbb{C}^{16}\}$; $\{F_4, B_4, D_5, \mathbb{C}^{16}\}$; $\{E_8, D_8, D_8, \mathbb{C}^{128}\}$; $\{E_6, D_5 + C + N, D_5 + C, \mathbb{C}^{16}\}$; $\{E_6, D_5 + C, D_6, \mathbb{C}^{32}\}$.

The proof proceeds in the same way as for Lemma 1.4.

LEMMA 3.4. *If* $P = sl(m, \mathbb{C})$, $V = S^k\mathbb{C}^m$, $k \geq 2$ *then* $k = 2$, $G = gl(m, \mathbb{C})$, $V = S^2\mathbb{C}^m$, $L = C_m$, $L_0 = gl(m) + N$, *where* $N \simeq S^2(\mathbb{C}^m)^*$ *is the nilradical of* L_0.

PROOF. If $k > 3$, then there are $\beta \in \Sigma$ and $t \in H$ such that $\beta, \beta - t, \beta - 2t, \beta - 3t, \beta - 4t$ are roots, but that is false. If $k = 3$, then we have two weights $\lambda_1 = Z + 3\delta_1$, $\lambda_2 = Z + 3\delta_2$ such that $(2\lambda_1 + \lambda_2)/3$, $(\lambda_1 + 2\lambda_2)/3$ are also weights. The corresponding property for roots is possible only if $L = G_2$. In this case $m = 2$ and Q contains four roots, so the system $\Sigma \backslash Q$ is not closed.

Let $G = gl(m, \mathbb{C})$, $V = S^2\mathbb{C}^m$. Then the weights are $\{\delta_i + \delta_j + Z \mid 1 \leq i, j \leq m\}$. Since $\frac{1}{2}\left[(2\delta_i + Z) + (2\delta_j + Z)\right] = Z + \delta_i + \delta_j$, there are roots $\gamma_1, \dots, \gamma_m$ such that $(\gamma_i + \gamma_j)/2$ are roots. This is obviously impossible if $L = A_l$ or $L = D_l$.

Let $L = C_l$. Then $\gamma_i \in \{2\varepsilon_1, \dots, 2\varepsilon_l\}$. The subset $\Sigma \backslash Q$ is closed if and only if $m = l$. So $Q = \{2\varepsilon_k, \varepsilon_i + \varepsilon_j \mid 1 \leq i < j \leq l, 1 \leq k \leq l\}$ and $L_0 = gl(m, \mathbb{C}) + N$, $N = \sum_{\beta \in R} \mathbb{C} e_\beta$, $R = \{-2\varepsilon_k, -\varepsilon_i - \varepsilon_j \mid 1 \leq k \leq l, 1 \leq i, j \leq l\}$.

Let $L = B_l$. Obviously, the γ_i must be long roots. Moreover, $m = 2$, $l = 2$. But this case may be included in the previous series because $B_2 \simeq C_2$.

If $L = E_i$, $i = 6, 7, 8$, then there are no roots γ_1, γ_2 such that $(\gamma_1 + \gamma_2)/2$ is also root.

If $L = F_4$, then $\gamma_i \in \{\varepsilon_1 \pm \varepsilon_2; \varepsilon_3 \pm \varepsilon_4\}$. But in every case the subsystem $\Sigma \backslash Q$ is not closed.

If $L = G_2$, then $(\alpha + \beta)/2 \in \Sigma$ if and only if $\langle \alpha, \beta \rangle = 0$. Thus $m = 2$, but $\Sigma \backslash Q$ is not closed. □

LEMMA 4.4. *The cases* $P = G_2$ *or* $P = F_4$ *are impossible.*

PROOF. Since every fundamental representation of F_4 has multiple weights, $P \neq F_4$. If $P = G_2$, then $V = V_{\omega_1} = \mathbb{C}^7$; thus, the subalgebra L_0 is not reductive. Let $\operatorname{rank} L = \operatorname{rank} G = 3$, so $k(H) = HG$. Then the weights of V are $\lambda_1 = Z + \delta_1 - \delta_2$; $\lambda_2 = Z + \delta_2 - \delta_1$; $\lambda_3 = Z + \delta_1 - \delta_3$; $\lambda_4 = Z + \delta_3 - \delta_1$; $\lambda_5 = Z + \delta_2 - \delta_3$; $\lambda_6 = Z + \delta_3 - \delta_2$; $\lambda_7 = Z$.

So we have $\lambda_1 + \lambda_2 = \lambda_3 + \lambda_4 = \lambda_5 + \lambda_6 = \lambda_7$. The same equality is true for the corresponding elements of Q. But this is impossible for $L = A_3$. If $L = B_3$ we obtain $Q = \{\varepsilon_1 + \varepsilon_2;\ -\varepsilon_2;\ \varepsilon_1 - \varepsilon_2;\ \varepsilon_2;\ \varepsilon_1 + \varepsilon_3;\ \varepsilon_1\}$, but in this case $\Sigma_0 = \Sigma \backslash Q$ is not closed. For $L = C_3$ we have $Q = \{2\varepsilon_1;\ -\varepsilon_1 + \varepsilon_2;\ \varepsilon_1 + \varepsilon_3;\ \varepsilon_1 + \varepsilon_2;\ \varepsilon_2 - \varepsilon_3;\ \varepsilon_1 - \varepsilon_3;\ \varepsilon_2 + \varepsilon_3\}$ but in this case Σ_0 is not closed also.

Let $\operatorname{rank} L = 2$. Then either $L = B_2$, $\dim L_0 = 3$ or $L = G_2$, $\dim L_0 = 7$. It is easy to check that the subsystem Σ_0 is not closed in either case. □

LEMMA 5.4. *If* $P = E_7$ *then* $G = E_7$, $V = V_{\omega_7} = \mathbb{C}^{56}$, $L = B_7$, $L_0 = gl(7, \mathbb{C})$.

PROOF. We have $V = V_{\omega_7}$, because other representations have multiple weights. Suppose $G = E_7 + Z$. The weights of V_{ω_7} are $Z \pm \delta_i \pm \frac{1}{2}(\delta_8 - \delta_7)$, $Z + \frac{1}{2}\sum_{i=1}^{6} c_i v_i$, $c_i = \pm 1$, $\prod_{i=1}^{6} c_i = 1$. For each weight $\lambda = Z + \widetilde{\lambda}$ there is a weight $\mu = Z - \widetilde{\lambda}$

such that $d(\lambda,\mu)=28$. Thus, either $L=E_8$ or Q is symmetric. If $L=E_8$, then $\dim L_0=192$, so the semisimple component P_0 of L_0 has at least 64 positive roots, but this is impossible. Thus Q and Σ_0 are symmetric, so we may assume that $G=E_7$. Further, $r\left(\frac{1}{2}(\delta_8-\delta_7)+\delta_i, \frac{1}{2}(\delta_8-\delta_7)+\delta_j\right)=12$. Thus if L is classical, then $S_k=\left\{z+\frac{k}{2}(\delta_8-\delta_7)+\delta_i \mid 1\le i\le 6,\ k=\pm 1\right\}$ are "stars" or "stars with vertex" and $L=B_7$ or C_7, $\dim L_0=49$. So $L_0=gl(7,\mathbb{C})$. If $L=B_7$, then the irreducible components of the isotropy representation are $\Lambda^2\mathbb{C}^7+\Lambda^2\mathbb{C}^{7*}+\mathbb{C}^7+\mathbb{C}^{7*}$. The components of the restriction to $gl(7,\mathbb{C})\subset E_7$, the representation V_{ω_7} would be the same. Thus, we have $L=B_7$, $L_0=gl(7,\mathbb{C})$. The case $L=C_7$ is impossible.

If L is exceptional, we must consider E_6, E_7. If $L=E_7$, then $\dim L_0=77$ but there is no such reductive subalgebra. Suppose $L=E_6$. Then $\dim L_0=22$. Let $\widetilde{L}_0$ be a maximal reductive subalgebra of maximal rank containing L_0. Then $\widetilde{L}_0=A_1+A_5$, $3A_2$, or D_5+C, so $L_0=2A_1+A_3+C$. Then the irreducible components of the isotropy representation are

$$\mathbb{C}^4\otimes\mathbb{C}^2\otimes\widetilde{\mathbb{C}}^2+\mathbb{C}^{4*}\otimes\mathbb{C}^{2*}\otimes\widetilde{\mathbb{C}}^{2*}+\mathbb{C}^6+\mathbb{C}^{6*}+\mathbb{C}^2+\mathbb{C}^{2*}+\widetilde{\mathbb{C}}^2+\widetilde{\mathbb{C}}^{2*}+4\mathbb{C}^1,$$

where $\mathbb{C}^2$, $\widetilde{\mathbb{C}}^2$ are the standard representations of two components of type A_1 of $L_0=A_1+A_1+A_3+C$. The restriction of the structure representation on $k(L_0)\subset E_7$ is $\mathbb{C}^4\otimes\mathbb{C}^2\otimes\widetilde{\mathbb{C}}^2+\mathbb{C}^{4*}\otimes\mathbb{C}^{2*}\otimes\widetilde{\mathbb{C}}^{2*}+\mathbb{C}^6\otimes\mathbb{C}^2\otimes\widetilde{\mathbb{C}}^2$. So $L\ne E_6$. □

LEMMA 6.4. *If $P=E_6$, then $V=V_{\omega_1}=\mathbb{C}^{27}$, $L=E_7$, $L_0=E_6+C+N$, $G=E_6+Z$, $N\simeq V_{\omega_6}$.*

PROOF. Obviously, V_{ω_1} is the only representation without multiple weights. The subalgebra L_0 is not reductive, because $\dim V$ is odd. So we take $G=E_6+Z$. The weights of V_{ω_1} are $Z+\dfrac{2\Delta}{3}$, $Z+\dfrac{\Delta}{6}-\varphi$, $Z+\dfrac{\Delta}{6}+\varphi-\delta_i$, $Z-\dfrac{\Delta}{3}\pm\delta_i$, $Z+\dfrac{\Delta}{6}-\varphi+\delta_i+\delta_j$, $1\le i\le 5$, $1\le i<j\le 5$, where $\varphi=\frac{1}{2}\sum_{i=1}^{5}\delta_i$, $\Delta=\delta_8-\delta_7-\delta_6$. The weights $\lambda_i=Z+\dfrac{\Delta}{6}+\varphi-\delta_i$ have the property that $\lambda_i-\varphi-\dfrac{\Delta}{2}$ are also weights. So there are roots $\widetilde{\lambda}_i=\left.\left(Z+\dfrac{\Delta}{6}+\varphi-\varepsilon_i\right)\right|_H$ such that $\widetilde{\lambda}_i-\left.\left(\varphi+\dfrac{\Delta}{2}\right)\right|_H$ are other roots. If L is classical, then $S_1=\{\widetilde{\lambda}_i : 1\le i\le 5\}$, $S_2=\left\{\lambda_i-\left.\left(\varphi+\dfrac{\Delta}{2}\right)\right|_H : 1\le i\le 5\right\}$ are "stars" or "stars with a vertex". We write $\widetilde{\lambda}_i=x-\varepsilon_i$, where $x=\varepsilon_6$ if $L=A_l, D_l$; $x=\varepsilon_6$ or $x=0$ if $L=B_l$; $x=\varepsilon_6$ or $x=\varepsilon_1$ if $L=C_l$. So $\delta_i|_H=\varepsilon_i$, $\left.\left(Z+\varphi+\dfrac{\Delta}{6}\right)\right|_H=x$. Further, $\left.\left(Z-\varphi+\dfrac{\Delta}{6}\right)\right|_H+\varepsilon_i+\varepsilon_j=\left.\left(Z+\delta_i+\delta_j+\dfrac{\Delta}{6}-\varphi\right)\right|_H$ are roots. Thus $\left.\left(Z+\dfrac{\Delta}{6}-\varphi\right)\right|_H=0$, but this is impossible because $Z+\dfrac{\Delta}{6}-\varphi$ is a weight. Therefore L is exceptional. Of course, $L\ne G_2, E_8$. If $L=F_4$, then $\dim L_0=25$ and such a subalgebra is not parabolic. If $L=E_6$ then $\dim L_0=51$ and its semisimple component P_0 has 9 positive roots. Since P_0 is contained in one of the maximal reductive subalgebras of the maximal rank which are A_1+A_5, $3A_2$, D_5+C we conclude that $P_0=3A_2$. In this case $L=3A_2+\mathscr{V}_1+\mathscr{V}_2$, where $\mathscr{V}_1=\mathbb{C}^3\otimes\mathbb{C}^3\otimes\mathbb{C}^3$, $\mathscr{V}_2=\mathbb{C}^{3*}\otimes\mathbb{C}^{3*}\otimes\mathbb{C}^{3*}$. But $P_0+\mathscr{V}_1$ would not be a subalgebra of L.

Let $L = E_7$. Since L_0 is a parabolic subalgebra, its semisimple component P_0 has 36 positive roots. Since $B_6 + C$, $C_6 + C$ are not imbeddable in E_7 we conclude that $P_0 = E_6$. Thus $L = E_7$, $L_0 = E_6 + C + \mathscr{V}_1$, $G = E_6 + Z$, $V = V_{\omega_1}$, where $L = E_6 + C + \mathscr{V}_1 + \mathscr{V}_2$, $\mathscr{V}_1, \mathscr{V}_2$ are conjugate E_6-submodules of L, $\dim \mathscr{V}_i = 27$. □

LEMMA 7.4. *If $P = sl(n, \mathbb{C})$, $V = \Lambda^3\mathbb{C}^n$, then $n = 6$ and either $L = B_4$, $L_0 = gl(4, \mathbb{C})$ or $L = D_5$, $L_0 = gl(5, \mathbb{C})$.*

LEMMA 8.4. *If $P = sl(n, \mathbb{C})$, $V = \Lambda^4\mathbb{C}^n$ then $L = E_7$, $L_0 = P = A_7$, $n = 8$.*

The proofs of these lemmas are rather sophisticated and they are based on the same ideas as in §3. See **[12]**.

LEMMA 9.4. *If $G = gl(n, \mathbb{C})$, $V = \Lambda^k\mathbb{C}^n$, then $k \leq 4$.*

PROOF. Let $k = 5$; thus $n \geq 10$, $\dim V \geq 252$, and so L is classical. If $\operatorname{rank} L < \operatorname{rank} G$, then $\binom{n}{5} \leq 2(n-1)^2$, but this is impossible. If $\operatorname{rank} L = \operatorname{rank} G$, then L is a parabolic subalgebra because the sum of two arbitrary weights is not zero. Therefore $\binom{n}{5} \leq n^2$, but that is false. Obviously, the case $k > 5$ is also impossible. □

LEMMA 10.4. *If $P = B_m, C_m, D_m$, then V is a spin representation of an orthogonal algebra.*

PROOF. The lemma is true because the multiplicity of all weights must be 1. □

Thus we have proved

THEOREM 1.4. *Let L_0 be a Lie subalgebra of maximal rank of a semisimple Lie algebra L over the complex number field and let there be an irreducible subalgebra $G \subset \operatorname{End} V$ containing the image $\mathscr{E}(L_0) \subset \operatorname{End} V$ of the isotropy representation. Then Table 3 (next page) contains all $\{L, L_0, G, V\}$ provided the representation V of G is not standard.*

§5. The case of real algebras

It is not difficult to describe every (L, L_0, G, V) over the real number field, provided the structure representation V of G is absolutely irreducible. In fact, the complexification $\{L^c, L_0^c, G^c, V^c\}$ is contained in our list, so we need to find every admissible real form.

There are the following cases:

1. L_0^c be reductive, $G^c = L_0^c$, (L^c, L_0^c) be a symmetric pair, so that $G = L_0$. Then (L, L_0) is contained in the list of Berger in **[3]** (pp.157–161). The irreducibility of the structure representation is equivalent to the semisimplicity of L_0.

Thus, we have proved

THEOREM 1.5. *If (L, L_0) is a symmetric pair and $G^c = L_0^c$, then $G = L_0$ and (L, L_0) are those pairs contained in the list of M. Berger such that L_0 is a semisimple subalgebra of maximal rank.*

2. Let (L^c, L_0^c) be a symmetric pair, $G^c \neq L_0^c$, P not simple. Suppose that $L_0 = L_{0,1} + L_{0,2}$, $G = G_1 + G_2$, $V^c = \mathbb{C}^{n_1} \otimes \mathbb{C}^{n_2}$, $L_{0,i} \subset G_i$, where $\mathbb{C}^{n_i}$ is a complex representation of the real algebra G_i. Then both $\mathbb{C}^{n_i}$ admit either real or quaternionic structure.

TABLE 3

L	L_0	G	V
$sl(p+q)$	$sl(p)+sl(q)+C+N$	$sl(p)+sl(q)+Z$	$\mathbb{C}^p\otimes\mathbb{C}^q$
C_{p+q}	C_p+C_q	$sl(2p)+sl(2q)$,	$\mathbb{C}^{2p}\otimes\mathbb{C}^{2q}$
		$sl(2p)+C_q,\ L_0$	$\mathbb{C}^{2p}\otimes\mathbb{C}^{2q}$
$so(p+q)$	$so(p)+so(q)$	$sl(p)+sl(q)$,	$\mathbb{C}^p\otimes\mathbb{C}^q$
		$sl(p)+so(q),\ L_0$	$\mathbb{C}^p\otimes\mathbb{C}^q$
C_{p+q+1}	C_p+C_q+C+N	$sl(2p+1)+$	
		$sl(2q+1)+Z$	$\mathbb{C}^{2p+1}\otimes\mathbb{C}^{2q+1}$
G_2	$2A_1$	$sl(4)+A_1$,	$\mathbb{C}^4\otimes\mathbb{C}^2$
		$sp(2)+A_1,\ L_0$	$\mathbb{C}^4\otimes\mathbb{C}^2$
G_2	$gl(2)$	$so(5)+sl(2)$,	$\mathbb{C}^5\otimes\mathbb{C}^2$
		$sl(5)+sl(2)$	$\mathbb{C}^5\otimes\mathbb{C}^2$
F_4	C_3+A_1	$sl(14)+A_1$,	$\mathbb{C}^{14}\otimes\mathbb{C}^2$
		$sp(7)+A_1,\ L_0$	$\mathbb{C}^{14}\otimes\mathbb{C}^2$
E_6	A_5+A_1	$sl(20)+A_1$,	$\mathbb{C}^{20}\otimes\mathbb{C}^2$
		$sp(10)+A_1,\ L_0$	$\mathbb{C}^{20}\otimes\mathbb{C}^2$
E_7	D_6+A_1	$sl(32)+A_1$,	$\mathbb{C}^{32}\otimes\mathbb{C}^2$
		$sp(16)+A_1,\ L_0$	$\mathbb{C}^{32}\otimes\mathbb{C}^2$
E_8	E_7+A_1	$sl(56)+A_1$,	$\mathbb{C}^{56}\otimes\mathbb{C}^2$
		$sp(28)+A_1,\ L_0$	$\mathbb{C}^{56}\otimes\mathbb{C}^2$
D_4	$4A_1$	$4A_1$	$\mathbb{C}^2\otimes\mathbb{C}^2\otimes\mathbb{C}^2\otimes\mathbb{C}^2$
$so(p+4)$	$so(p)+so(4)$	$so(2p)+A_1$,	$\mathbb{C}^{2p}\otimes\mathbb{C}^2$
		$sl(2p)+A_1$	$\mathbb{C}^{2p}\otimes\mathbb{C}^2$
		$so(p)+2A_1$	$\mathbb{C}^p\otimes\mathbb{C}^2\otimes\mathbb{C}^2$
		$sl(p)+2A_1$	$\mathbb{C}^p\otimes\mathbb{C}^2\otimes\mathbb{C}^2$
E_8	D_8	D_8	V_{ω_7}
E_7	A_7	A_7	V_{ω_3}
E_6	D_5+C	D_6	V_{ω_5}
F_4	B_4	B_4	V_{ω_4}
F_4	B_4	D_5	V_{ω_4}
B_7	A_6+C	E_7	V_{ω_7}
B_4	A_3+C	A_5	V_{ω_3}
D_5	A_4+C	A_5	V_{ω_3}
B_l	$A_{l-1}+C+N$	A_l+Z	V_{ω_2}
D_l	$A_{l-1}+C+N$	$A_{l-1}+Z$	V_{ω_2}
C_l	$A_{l-1}+C+N$	$A_{l-1}+Z$	$V_{2\omega_1}$
E_6	D_5+C+N	D_5+Z	V_{ω_4}
E_7	E_6+C+N	E_6+Z	V_{ω_1}

Let $n_2 = 2$; then $G_1^c = sl(n,\mathbb{C})$ or $G_1^c = Sp(n_1/2,\mathbb{C})$. If $L_{0,2} = sl(2,\mathbb{R})$, then $G = sl(n,\mathbb{R}) + sl(2,\mathbb{R})$ or $G = Sp(n_1/2,\mathbb{R}) + sl(2,\mathbb{R})$. If $L_{0,2} = su(2)$, then $G = su^*(n_1) + su(2)$ or $G = Sp(n_1/2) + su(2)$.

Let f be an involutive automorphism of the simple algebra L^c, $P + Z = L^+ = \{x \in L^c : fx = x\}$, where P is a semisimple subalgebra, Z is the one-dimensional center, $L^- = \{x \in L : fx = -x\} = \mathscr{V}_1^c + \mathscr{V}_2^c$, where the $\mathscr{V}_i^c$ are irreducible P-submodules, $\mathscr{V}_1^c = \ker k$, $k(L_0) = Z + P = G^c$, $V \simeq \mathscr{V}_2^c$. Then any involution φ of L^c such that $\varphi L_0^c \subset L_0^c$ must preserve its nilradical $\mathscr{V}_1^c$, center Z, and semisimple component P. Therefore the pair (L, L_0'), where $L_0' = Z + P$, is contained in the list of M. Berger. Moreover, the isotropy representation of L_0' on L/L_0' must be reducible. For example, if $L^c = E_7$, $G = E_6^c + Z$, $L_0^c = E_6^c + Z + \mathscr{V}_1$, $V^c = V_{\omega_1} = \mathbb{C}^{27}$, then the admissible real forms of (L^c, L_0^c) are $(E_7^1, E_6^1 + \mathscr{V}_1)$ and $(E_7^3, E_6^4 + R + \mathscr{V}_1)$.

Let (L^c, L_0^c) be a symmetric pair, $G \neq k(L_0)$, $G = Z + P$, where P is simple, Z is one-dimensional center, or $Z = 0$. For example, if $L^c = F_4$, $L_0^c = B_4$, $G^c = D_5$, $V = \mathbb{C}^{16}$, then there is a real structure on V if and only if $G = so(2,8)$, $so(4,6)$, $so(5,5)$, $so^*(10)$. The real forms of (L^c, L_0^c) are $(F_4, so(9))$, $(F_4^2, so(9))$, $(F_4^1, so(4,5))$, $(F_4^2, so(1,8))$.

The inclusion $k(L_0) \subset G$ is possible only if $L_0 = so(4,5)$, $G = so(4,6)$, $so(5,5)$, $L_0 = so(1,8)$, $G = so(2,8)$.

THEOREM 2.5. *Table* 4 *contains every* (L, L_0, G), *where* (L, L_0) *is not a symmetric pair.*

TABLE 4

L	L_0	G
$sl(p+q, \mathbb{R})$	$sl(p, \mathbb{R}) + sl(q, \mathbb{R}) + Z + N$	$sl(p, \mathbb{R}) + sl(q, \mathbb{R}) + Z$
$su(2(p+q))$	$su(2p) + su(2q) + Z + N$	$su(2p) + su(2q) + Z$
$sp(p+q+1, \mathbb{R})$	$sp(p, \mathbb{R}) + sp(q, \mathbb{R}) + Z + N$	$sl(2p+1, \mathbb{R}) + sl(2q+1, \mathbb{R})$
$so(N, N+1)$	$gl(N, \mathbb{R}) + N$	$gl(N+1, \mathbb{R})$
$sp(N, \mathbb{R})$	$gl(N, \mathbb{R}) + N$	$gl(N, \mathbb{R})$
$sp(N, N)$	$su(2N) + Z + N$	$su(2N) + Z$
$so(N, N)$	$gl(N, \mathbb{R}) + N$	$gl(N, \mathbb{R})$
$so^*(4N)$	$su(2N) + Z + N$	$su(2N) + Z$
G_2^*	$gl(2, \mathbb{R})$	$sl(5, \mathbb{R}) + sl(2, \mathbb{R})$ $so(2,3) + sl(2, \mathbb{R})$
E_6^1	$so(5,5) + Z + N$	$so(5,5)$
E_6^4	$so(1,9) + Z + N$	$so(1,9)$
E_7^1	$E_6^1 + Z + N$	$E_6^1 + Z$
E_7^3	$E_6^4 + Z + N$	$E_6^4 + Z$
$so(15)$ $so(1,14)$	$u(7)$	E_7^1
$so(4,5)$	$gl(4, \mathbb{R})$	$sl(6, \mathbb{R})$
$so(4,5)$	$u(2,2)$	$su(3,3)$
$so(3,6)$ $so(2,7)$	$u(1,3)$	$su(3,3)$ $su(1,5)$
$so(9)$ $so(1,8)$	$u(4)$	$su(1,5)$

THEOREM 3.5. *Every* (L, L_0, G) *such that the pair* (L, L_0) *is symmetric,* $k(L_0) \neq G$*,* P *is simple, is listed in Table* 5.

TABLE 5

L	L_0	G
$so(5,5)$	$gl(5,\mathbb{R})$	$sl(6,\mathbb{R})$
$so(10)$	$u(5)$	$su(1,5)$
$so(2,8)$	$u(1,4)$	$su(1,5)$
$so(4,6)$	$u(2,3)$	$su(3,3)$
$so(10)$	$u(5)$	$su(1,5)$
	$u(1,4)$	$su(1,5)$
	$u(2,3)$	$su(3,3)$
F_4^1	$so(4,5)$	$so(5,5)$
F_4^2	$so(1,8)$	$so(1,9)$
E_6^1	$so(5,5)+Z$	$so(5,7)$ $so(6,6)$
E_6^2	$so(4,6)+Z$	$so(5,7)$ $so(6,6)$
E_6^3	$so(2,8)+Z$	$so(2,10)$ $so(3,9)$
E_6^4	$so(1,9)+Z$	$so(2,10)$ $so(1,11)$ $so(3,9)$
E_6^4 E_6	$so(10)+Z$	$so(2,10)$ $so(1,11)$

References

1. N. Bourbaki, Groupes et algèbres de Lie, Chapters 4–6, Hermann, Paris, 1968; Chapters 7–8, Hermann, Paris, 1975.
2. M. Goto and F. Grosshans, *Semisimple Lie algebras*, Lecture Notes in Pure and Appl. Math., vol. 38, Marcel Dekker, New York and Basel, 1978.
3. M. Berger, *Les éspaces symetriques noncompacts*, Ann. Sci. École Norm. Sup. (3) **74** (1957), 85–177.
4. E. B. Dynkin, *Semisimple subalgebras of semisimple Lie algebras*, Mat. Sb. **30** (1952), no. 2, 349–462; English transl. in Amer. Math. Soc. Transl. Ser. 2 **6** (1957).
5. A. I. Mal′cev, *On semisimple subgroups of Lie groups*, Izv. Akad. Nauk SSSR Ser. Mat. **8** (1944), no. 4, 143–174; English transl. in Amer. Math. Soc. Transl. Ser. 2 **9** (1962).
6. R. Hermann, *Compact homogeneous almost complex spaces of positive characteristics*, Trans. Amer. Math. Soc. **83** (1956), 471–481.
7. H.-C. Wang, *Closed manifolds with homogeneous complex structure*, Amer. J. Math. **76** (1954), 1–32.
8. J. Wolf, *Complex homogeneous contact manifolds and quaternionic symmetric spaces*, J. Math. Mech. **14** (1965), 1033–1047.
9. E. B. Vinberg and A. L. Onishchik, *Seminar algebraic groups and Lie groups*, "Nauka", Moscow, 1988; English transl., *Lie groups and algebraic groups*, Springer-Verlag, Berlin and New York, 1990.
10. P. Ya. Grushko, *On some exceptional homogeneous spaces*, Problems in Group Theory and Homological Algebra, Yaroslavl. Gos. Univ., Yaroslavl′, 1985, pp. 153–154. (Russian)
11. ———, *Special systems*, Problems in Group Theory and Homological Algebra, Yaroslavl. Gos. Univ., Yaroslavl′, 1983, pp. 43–49. (Russian)
12. L. A. Osipenko, *On one class of nonreductive maximal rank subalgebras in semisimple complex Lie algebras*, Preprint No. 475384, Irkutsk. Gos. Univ., Irkutsk, 1984. (Russian)
13. ———, *Special nonreductive systems*, Differential Geometry in Homogeneous Spaces, Irkutsk. Gos. Univ., Irkutsk, 1988. (Russian)

Amer. Math. Soc. Transl.
(2) Vol. **163**, 1995

Algebraic Principles of Building Mathematical Structures

V. K. Ionin

ABSTRACT. This paper is an overview of a series of author's papers describing an algebraic method for producing mathematical structures, including a number of known structures of geometry, algebra, topology, etc. The notion of structure in the sense of Bourbaki does not distinguish structures of interest. Our algebraic approach is like a mechanism for enriching structures: it is fed by something simpler and produces new, richer ones.

§1. Preliminary comments on categories

To fix notation, we begin with some well-known definitions concerning categories (cf. [**1–5**]), with some minor modifications.

1.1. A *category* $\mathscr{K}$ consists of a class $\operatorname{Ob}\mathscr{K}$ of *$\mathscr{K}$-objects* and a class $\operatorname{Hom}\mathscr{K}$ of *$\mathscr{K}$-morphisms* satisfying the following conditions:

C1. To each ordered pair of $\mathscr{K}$-objects A, B there corresponds a set of $\mathscr{K}$-morphisms $\operatorname{Hom}(A, B)$ so that each $\mathscr{K}$-morphism belongs to $\operatorname{Hom}(A, B)$ for one and only one pair of objects A, B.

C2. If $f \in \operatorname{Hom}(A, B)$ and $g \in \operatorname{Hom}(B, C)$, there exists one and only one element of the set $\operatorname{Hom}(A, C)$ called the *composition* of f, g and denoted by gf.

C3. If $f \in \operatorname{Hom}(A, B)$, $g \in \operatorname{Hom}(B, C)$, $h \in \operatorname{Hom}(C, D)$, then $h(gf) = (hg)f$.

C4. To each $\mathscr{K}$-object A there corresponds one and only one $\mathscr{K}$-morphism $1_A \in \operatorname{Hom}(A, A)$ such that for all $f \in \operatorname{Hom}(B, A)$ and $g \in \operatorname{Hom}(A, C)$ $1_A f = f$ and $g 1_A = g$.

1.2. Instead of $f \in \operatorname{Hom}(A, B)$, we write also $f: A \to B$ and say that "f is directed from A to B". Where several categories are considered, to avoid misunderstanding, we write $\operatorname{Hom}_{\mathscr{K}}(A, B)$ instead of $\operatorname{Hom}(A, B)$.

1.3. A morphism $u: A \to B$ is called an *isomorphism* if there exists a $v: B \to A$ such that $vu = 1_A$, $uv = 1_B$. An isomorphism from A to A is called an *automorphism* of A; the set of all such automorphisms is denoted by $\operatorname{Aut} A$. Any morphism from A to A is called an *endomorphism* of A; the set of all such endomorphisms is denoted by $\operatorname{End} A$.

1991 *Mathematics Subject Classification*. Primary 18Bxx.

1.4. A one-to-one correspondence $f \to f'$ between $\operatorname{Hom} \mathscr{K}$ and $\operatorname{Hom} \mathscr{K}'$ is called an *isomorphism* between the categories $\mathscr{K}$ and $\mathscr{K}'$ if it satisfies the following condition: gf is defined if and only if $g'f'$ is defined, and then $(gf)' = g'f'$.

As the objects are in one-to-one correspondence with identity morphisms, it follows that $\operatorname{Ob} \mathscr{K}$ is in a one-to-one correspondence with $\operatorname{Ob} \mathscr{K}'$. Two categories are called *equivalent* or *isomorphic* if there exists an isomorphism between them.

1.5. An isomorphism is a particular case of a *functor*. A functor from a category $\mathscr{K}$ to a category $\mathscr{K}'$ is a pair of maps $\operatorname{Ob} \mathscr{K} \to \operatorname{Ob} \mathscr{K}'$ and $\operatorname{Hom} \mathscr{K} \to \operatorname{Hom} \mathscr{K}'$ (denoted by the same symbol, e.g., T), such that (a) $T(1_A) = 1_{T(A)}$ for all A; (b) if gf is defined in $\mathscr{K}$, then $T(g)T(f)$ is defined in $\mathscr{K}'$, and $T(gf) = T(g)T(f)$.

$T : \mathscr{K} \to \mathscr{K}'$ means that T is a functor from $\mathscr{K}$ to $\mathscr{K}'$. To be precise, this is a covariant functor, which is needed in the sequel. To get the definition of a contravariant functor, one has to change $T(g)T(f)$ in the condition (b) to $T(f)T(g)$.

1.6. A category $\mathscr{K}'$ is called a *subcategory* of a category $\mathscr{K}$ (written $\mathscr{K}' \subset \mathscr{K}$), if the following conditions are satisfied:

(a) $\operatorname{Ob} \mathscr{K}' \subset \operatorname{Ob} \mathscr{K}$, $\operatorname{Hom} \mathscr{K}' \subset \operatorname{Hom} \mathscr{K}$.

(b) If $A, B \in \operatorname{Ob} \mathscr{K}'$ then $\operatorname{Hom}_{\mathscr{K}'}(A, B) \subset \operatorname{Hom}_{\mathscr{K}}(A, B)$.

(c) If $f, g \in \operatorname{Hom} \mathscr{K}'$ and gf is defined in $\mathscr{K}$, then gf is the composition of f and g in $\mathscr{K}'$.

(d) Each identity morphism of $\mathscr{K}'$ is an identity morphism in $\mathscr{K}$.

A subcategory $\mathscr{K}'$ of a category $\mathscr{K}$ is called *complete* if, instead of (b), a stronger condition is satisfied:

(b′) If $A, B \in \operatorname{Ob} \mathscr{K}'$, then $\operatorname{Hom}_{\mathscr{K}'}(A, B) = \operatorname{Hom}_{\mathscr{K}}(A, B)$.

To define a complete subcategory, it is sufficient to specify the corresponding subclass of objects of the given category.

1.7. The *subcategory of isomorphisms* $B_{ij}\mathscr{K}$ of a category $\mathscr{K}$ is defined as follows: (a) $\operatorname{Ob} B_{ij}\mathscr{K} = \operatorname{Ob} \mathscr{K}$; (b) The class $\operatorname{Hom} B_{ij}\mathscr{K}$ consists of all isomorphisms of $\mathscr{K}$.

1.8. The *genus of a structure* is a covariant functor $T : B_{ij}\mathscr{K} \to B_{ij}\,\mathrm{Ens}$, where Ens is the category of all sets and maps.

EXAMPLES. 1. Let M be a set, and $T(M)$ the set of all maps of $M \times M$ into the real line R satisfying the axioms of a metric space. Then we have an obvious functor $T : B_{ij}\,\mathrm{Ens} \to B_{ij}\,\mathrm{Ens}$, the genus of metric space structure.

2. Let X be a topological space, that is, an object of the category Top of topological spaces and continuous maps, and $T(X)$ the set of all continuous binary operations $X \times X \to X$ satisfying the group axioms, so that inversion is continuous. Then we have the functor $T : B_{ij}\,\mathrm{Top} \to B_{ij}\,\mathrm{Ens}$, the genus of continuous group structure.

Taking, instead of Top, the category of differentiable manifolds and maps, we get the genus of Lie group structures.

3. Let X be a set, $T(X)$ the set of all reflexive, antisymmetric, and transitive binary relations on X. In this case we get the genus of (partial) order structure.

1.9. Some general structures are *equivalent*, in an intuitively evident sense, of which we shall have examples in the sequel. The precise definition, for general functors, is as follows.

A *functorial morphism* $h : T \to S$, where T and S are covariant functors from a category $\mathscr{K}_1$ to a category $\mathscr{K}_2$, is given if for each object $A \in \operatorname{Ob} \mathscr{K}_1$ a morphism

$h(A)\colon T(A) \to S(A)$ is defined such that for all morphisms $u\colon A \to B$ of $\mathscr{K}_1$ the condition $S(u)h(A) = h(B)T(u)$ is satisfied. This means that the diagram

$$\begin{array}{ccc} T(A) & \xrightarrow{h(A)} & S(A) \\ {\scriptstyle T(u)}\downarrow & & \downarrow{\scriptstyle S(u)} \\ T(B) & \xrightarrow{h(B)} & S(B) \end{array}$$

is commutative.

§2. The category $\mathscr{K}(\Gamma)$

Fix a category $\mathscr{K}$, a pair of its objects (A, B), and a set $\Gamma \subset \mathrm{Hom}(A, B)$. Proceeding from $(\mathscr{K}, A, B, \Gamma)$ we define in a natural way a new category $\mathscr{K}(A, B, \Gamma)$ underlying our process of "enriching mathematical structures". As Γ defines the pair (A, B) (with the exception of $\Gamma = \varnothing$), the notation may be simplified to $\mathscr{K}(\Gamma)$.

2.1. Let n be a positive integer and let $X_1, X_2, \dots, X_n, X_{n+1}$ be objects of the category $\mathscr{K}$. For each $i \in \{1, \dots, n\}$ choose an arbitrary set of morphisms $H_i \subset \mathrm{Hom}(X_i, X_{i+1})$. The set of all compositions $h_n \dots h_2 h_1 \in \mathrm{Hom}(X_1, X_{n+1})$ is denoted by $H_n \dots H_2 H_1$ and is called the composition of the sets $H_1, H_2, \dots, H_n$. Then, by definition, the notation

$$H_n \dots \dot{H}_i \dots H_1 \subset H,$$

where $H \subset \mathrm{Hom}(X_1, X_{n+1})$, means that the inclusion $H_n \dots H_i \dots H_1 \subset H$ holds in which H_i cannot be extended; that is, if $H_n \dots H'_i \dots H_1 \subset H$ for any $H'_i \supset H_i$, $H'_i \subset \mathrm{Hom}(X_i, X_{i+1})$, then $H'_i = H_i$. This notation is obviously generalized to the case of dots over several of the H_i's.

2.2. A triple (X, F, Φ), where

$$X \in \mathrm{Ob}\,\mathscr{K}, \qquad F \subset \mathrm{Hom}(A, X), \qquad \Phi \subset \mathrm{Hom}(X, B),$$

is called a Γ-*space*, and the pair (F, Φ) a Γ-*structure*, if the following axiom is satisfied:

$$\dot{\Phi}\dot{F} \subset \Gamma.$$

Later on, we write simply X instead of (X, F, Φ), if it does not give rise to misunderstanding. Each morphism from F (respectively, from Φ) is called the *input morphism* (respectively, the *output morphism*). Γ defines in an obvious way a genus of structure $T(\Gamma)\colon B_{ij}\mathscr{K} \to B_{ij}\,\mathrm{Ens}$ associating to each object of $\mathscr{K}$ the set of all its Γ-structures.

2.3. As is easily seen, any set $F' \subset \mathrm{Hom}(A, X)$ defines a Γ-structure (F, Φ) on X, in the following way. There exists one and only one pair of sets (F, Φ), $F \subset \mathrm{Hom}(A, X)$, $\Phi \subset \mathrm{Hom}(X, B)$, such that $\dot{\Phi}F' \subset \Gamma$, $\Phi\dot{F} \subset \Gamma$, and for these $\dot{\Phi}\dot{F} \subset \Gamma$ holds. Then F' is called the *input base* of the Γ-structure (F, Φ). Similarly, any set $\Phi' \subset \mathrm{Hom}(X, B)$ defines through relations $\Phi'\dot{F} \subset \Gamma$, $\dot{\Phi}F \subset \Gamma$, a Γ-structure (F, Φ), of which it is the *output base*.

The fixed objects A, B get themselves natural Γ-structures (F_A, Φ_A) (respectively, (F_B, Φ_B)), Γ being the output base for (F_A, Φ_A) (respectively, the input base for (F_B, Φ_B)). Obviously, $\Phi_A = F_B = \Gamma$. Later on we shall assume, noting the fact explicitly, that A and B are endowed with their natural Γ-structures. A is then called the *input* Γ*-space*, and B the *output* Γ*-space*.

2.4. PROPOSITION. *Let F' be the input base of the Γ-space (X, F, Φ), Ψ' the output base of the Γ-space (Y, G, Ψ). Then for any set $H \subset \mathrm{Hom}(X, Y)$ the following six statements are equivalent*: (a) $\Psi \dot{H} F \subset \Gamma$; (b) $\dot{H} F \subset G$; (c) $\Psi \dot{H} \subset \Phi$; (d) $\Psi' \dot{H} F' \subset \Gamma$; (e) $\dot{H} F' \subset G$; (f) $\Psi' \dot{H} \subset \Phi$.

DEFINITION. A morphism of the set H is called a Γ-morphism of the Γ-space X into the Γ-space Y if one of the statements (a)–(f) of Proposition 2.4 holds for H (and thus all other statements).

The class of Γ-spaces, the class of Γ-morphisms, and the composition of Γ-morphisms induced from the category $\mathscr{K}$ constitute a category called $\mathscr{K}(\Gamma)$.

2.5. Comparison of Γ-structures. Let $(X_i, F_i, \Phi_i) \in \mathrm{Ob}\,\mathscr{K}(\Gamma)$, $\gamma_i = (F_i, \Phi_i)$, $i = 1, 2$. The Γ-structure γ_1 *majorizes* the Γ-structure γ_2 (or γ_2 is *majorized* by γ_1), if $X = X_1 = X_2$ is in $\mathscr{K}$, and the identity morphism $1_X : X_1 \to X_2$ is a Γ-morphism. If, furthermore, $\gamma_1 \neq \gamma_2$, then γ_1 is *stronger* than γ_2 (or γ_2 is *weaker* than γ_1).

The following statements are obviously equivalent: (a) γ_1 majorizes γ_2; (b) γ_2 is majorized by γ_1; (c) $F_1 \subset F_2$; (d) $\Phi_2 \subset \Phi_1$. Similarly, four other statements are equivalent: (e) γ_1 is stronger than γ_2; (f) γ_2 is weaker than γ_1; (g) F_1 is a proper part of F_2; (h) Φ_2 is a proper part of Φ_1.

(F, Φ) is the strongest (respectively, the weakest) Γ-structure in $\mathscr{K}(\Gamma)$ if and only if $\Phi = \mathrm{Hom}(X, B)$ (respectively, $F = \mathrm{Hom}(A, X)$).

2.6. Let I be a set of indices and $\{H_i\}_{i\in I}$ a family of sets of morphisms $H_i \subset \mathrm{Hom}(X, X_i)$ in the category $\mathscr{K}$. Suppose each object X_i, $i \in I$, is endowed with a Γ-structure (F_i, Φ_i). It can be shown that on the object X there exists a strongest Γ-structure (F, Φ) for which each H_i, $i \in I$, contains only Γ-morphisms. This Γ-structure is called *initial* with respect to the family $\{H_i\}_{i\in I}$. Suppose X is a Γ-space, which is a set, and $X' \subset X$. Then we get the initial Γ-structure on X' with respect to the family $\{h\}$ containing the single map h, the inclusion of X' into X. X' is called a Γ-*subspace* of X.

2.7. Let I be again a set of indices and $\{H_i\}_{i\in I}$ a family of sets of morphisms $H_i \subset \mathrm{Hom}(X_i, X)$ in the category $\mathscr{K}$. Suppose each X_i, $i \in I$, is endowed with a Γ-structure. It can be shown that on the object X there exists a weakest Γ-structure (F, Φ) for which each $\{H_i\}_{i\in I}$ contains only Γ-morphisms. This Γ-structure is called *final* with respect to the family $\{H_i\}_{i\in I}$.

Suppose X is a Γ-space, which is a set, and X^* the quotient set of X with respect to some equivalence relation on X. Then we get a final Γ-structure on X^* with respect to the family $\{h\}$ containing a single map, the canonical projection $X \to X^*$. X^* is called the Γ-*quotient space* of X.

Note that the natural Γ-structures of the objects A and B are, respectively, initial and final with respect to the family containing a single set Γ.

2.8. Suppose that the objects of the category $\mathscr{K}$ are sets, perhaps endowed with some structures. In this case we need the following definitions for the sequel.

2.8.1. DEFINITION. A Γ-space (X, F, Φ) is said to be *separable* if for any two different points $x_1, x_2 \in X$ there exists a function $\varphi \in \Phi$ such that $\varphi(x_1) \neq \varphi(x_2)$.

2.8.2. Definition. A Γ-space (X, F, Φ) is said to be *connected* if for any two points $x_1, x_2 \in X$ there exists a function $f \in F$ such that $x_1, x_2 \in f(A)$.

§3. Affine spaces

Consider the category $\mathscr{K} = \text{Ens}$, the objects $A = B = \mathscr{P}$, where $\mathscr{P}$ is a field, and the set $\Gamma = \Lambda(\mathscr{P})$ of affine maps of $\mathscr{P}$ into $\mathscr{P}$, that is, the set of all maps $\lambda\colon \mathscr{P} \to \mathscr{P}$ of the form $\lambda(t) = at + b$ with two fixed elements a, b of $\mathscr{P}$.

3.1. Definition. Let X be a nonempty set and $\Gamma = \Lambda(\mathscr{P})$. Then a connected and separable Γ-space (X, F, Φ) is called an *affine space* over $\mathscr{P}$. The Γ-structure (F, Φ) is called an *affine structure*. The morphisms of F, respectively Φ, are called *affine maps* from $\mathscr{P}$ to X (respectively, from X to $\mathscr{P}$). A Γ-morphism $X \to Y$ of affine spaces is called an *affine map* from X to Y.

3.2. The field $\mathscr{P}$ may be considered as an input and an output Γ-space (cf. 2.3). It is easily seen that these Γ-spaces coincide. Ascribe to any Γ-space isomorphic to $\mathscr{P}$, by definition, *dimension* 1 (written $\dim X = 1$). Any set X consisting of a single element can be endowed by one and only one Γ-structure. Ascribe to the resulting Γ-space dimension 0 ($\dim X = 0$).

3.3. Let $\mathscr{E}$ be any nonempty set, and Ψ a nonempty set of functions $\Psi\colon X \to \mathscr{P}$ such that Ψ distinguishes points of $\mathscr{E}$; that is, for any pair of different points $p, q \in \mathscr{E}$ there exists a function $\psi \in \Psi$ for which $\psi(p) \neq \psi(q)$. For the set $\mathscr{E}$ with the set of functions Ψ, we define the *canonical inclusion* into the affine space (X, F, Φ).

Proceeding from $\mathscr{E}$, construct a sequence of sets $\mathscr{E}_0, \dots, \mathscr{E}_m, \dots$ as follows:

$$\mathscr{E}_0 = \mathscr{E}, \dots, \mathscr{E}_{m+1} = \mathscr{E}_m \bigcup (\mathscr{E}_m \times \mathscr{E}_m \times \mathscr{P}), \dots$$

The following recurrent rule assigns to any function $\psi \in \Psi$ a function $\psi_*\colon \mathscr{E}_* \to \mathscr{P}$, where $\mathscr{E}_* = \bigcup_{m=0}^{\infty} \mathscr{E}_m$: (a) $\psi_*(p) = \psi(p)$ if $p \in \mathscr{E}_0$; (b) $\psi_*(p, q, t) = (1-t)\psi_*(p) + t\psi_*(q)$ if $(p, q, t) \in \mathscr{E}_m$ $(m > 0)$. Denote by Ψ_* the set of all such functions ψ_*. On the set $\mathscr{E}_*$ define the Γ-structure (F_*, Φ_*) with the output base Ψ_*. Introduce an equivalence relation on $\mathscr{E}$ assuming $p \sim q$ if $\varphi(p) = \varphi(q)$ for each $\varphi \in \Phi_*$; let X be the quotient-set of $\mathscr{E}$ with respect to this equivalence, and $\pi\colon \mathscr{E}_* \to X$ the corresponding canonical projection. Take the final Γ-structure (F, Φ) of the projection π. Then it can be shown that (X, F, Φ) is an affine space.

The inclusion $j\colon \mathscr{E} \to \mathscr{E}_*$ engenders the injective map $\pi j\colon \mathscr{E} \to X$, which is also called an inclusion. With some licence of speech, we shall not distinguish the set $\mathscr{E}$ from $\pi j(\mathscr{E})$. Thus, we may suppose that $\mathscr{E} \in X$.

If Ψ consists of *all* maps of $\mathscr{E}$ into $\mathscr{P}$, the affine space (X, F, Φ) is called the *affine hull* of the (abstract) set $\mathscr{E}$. Any affine space can be obtained as the hull of some set. Two affine spaces are isomorphic if and only if they are affine hulls of sets of the same cardinality. This leads to the following

Definition. If an affine space X is the affine hull of a set $\mathscr{E}$ then the cardinality of $\mathscr{E}$ less one is called the *dimension* of X (and denoted $\dim X$).

Note that this definition is in agreement with 3.2, and that for an infinite set $\mathscr{E}$ the cardinality is not changed by subtracting one.

3.4. Suppose a set X is endowed with an affine structure (F, Φ) and $X' \subset X$. Then X' acquires a Γ-structure (F', Φ'), initial with respect to the inclusion $X' \to X$.

If this structure is affine, X' is called an *affine subspace* of X. When $\dim X' = 1$, X' is called a *straight line*. The following two statements are equivalent: (a) X' is a straight line of the affine space (X, F, Φ); (b) there exists a function $f \in F$ such that $f(\mathscr{P}) = X'$.

3.5. Introduce an equivalence relation on the set $X \times X$, where X is a space with an affine structure (F, Φ), assuming $(p, q) \sim (p', q')$ when $\varphi(q) - \varphi(p) = \varphi(q') - \varphi(p')$ for all $\varphi \in \Phi$. Denote by V the corresponding quotient-set. The elements of V are called *vectors*. Later on, we shall endow V with a vector space structure.

3.6. It can be proved that for any triple (p, q, u) of the space X there exists one and only one point $v \in X$ such that $(p, q) \sim (u, v)$. Therefore, the following definition is correct. Denote by $[p, q]$ the equivalence class of (p, q); then the sum $[p, q] + [u, v]$ is the vector $[p, r]$, r being that point for which $(q, r) \sim (u, v)$.

The product of a vector $[p, q]$ and an element $\lambda \in \mathscr{P}$ (denoted $\lambda[p, q]$) is, by definition, the vector $[p, r]$, r being that point for which $\varphi(r) - \varphi(p) = \lambda(\varphi(q) - \varphi(p))$ for all $\varphi \in \Phi$.

It can be shown that V, with the two operations just defined, satisfies the axioms of a vector space.

3.7. Now define addition of a point $u \in X$ and a vector $[p, q] \in V$ (denoted $v = u + [p, q]$) as follows: v is the sum of u and $[p, q]$ if $(u, v) \sim (p, q)$. This operation can be considered as an action of the vector space V on the set X. As is easily verified, this action is effective and transitive or, which amounts to the same, the triple consisting of the set X, the vector space V, and the action of V on X is an affine space in the sense of the usual definition (cf., e.g., [**6**, Chapter 2]).

3.8. It can be seen that our definition of an affine space is essentially equivalent to the usual definition. The precise meaning of this is as follows. According to [**6**], an affine space is defined as a triple (X, V, D), X being a set, V a vector space over a field $\mathscr{P}$, and D an action of V on X. Thus, the class of all affine structures $S(X)$ on a given set X consists of all pairs (V, D), and thus is *not a set*. Let us modify this definition, without changing the essential contents. Consider an equivalence relation $\sim$ on $X \times X$ such that for any three points p, q, $r \in X$ there exists one and only one point $s \in X$ for which $(p, q) \sim (r, s)$. Denote by V the quotient set corresponding to this equivalence. Endow V with some vector space structure so that $[p, q] + [q, r] = [p, r]$, where $[p, q]$ is the equivalence class containing the pair (p, q). Finally, define the action of V on X by the relation $D([p, q], p) = q$. Then the modified definition (cf. [**6**]) is as follows: an affine space is a triple (X, V, D) with a pair (V, D) just defined. Now all affine structures on X form a *set* $S(X)$, and thus the genus of the structure $S\colon B_{ij}\,\mathrm{Ens} \to B_{ij}\,\mathrm{Ens}$ is obtained. It can be shown that this functor S is equivalent to the functor T considered above. For details of the proof see [**7**], pp. 3–16. V is called the *adjoint vector space* of the affine space X.

3.9. Let (X, V, D) be an affine space over the field of real numbers R (or the field of complex numbers C), that is, $\mathscr{P} = R$ or C. This space is called a *normed affine space* if V is a normed vector space. It is clear that each normed affine space is a metric space with an interior metric. So we get the Minkovskian geometries.

Now we define a normed affine space as a Γ-space.

DEFINITION. Let X be a nonempty set and $\Gamma \subset \Lambda(R)$, where $\Lambda(R)$ is the set of all functions $\lambda\colon R \to R$ of the form $\lambda(t) = at + b$, a and b being elements of R such that $|a| \leq 1$. Then a connected and separable Γ-space (X, F, Φ) is called a *normed affine space*. (For $\mathscr{P} = C$, the definition is similar.)

This definition can be shown to be equivalent to the usual definition cited above.

3.10. Denote by $\operatorname{Aut}\mathscr{P}$ the set of all automorphisms of the field $\mathscr{P}$. Fix an arbitrary set $Q \subset \operatorname{Aut}\mathscr{P}$, $1_{\mathscr{P}} \in Q$, and define Γ as follows: Γ consists of all functions $\lambda\colon \mathscr{P} \to \mathscr{P}$ of the form $\lambda(t) = a\sigma(t) + b$, $t \in \mathscr{P}$, where $\sigma \in Q$. Note that $\big(A(\mathscr{P}) \cup Q\big) \subset \Gamma$.

DEFINITION. Let X be a nonempty set. Then a connected and separable Γ-space (X, F, Φ), with Γ defined above, is called a *semiaffine space* with respect to the set Q of automorphisms of $\mathscr{P}$ (cf. **[6]**, Chapter 2). If $Q = \operatorname{Aut}\mathscr{P}$, X is called a *complete* semiaffine space (or simply a semiaffine space). If Q consists of a single element, the identity map, then X is an affine space in the sense of 3.1.

3.11. Let A and B be affine spaces, $\dim A > 0$, $\dim B > 0$, and Γ the set of all affine maps from A to B. Definition 3.1 can be generalized as follows.

DEFINITION. A connected and separable Γ-space (X, F, Φ), with $X \neq 0$, is called an *affine space*, and the Γ-structure (F, Φ) an *affine structure*. A Γ-morphism from an affine space to another affine space is called an *affine map*.

As is easily seen, this is equivalent to Definition 3.1. For various purposes it is advantageous to have a large choice of equivalent definitions; in our case, we have obtained a definition for every pair (A, B).

§4. Vector spaces

In this section, unless stated otherwise, $\mathscr{K} = \text{Ens}$, $A = \mathscr{P}^2$, $B = \mathscr{P}$, where $\mathscr{P}$ is a field, and Γ is the set of all linear maps of $\mathscr{P}^2$ into $\mathscr{P}$, that is, maps of the form $\gamma(s, t) = as + bt$, $(s, t) \in \mathscr{P}^2$, where a, b are arbitrary elements of $\mathscr{P}$.

4.1. DEFINITION. A connected and separable Γ-space (X, F, Φ), $X \neq \varnothing$, is called a *vector space*. The Γ-structure (F, Φ) is called a *vector structure*. A Γ-morphism of a vector space into another vector space is called a *vector map* (or a *linear map*).

4.2. Inclusion of a set $\mathscr{E}$ into a vector space (X, F, Φ). Construct a sequence of sets $\mathscr{E}_0, \mathscr{E}_1, \dots, \mathscr{E}_m, \dots$ as follows. Set $\mathscr{E}_0 = \mathscr{E} \cup \{\theta\}$, where θ is an element not belonging to $\mathscr{E}$. For each $m > 0$ define $\mathscr{E}_m = \mathscr{E}_{m-1} \bigcup \big(\mathscr{E}_{m-1}^2 \times \mathscr{P}_2\big)$. Let Ψ be a set of functions $\psi\colon \mathscr{E} \to \mathscr{P}$. Extend these functions to $\mathscr{E}_0$ by putting $\psi(\theta) = 0$ and assign to each ψ a function $\psi_*\colon \mathscr{E}_* \to \mathscr{P}$ defined by the recurrent rule: (a) $\psi_*(p) = \psi(p)$ for all $p \in \mathscr{E}_0$; (b) $\psi_*(p, q, s, t) = s\psi_*(p) + t\psi_*(q)$ for $p, q \in \mathscr{E}_{m-1}$, $s, t \in \mathscr{P}$. Denote by Ψ_* the set of all such functions ψ_*, for $\psi \in \Psi$. Define on $\mathscr{E}_*$ a Γ-structure (F_*, Φ_*) with the input base Ψ_*. Introduce an equivalence relation $\sim$ on $\mathscr{E}_*$ assuming $p \sim q$ if $\varphi(p) = \varphi(q)$ for all $\varphi \in \Phi_*$. Let X be the quotient-set of $\mathscr{E}$ with respect to this equivalence, and $\pi\colon \mathscr{E}_* \to X$ the corresponding canonical projection. Take the final Γ-structure (F, Φ) on X for the projection π. It can be shown that (X, F, Φ) is a vector space.

The inclusion $j\colon \mathscr{E} \to \mathscr{E}_*$ engenders an injective map $\pi j\colon \mathscr{E} \to X$, which is also called inclusion. In the sequel, we shall not distinguish the set $\mathscr{E}$ from $\pi j(\mathscr{E})$. Thus, each function $\psi \in \Psi$ turns out to be a restriction of some $\varphi \in \Phi$. If Ψ contains *all* maps $\psi \colon \mathscr{E} \to \mathscr{P}$, the vector space (X, F, Φ) is called *vector* (or linear) *hull* of the (abstract) set $\mathscr{E}$. As for affine spaces, we have the following

DEFINITION. The *dimension* of the vector space X (written $\dim X$) is the cardinality of the set $\mathscr{E}$, of which it is the vector hull.

4.3. For any subset X' of the vector space (X, F, Φ) we get a Γ-structure (F', Φ'), initial with respect to the inclusion $X' \to X$. If this structure is a vector structure, then (X', F', Φ') is called a *vector subspace* of the vector space (X, F, Φ).

4.4. Given a vector space X, introduce on X a usual vector structure as follows. Assign to a linear combination $sp + tq$ $(p, q \in X;\ s, t \in \mathscr{P})$ the vector r for which $\varphi(r) = s\varphi(p) + t\varphi(q)$ for any $\varphi \in \Phi$. This definition is shown to be correct and satisfies the usual axioms of vector space. It can be shown that Definition 4.1 is equivalent to the usual definition of a vector space.

4.5. Now fix two vector spaces A, B with $\dim A > 1$, $\dim B > 0$, and let Γ be the set of all linear maps from A to B. Definition 4.1 can be generalized as follows.

DEFINITION. A connected and separable Γ-space (X, F, Φ), with $X \neq 0$, is called a Γ-*space*, and (F, Φ) its vector structure. A Γ-morphism from a vector space into another vector space is called a *vector map* (or *linear map*).

4.6. It is easily shown that Definitions 4.1 and 4.5 are equivalent. Definition 4.5 is useful in the proof of the following theorem, which characterizes a vector structure in terms of its linear maps.

THEOREM. *Consider vector spaces over a field $\mathscr{P}$ such that the only automorphism of $\mathscr{P}$ is the identity map. (This is true, in particular, for the field of reals R.) Let*

$$(X, F, \Phi), \quad (Y, G, \Psi), \quad (X, F', \Phi'), \quad (Y, G', \Psi')$$

be vector spaces, with $\dim (X, F, \Phi) > 1$, $\dim (Y, G, \Psi) > 0$. *If the set of all linear maps from (X, F, Φ) to (Y, G, Ψ) coincides with the set of all linear maps from (X, F', Φ') to (Y, G', Ψ'), then*

$$F = F', \quad \Phi = \Phi', \quad G = G', \quad \Psi = \Psi'.$$

For the proof see **[8]**.

4.7. The condition on $\dim (X, F, \Phi)$ in Theorem 4.6 is essential, as is seen from the following example. Take $\mathscr{P} = R$. The usual addition and multiplication define a vector structure on R, which we denote simply by R. Now let $\lambda\colon R \to R$ be a bijective map such that $\lambda(0) = 0$, $\lambda(-1) = -1$, $\lambda(1) = 1$, $\lambda(st) = \lambda(s)\lambda(t)$ for all s, $t \in R$ (one may take, e.g., $\lambda(t) = \sqrt[3]{t}$). Consider a new vector structure R' on the same set R with addition

$$u \oplus v = \lambda(\lambda^{-1}(u) + \lambda^{-1}(v)), \qquad u, v \in R',$$

and multiplication $t * u = \lambda(t)u$ of $u \in R'$ and a real t. Then the sets of all linear maps of R into R and of R' into R' coincide, whereas the vector structures for $\lambda(t) \not\equiv t$ are different.

4.8. To extend Theorem 4.6 to an arbitrary field $\mathscr{P}$, define the notion of a *semivector space*. Fix a set $Q \subset \operatorname{Aut}\mathscr{P}$, and consider the set Γ_1 of all maps $\gamma : \mathscr{P}^2 \to \mathscr{P}$ of the form $\gamma(s,t) = a\sigma(s) + b\sigma(t)$, where $\sigma \in Q$ and $a, b \in \mathscr{P}$.

DEFINITION. A connected and separable Γ_1-space (X, F, Φ) with $X \neq \varnothing$ is called a *semivector space* with respect to Q. When $Q = \operatorname{Aut}\mathscr{P}$, X is called a *complete* semivector space (or, simply, semivector space). If Q contains only the identity map, X is a vector space in the sense of 4.1.

For any vector space (X, F, Φ) in the sense of 4.1, over an arbitrary field $\mathscr{P}$, extend the set Γ of linear maps $\mathscr{P}^2 \to \mathscr{P}$ to the set Γ_1 described in 4.8 with *all* automorphisms σ, that is, with $Q = \operatorname{Aut}\mathscr{P}$. Then, define the complete semivector space (X, F_1, Φ_1) on X corresponding to the extended set Γ_1. Its structure satisfies the conditions $\dot{\Phi}_1 F \subset \Gamma$, $\Phi \dot{F}_1 \subset \Gamma(X, F_1, \Phi_1)$ and it is called the complete semivector space *induced* by the vector space (X, F, Φ). Now, suppose again all linear maps of a vector space into another vector space are known. For a general field $\mathscr{P}$, this is not sufficient to recover the vector structure, but the induced semivector structure can be recovered, with the same dimensional restrictions as in Theorem 4.6.

THEOREM. *Consider vector spaces over an arbitrary field* $\mathscr{P}$. *Let*

$$(X, F, \Phi), \qquad (Y, G, \Psi), \qquad (X, F', \Phi'), \qquad (Y, G', \Psi')$$

be vector spaces, and

$$(X, F_1, \Phi_1), \qquad (Y, G_1, \Psi_1), \qquad (X, F_1', \Phi_1'), \qquad (Y, G_1', \Psi_1')$$

the corresponding induced semivector spaces, where

$$\dim (X, F, \Phi) > 1, \qquad \dim (Y, G, \Psi) > 0.$$

If the set of all linear maps from (X, F, Φ) *to* (Y, G, Ψ) *coincides with the set of all linear maps from* (X, F', Φ') *to* (Y, G', Ψ'), *then*

$$F_1 = F_1', \qquad \Phi_1 = \Phi_1', \qquad G_1 = G_1', \qquad \Psi_1 = \Psi_1'.$$

§5. Special relativity

In this section $\mathscr{K} = \text{Ens}$, $A = B = R$, and Γ is the set of all nondecreasing affine maps from R to R, that is, maps of the form $\gamma(t) = at + b$, $t \in R$, where $a \geq 0$ and b are elements of R.

5.1. We shall define by some axioms a type of Γ-spaces called *space-time* (or "universe"). The points of these spaces are called *events*. Suppose $X \neq \varnothing$ and introduce the following definitions concerning Γ-spaces (X, F, Φ).

5.2. A subset $\alpha \in X$ is called a *particle* if it contains more than one event, and if there exists a function $f \in F$ for which $\alpha = f(R)$. The set of all particles is denoted by H.

5.3. A particle α is said to be *massive* (or *time-like*) if it satisfies the following condition: for any nonconstant functions $f \in F$, $\varphi \in \Phi$ such that $f(R) = \alpha$, φf is nonconstant. The set of all massive particles is denoted by T. A particle α is said to be *massless* (or isotropic) if it belongs to the set $I = H \backslash T$.

The automorphisms of the Γ-space X map time-like particles into time-like, and isotropic particles into isotropic.

5.4. Let Γ' be the set of *all* affine maps from R to R; $\Gamma \subset \Gamma'$. Consider the Γ'-space (X, F', Φ') with the output base Φ. A subset $\alpha \subset X$ is called a *space-like straight line* if it contains more than one event, and if there exists a function $f \in F' \backslash F$ such that $\alpha = f(R)$. The set of all space-like straight lines is denoted by S. For the sake of brevity, we sometimes call a space-like straight line a *tachyon*.

5.5. A space-time X is said to be *connected* if for any two events $x, y \in X$ there exists a finite sequence of events $x_1, x_2, \dots, x_n$ such that $x_1 = x$, $x_n = y$ and for each $i = 1, \dots, n-1$ the events x_i, x_{i+1} belong to some particle, or to some tachyon.

5.6. Let $\mathscr{E} \subset X$. Denote by $G(\mathscr{E})$ the set of all automorphisms g of X such that $g(\mathscr{E}) = \mathscr{E}$. Then $G(\mathscr{E})$ is a subgroup of the group of all automorphisms $\operatorname{Aut} X$. Obviously, $G(\varnothing) = G(X) = \operatorname{Aut} X$. It is easily proved that $G(T) = G(I) = G(S) = \operatorname{Aut} X$.

5.7. Define the *dimension* of the space-time X. Take any set $\mathscr{E}_0 \subset X$ and define a sequence of sets $\mathscr{E}_k$ $(k = 1, 2, \dots)$ by the following recurrent rule: an event x belongs to $\mathscr{E}_k$ if and only if there exists a set $\alpha \in H \cup S$ such that $x \in \alpha$, and $\alpha \cap \mathscr{E}_{k-1}$ contains at least two events. The dimension of X (written $\dim X$) is the least integer n for which the following condition holds: there exists a set $\mathscr{E}_0$ consisting of $n+1$ events such that $X = \bigcup_{k=0}^{\infty} \mathscr{E}_k$. If no such integer exists, we put $\dim X = \infty$.

5.8. Now we formulate the axioms for a space-time X.

1) The Γ-space X is separable.
2) The Γ-space X is connected (in the sense of Definition 5.5).
3) The dimension of X is finite.
4) The group $\operatorname{Aut} X$ acts transitively on the set I.
5) For each $\alpha \in I$, the group $G(\alpha)$ acts transitively on α.

Note that in Axioms 4 and 5 one may change I to T or S. It seems probable that these two axioms can be weakened.

It can be shown that (X, F', Φ'), considered as an affine space, has the same dimension as a space-time.

Denote by C_x the union of all isotropic particles containing an event x. It turns out that the family $\{C_x\}_{x \in X}$ consists of parallel elliptic cones. In a suitable affine coordinate system $(x_1, x_2, \dots, x_n)$ a cone of this family with its apex in the origin has the equation $x_1^2 + x_2^2 + \dots + x_{n-1}^2 - x_n^2 = 0$. Thus, it is possible to distinguish "the future" from "the past" by the sign of x_n.

The group $\operatorname{Aut} X$ is completely characterized as the maximal group acting effectively on the affine space X and preserving the family of cones $\{C_x\}_{x \in X}$. A. D. Aleksandrov calls this group the general Lorentz group **[9]**. To exclude homotheties, we could require in the definition of space-time that $0 \leq a \leq 1$, instead of $a \geq 0$; then $\operatorname{Aut} X$ would be, in dimension four, the usual inhomogeneous Lorentz group, called also the Poincaré group.

§6. Metric spaces

In this section we consider two categories. Besides the category $\mathscr{K} = \mathrm{Ens}$ of all sets and all maps, we need another category L, with $\operatorname{Ob} L = \operatorname{Ob} \mathrm{Ens}$, that is, with all sets as objects, but with morphisms which are not maps: these are special binary relations called partial functions.

6.1. For any $X, Y \in \operatorname{Ob} L$ define $\operatorname{Hom}_L(X, Y)$ as the set of all subsets h_{XY} of $X \times Y$ such that for each $x \in X$ there exists at most one $y \in Y$ for which $(x, y) \in h_{XY}$. Each of the binary relations h_{XY} is identified with the corresponding function $h\colon \operatorname{pr}_X(h_{XY}) \to Y$ called a *partial function* from X to Y, assigning to any $x \in \operatorname{pr}_X(h_{XY})$ that $y \in Y$ for which $(x, y) \in h_{XY}$. Denote $h(X) = \operatorname{pr}_Y(h_{XY})$, $h^{-1}(Y) = \operatorname{pr}_X(h_{XY})$. The composition kh of morphisms $h \in \operatorname{Hom}(X, Y)$ and $k \in \operatorname{Hom}(Y, Z)$ is defined as a morphism $l \in \operatorname{Hom}(X, Z)$ with $l^{-1}(Z) = k^{-1}(h^{-1}(Z))$ and $l(x) = k(h(x))$. To complete the definition of the category L, define the identity morphism 1_X for any $X \in \operatorname{Ob} L$ as the partial function with $1_X(x) = x$ and $1_X^{-1}(x) = x$ for all $x \in X$.

For $h, k \in \operatorname{Hom}_L(X, Y)$, k is called an extension of h if $h^{-1}(Y) \subset k^{-1}(Y)$ and $k(x) = h(x)$ for $x \in h^{-1}(Y)$.

6.2. Fix two sets A, B and a set $\Gamma \subset \operatorname{Hom}_L(A, B)$ defining a category $L(\Gamma)$. Since now our morphisms are not maps, we need a new version of two definitions. A Γ-space (X, F, Φ) is said to be *connected* if for any points $p, q \in X$ there exists an $f \in F$ such that $p \in f(A)$, $q \in f(A)$. A Γ-space (X, F, Φ) is said to be *separable* if for any points $p, q \in X$, $p \neq q$, there exists $\varphi \in \Phi$ such that $p \in \varphi^{-1}(B)$, $q \in \varphi^{-1}(B)$, and $\varphi(p) \neq \varphi(q)$.

6.3. Now we define a metric space as a Γ-space of the category L. Let $A = B = R$, and let Γ consist of all morphisms $\gamma\colon A \to B$ such that

$$|\gamma(p) - \gamma(q)| \leq |p - q|$$

for any $p, q \in \gamma^{-1}(B)$.

DEFINITION. A connected and separable Γ-space (X, F, Φ) over the category L, with a set Γ as described above, is called a *metric space* if the following two conditions are satisfied:

(a) If $f \in F$ and $f(R)$ contains at most two points, then for any $p \in X$ the partial function f can be extended to a $g \in F$ such that $p \in g(R)$.

(b) If $\varphi \in \Phi$ and $\varphi^{-1}(B)$ contains at most two points, then for any $q \in X$ the partial function φ can be extended to a $\psi \in \Phi$ such that $q \in \Psi^{-1}(R)$.

The pair (F, Φ) is called a *metric structure* on X.

6.4. Let X be a metric space in the usual sense with a metric $\rho\colon X \times X \to R$. This metric engenders in a natural way a Γ-structure (F, Φ) that is a metric structure in the sense of 6.3. The sets F, Φ are defined by the following characteristic conditions:

(a) If $f \in F$ and $s, t \in f^{-1}(X)$, then

$$\rho\big(f(s), f(t)\big) \leq |t - s|.$$

(b) If $\varphi \in \Phi$ and $p, q \in \varphi^{-1}(R)$, then

$$|\varphi(p) - \varphi(q)| \leq \rho(p, q).$$

As is easily shown, (X, F, Φ) is a metric space in the sense of Definition 6.3.

6.5. Now let (X, F, Φ) be a metric space in the sense of 6.3. Define a usual metric ρ on X by the rule

$$\rho(p, q) = \sup |\varphi(p) - \varphi(q)|,$$

where sup is taken over all L-morphisms $\varphi \in \Phi$ such that $p, q \in \varphi^{-1}(B)$.

The same distance can be defined otherwise:

$$\rho(p,q) = \inf |f^{-1}(p) - f^{-1}(q)|,$$

where inf is taken over all L-morphisms $f \in F$ for which the sets $f^{-1}(p)$ and $f^{-1}(q)$ are singletons.

It can be easily proved that Definition 6.3 and the usual definition of a metric space are equivalent, that is, that the genus of a metric space structure in the sense of 6.3 is equivalent to the genus of the usual metric space structure.

6.6. Suppose A, B are arbitrary metric spaces, and $\Gamma \subset \mathrm{Hom}_L(A, B)$ is defined in the same way as in 6.3; that is, Γ consists of all morphisms $\gamma\colon A \to B$ such that

$$\rho_B\big(\gamma(p), \gamma(q)\big) \leq \rho_A(p,q)$$

for any p, $q \in \gamma^{-1}(B)$. An (A, B)-space is any connected and separable Γ-space (X, F, Φ), with Γ just described, satisfying conditions analogous to (a), (b) of 6.3. Then each (A, B)-space is a metric space. Moreover, if the sets of all distances in both A and B are dense in the space of all nonnegative reals, then any metric space can be represented as an (A, B)-space.

6.7. Starting with this subsection, we shall work in the category $\mathscr{K}$. Let $A = B = R$, and let Γ be the set of all contractions from R to R, that is, maps $\gamma\colon R \to R$ such that

$$|\gamma(s) - \gamma(t)| \leq |s - t|$$

for all s, $t \in R$.

DEFINITION. A connected and separable Γ-space (X, F, Φ), with Γ as just described, is called *metric space with intrinsic metric*.

6.8. The usual definition of a space with an instrinsic metric, in the sense of A. D. Aleksandrov, is as follows **[10]**: a metric on a set is called *intrinsic* if the distance between any two points of this set equals the exact lower bound of lengths of all curves connecting these points. For brevity, we call these spaces Aleksandrov spaces.

On each Aleksandrov space a Γ-structure (F, Φ) is defined. The sets F and Φ are characterized by the following conditions:

(a) If $f \in F$, then

$$\rho\big(f(s), f(t)\big) \leq |s - t|$$

for any s, $t \in R$.

(b) If $\varphi \in \Phi$, then

$$|\varphi(p) - \varphi(q)| \leq \rho(p,q)$$

for any p, $q \in X$.

It can be shown that (X, F, Φ) is a metric space with an intrinsic metric, in the sense of 6.7.

6.9. Conversely, on a connected and separable Γ-space (X, F, Φ) as defined in 6.7 the structure of Aleksandrov space is obtained as follows: the distance between points $p, q \in X$ is the number

$$\rho(p, q) = \sup |\varphi(p) - \varphi(q)|,$$

where sup is taken over all $\varphi \in \Phi$. There is an alternative definition:

$$\rho(p, q) = \inf |f^{-1}(p) - f^{-1}(q)|,$$

where inf is taken over all $f \in F$ such that the sets $f^{-1}(p)$, $f^{-1}(q)$ are singletons. Thus, Definition 6.7 is equivalent to the definition of an Aleksandrov space.

6.10. Euclidean spaces are characterized as Γ-spaces within the category $\mathscr{K}$ by means of a particularly simple Γ. Take $A = B = \{0, 1\}$; that is, let A and B each consist of two elements realized as the integers 0, 1. There are four maps $\gamma: A \to B$. Γ is obtained by excluding one of the constant maps; thus $\Gamma = \{\gamma_1, \gamma_2, \gamma_3\}$, where

$$\gamma_1(0) = 0, \quad \gamma_1(1) = 1; \quad \gamma_2(0) = 1; \quad \gamma_2(1) = 0; \quad \gamma_3(0) = \gamma_3(1) = 0.$$

Let X be a metric space with metric ρ. Define on X a Γ-structure (F, Φ) as follows:

(a) $f: A \to X$ belongs to F if and only if $\rho(f(0), f(1)) > 1$.

(b) $\varphi: X \to B$ belongs to Φ if and only if the diameter of the set $\varphi^{-1}(1)$ does not exceed 1.

Denote this Γ-structure by $\sigma(X, \rho)$. We say that Γ remembers the metric structure on X if $\sigma(X, \rho) = \sigma(X, \rho')$ implies $\rho = \rho'$. As is easily seen, Γ cannot remember the Euclidean metric on R. But for any dimension $n > 1$, Γ remembers the metric of the Euclidean space R^n, as can be inferred from a theorem by Beckman and Quarles **[11]**. The result is as follows:

THEOREM. *Let ρ and ρ' be Euclidean metrics on a set X, both of dimensions* > 1. *Then, if $\sigma(X, \rho) = \sigma(X, \rho')$ the metrics ρ, ρ' coincide.*

It would be interesting to know whether this theorem can be generalized to broader classes of metric spaces X.

§7. Topological spaces

In this section $\mathscr{K} = \mathrm{Ens}$, $A = \{0, 1, 2, \dots\}$, $B = \{0, 1\}$. The set Γ, unless stated otherwise, is defined as follows: a map $\gamma: A \to B$ belongs to Γ if and only if $\gamma^{-1}(0)$ is a finite set when $\gamma(0) = 1$. Thus, Γ contains all maps γ for which $\gamma(0) = 0$, and some of those for which $\gamma(0) = 1$, namely the maps for which $\gamma^{-1}(0)$ is finite.

7.1. Let (X, F, Φ) be a Γ-space. It can be shown that all sets of the form $\varphi^{-1}(1)$, considered as open sets, define a topology on X. This topology is called *engendered* by the Γ-structure (F, Φ). We suppose without further notice that any Γ-space is endowed with a topology engendered by its Γ-structure.

7.2. The following two statements hold for any Γ-space (X, F, Φ):

(a) If a sequence of points $(x_1, x_2, \dots, x_n, \dots)$ converges in X to a point x_0, then the function $f: A \to X$ defined by $f(m) = x_m$ belongs to F.

(b) If $f \in F$, then the sequence $(f(1), f(2), \dots)$ converges to $f(0)$.

The procedure described in 7.1 leads to a broad class of topological spaces, though, as we shall see in 7.3, does not yield all of them. For example, we have the following

THEOREM. *Any topological space with a countable base can be obtained as a* Γ-*space with* Γ *as described above.*

7.3. Let X be any noncountable set. Denote by X_1 the topological space supported by X with the strongest topology, that is, the space in which all sets are open. Consider another topological space X_2 on X, characterized by the condition: a set $U \subset X$ is open if and only if $X \backslash U$ is at most countable. As is easily seen, in both spaces X_1 and X_2 a sequence $(x_1, x_2, \dots)$ converges to x_0 if and only if $x_m = x_0$ beginning from some number m_0. It is obvious that only X_1 can be represented as a Γ-space, but not X_2.

Note that the set Φ' of all maps $\varphi\colon X \to B$ for which $\varphi^{-1}(1)$ is open in X_2 is an output base for the Γ-structure of X_1.

7.4. Let T be the set of all topologies on a set X. Consider two topologies of T equivalent if any sequence converging to a point in one of them converges to the same point in the other. As shown by Y. Nikanorov, a student at the Novosibirsk University, in each of the equivalence classes only the strongest topology is engendered by some Γ-structure. Thus, for a given topology on X one can find one and only one equivalent topology, engendered by a Γ-structure.

7.5. Now let A be an arbitrary set with a filter Σ, $B = \{0, 1\}$ and Γ the set of all maps $\gamma\colon A \to B$ such that $\gamma^{-1}(1) \in \Sigma$.

With these A, B, and Γ, consider a Γ-space (X, F, Φ). It can be shown that the sets $\varphi^{-1}(1)$, for all $\varphi \in \Phi$, define a topology on X, of which they are the open sets. Call this topology *engendered* by the Γ-structure (F, Φ). Then for each topological space X a set A with a filter Σ can be found such that the topology of X is engendered by the Γ-structure just described. This procedure defines the corresponding spaces for each pair (A, Σ).

7.6. Let, again, $A = \{0, 1, 2, \dots\}$, $B = \{0, 1\}$ and let Γ be the set of all maps $\gamma\colon A \to B$ such that $\gamma^{-1}(1)$ is *infinite* when $\gamma(0) = 1$.

DEFINITION. A Γ-space (X, F, Φ) is called a *topological* Γ-*space* if the family of all sets $\varphi^{-1}(1)$, with $\varphi \in \Phi$, satisfies the axioms for open sets of a topological space.

It can be proved that the class of topological Γ-spaces coincides with the class of topological spaces defined in 7.1.

As will be shown by examples in 7.7–7.9, there exist Γ-spaces in the sense of 7.6 that *are not topological* Γ-spaces, that is, for which the corresponding families of sets $\varphi^{-1}(1)$ do not satisfy the axioms for open sets. So, we have a generalization of the concept of topological space deserving further investigation.

7.7. Let X be the real line R. Define the set F' as follows: a function $f\colon A \to R$ belongs to F' if and only if the following two conditions are satisfied:

(a) The sequence $(f(1), f(2), \dots)$ converges in the Euclidean topology of R.

(b) $f(2n) \le f(0) \le f(2n+1)$ for each $n \in A$.

R is endowed with a Γ-structure (F, Φ) for which F' is an input base. This Γ-structure is not topological.

7.8. Let X be the plane R^2. Define the set F' as follows: a function $f: A \to R^2$ belongs to F' if and only if the following two conditions are satisfied:

(a) $\lim_{n\to\infty} f(n) = f(0)$ in the Euclidean topology of R^2.

(b) $f(n) \neq f(0)$ for $n \neq 0$, and the union of straight lines connecting $f(0)$ with $f(n)$ is dense in R^2, in the Euclidean topology of R^2.

R^2 is endowed with a Γ-structure (F, Φ) for which F' is an input base. This Γ-structure is not topological.

7.9. Let X be the m-dimensional Euclidean space R^m ($m \geq 2$). Define F' as follows: a function $f: A \to R^m$ belongs to F' if and only if there exists an infinite number of indices n for which the Euclidean distance between $f(0)$ and $f(n)$ does not exceed 1. Take this F' as an input base for a Γ-structure (F, Φ). It can be shown that each Γ-automorphism of the space (X, F, Φ) is an orthogonal transformation. Moreover, each Γ-endomorphism of this space is a contraction. This is not a topological space, either.

References

1. R. Goldblatt, *Topoi. The categorial analysis of logic*, North-Holland, Amsterdam and New York, 1979.
2. S. Lang, *Algebra*, Addison-Wesley, Reading, MA, 1965.
3. I. Bucur and A. Deleanu, *Introduction to the theory of categories and functors*, Interscience, New York, 1968.
4. S. Mac Lane, *Categories for the working mathematician*, Springer-Verlag, New York and Berlin, 1971.
5. P. M. Cohn, *Universal algebra*, 2nd ed., Reidel, Boston, 1981.
6. M. Berger, *Geometrie*, CEDIC, Paris, 1977.
7. V. K. Ionin, *One way to define affine structure*, Geometritcheskii Sb. **1982**, no. 23, 3–16. (Russian)
8. V. K. Ionin and E. K. Suleimenov, *Characterization of vector structures by linear maps*, Dokl. Akad. Nauk SSSR **268** (1983), no. 4, 781–784; English transl. in Soviet Math. Dokl. **27** (1983).
9. A. D. Aleksandrov, *Cones with transitive group*, Dokl. Akad. Nauk SSSR **189** (1969), no. 4, 695–698; English transl. in Soviet Math. Dokl. **10** (1969).
10. ———, *Intrinsic geometry of convex surfaces*, OGIZ, Moscow, 1948. (Russian)
11. F. S. Beckman and D. A. Quarles, Jr., *On isometries of Euclidean spaces*, Proc. Amer. Math. Soc. **4** (1953), 810–815.

Amer. Math. Soc. Transl.
(2) Vol. **163**, 1995

On the Absence of Sullivan's Cusp Finiteness Theorem in Higher Dimensions

Michael Kapovich

ABSTRACT. We prove the existence of a discontinuous finitely generated free conformal group K_3 acting on $\mathbb{R}^3$ such that the number of rank 1 cusps of K_3 is infinite. Small deformations of K_3 provide a finitely generated Kleinian group with infinitely many conjugacy classes of finite order elements.

§1. Introduction

1.1. In this paper we continue the discussion **[K-P 1]** of the failure of Ahlfors's finiteness theorem for Kleinian groups in dimensions ≥ 3. Here we establish the failure of Sullivan's cusp finiteness theorem. Namely

THEOREM. *There exists a finitely generated free Kleinian group $K_3 \subset \mathrm{Mob}(S^3)$ such that*

(a) *The number of conjugacy classes of maximal parabolic subgroups of K_3 is infinite.*

(b) *The fixed points of these parabolic subgroups $\langle u_i \rangle \subset K_3$ have pairwise disjoint horoball neighborhoods $U_i \subset H^4$ which are precisely invariant under $\langle u_i \rangle$ in K_3.*

(c) *If $K_n \subset \mathrm{Mob}(S^n)$ is the conformal (Poincaré) extension of K_3 to S^n $(n \geq 3)$, then*

$$\mathrm{rank}\Big(H_{n-1}\big(\Omega(K_n)/K_n, \mathbb{Q}\big)\Big) = \infty.$$

Here and below $\Omega(\ )$ denotes the discontinuity domain for a Kleinian group.

COROLLARY 1. *For each $n \geq 3$ there exists a finitely generated Kleinian group $K_n \subset \mathrm{Mob}(S^n)$ such that the quotient manifold $M(K_n) = \Omega(K_n)/K_n$ has infinite homotopy type.*

So, the analog of Ahlfors's finiteness theorem fails for every dimension $n \geq 3$.

COROLLARY 2. *For all but finitely many $q \in \mathbb{Z}$ there exist:*
1) *an integer r, the rank of free group $F_r = \langle x_1, \ldots, x_r \rangle$,*

1991 *Mathematics Subject Classification*. Primary 57S30; Secondary 20H10, 30F40.

2) *an automorphism* $\theta\colon F_r \to F_r$ *such that the group*

$$F_{r,q} = \left\langle x_1, \dots, x_r : \ (\theta^n(x_1))^q = 1, \ n = \pm 1, \pm 2, \dots \right\rangle$$

has infinitely many conjugacy classes of finite order elements $[\theta^n(x_1)]$ *and* $F_{r,q}$ *admits a discrete faithful representation in* $\mathrm{Mob}(S^3)$.

REMARK. According to Selberg's lemma **[Se]** $F_{r,q}$ has a torsion-free finite index subgroup (as a finitely generated linear group).

The manifold $\Omega(K_3)/K_3$ is homeomorphic to an open handlebody X_r from which some ∞-component link L_∞ is removed. Each component of the link is a knot representing a free generator of $\pi_1(X_r)$.

The underlying space of the orbifold $M(F_{r,q})$ is homeomorphic to X_r, where the singular set is L_∞(see above).

Historical remarks.

1.2. The first version of the cusp finiteness theorem for Kleinian subgroups of $PSL(2, \mathbb{C})$ was given by Ahlfors **[Ah]** (see also **[Kr 1]**). In particular, he proved that for a finitely generated Kleinian group $K \subset PSL(2, \mathbb{C})$ the surface $\Omega(K)/K$ has only finitely many punctures. Combined with the Leutbecher-Shimizu lemma, this implies that K can have only finitely many conjugacy classes of maximal parabolic subgroups of rank 1. Ahlfors's proof was based on analytic function theory and fails in higher dimensions. Fifteen years later Sullivan **[Su]** proved that any finitely generated discrete subgroup of $PSL(2, \mathbb{C})$ has only finitely many cusps (i.e., conjugacy classes of maximal parabolic subgroups). Sullivan also gave a numerical estimate for the number of cusps (in terms of the number of generators of the group K). That estimate was improved by Kulkarni and Shalen **[K-S, K]** and by Abikoff **[Ab 1]**, who used topological considerations based on existence of a compact "Scott core" for 3-manifolds with finitely generated fundamental group. Certainly, a "Scott core" does not exist in higher dimensions. In contrast, Sullivan exploited Eisenstein series associated with cusps of K and elements of the finite-dimensional space $H^1(K, sl(2, \mathbb{C}))$ (these arguments were elaborated in **[Kr 2]**). The idea of Eisenstein series remains meaningful in higher-dimensional hyperbolic spaces; hence there was a hope of generalizing Sullivan's proof to the case of discrete subgroups in $\mathrm{Mob}(S^n)$ ($n > 2$). This is probably possible under some restrictions on K (see the Conjecture below). However, the main theorem above shows that in general the cusp finiteness theorem fails in higher dimensions.

CONJECTURE. *Let* Γ *be a finitely generated discrete subgroup of* $\mathrm{Mob}(S^n)$ *such that*:

(1) *any cusp of* Γ *has "maximal" rank* $= n$;

(2) *every cusp of* Γ *corresponds to a "cusp end" of* H^{n+1}/Γ.

Then Γ *has only finitely many cusps.*

1.3. Evidently, every geometrically finite discrete subgroup of $\mathrm{Mob}(S^n)$ has only finitely many cusps.

1.4. We now discuss Corollary 2. Every finitely generated discrete subgroup of $\mathrm{Mob}(S^2)$ has only finitely many conjugacy classes of finite order elements (briefly CCFE) (the author could not find an exact reference; this statement will be proved in §7). Every geometrically finite subgroup of $\mathrm{Mob}(S^n)$ has only a finite number of

CCFE. As a generalization of this fact, Gromov **[Gr]** proved that any word hyperbolic group has only finitely many CCFE. There is also an old lemma of Selberg **[Se]** on this matter. Selberg showed that if in every finitely generated subgroup of $SL(n, \mathbb{C})$ the finite subgroups have bounded order, then there are only finitely many $SL(n, \mathbb{C})$–conjugacy classes of finite order elements. Probably, examples like Corollary 2 are known to algebraists for the case of arbitrary finitely generated subgroups of $SL(n, \mathbb{C})$.

1.5. Corollary 1 for $n = 3$ was proved by a different example in the joint paper of the author and Potyagailo **[K-P 1]**. For $n \geq 5$, Corollary 1 cannot be deduced from **[K-P 1]**, in view of Potyagailo's remark that the Kleinian group constructed there is not finitely presented.

1.6. The results of the present paper were announced in **[Ka 1, Ka 2]**. Corollary 2 was announced in **[K-P 2]**. In this current article we simplify the original proof of the main theorem first given in **[K-P 2]**.

§2. Outline of the proofs

2.1. Consider the configuration of four Euclidean spheres $\Sigma_1, \Theta_2, \Theta_3, \Sigma_4 \subset \mathbb{R}^3$ drawn in Figure 1. We shall construct Kleinian groups $\Gamma'_1, \Gamma'_2, \Gamma'_3, \Gamma'_4$ whose limit sets are the spheres $\Sigma_1, \Theta_2, \Theta_3, \Sigma_4$, respectively. All the groups Γ'_i (up to finite index) are conjugate in $\mathrm{Mob}(S^3)$ to the groups Γ_i of our previous paper **[K-P 1]**. The groups Γ'_i possess the following properties:

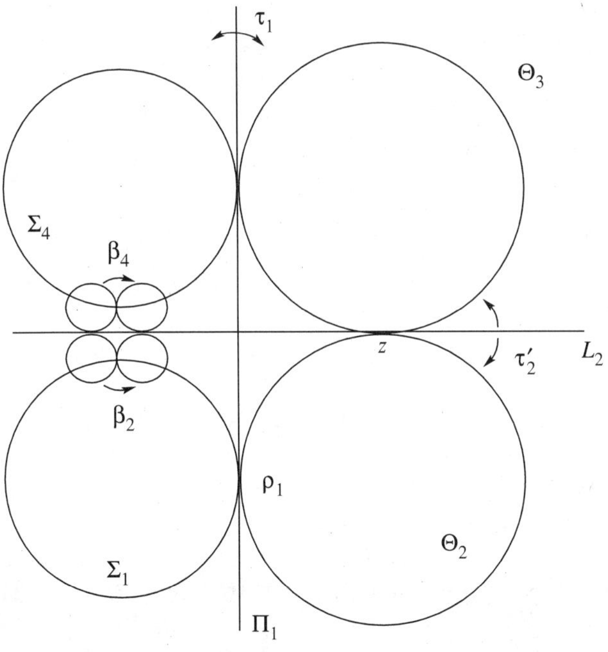

FIGURE 1

(1) The group $G' = \langle \Gamma'_1, \Gamma'_2, \Gamma'_3, \Gamma'_4 \rangle$ is Kleinian and contains a finitely generated free normal subgroup F' such that $G'/F' \simeq \mathbb{Z}$.
(2) If τ'_2 is the reflection in the symmetry plane L_2, then $\tau'_2 \Gamma'_1 \tau'_2 = \Gamma'_4$, $\tau'_2 \Gamma'_2 \tau'_2 = \Gamma'_3$.
(3) The groups $\Gamma'_1 \cap F'$ and $\Gamma'_4 \cap F'$ contain parabolic elements β_2 and $\beta_4 = \tau'_2 \beta_2 \tau'_2$, respectively, such that the isometric spheres $I(\beta_2)$, $I(\beta_2^{-1})$ touch L_2 at some point $x = I(\beta_2) \cap L_2$.
(4) The point x is a fixed point of the parabolic transformation $u = (\beta_4)^{-1} \circ \beta_2$. This point has a cusp neighborhood, precisely invariant under $\langle u \rangle$ in G'.

Then the group F' is the group K_3 we need. Indeed, consider an element $t_2 \in G'$ such that $\langle F', t_2 \rangle = G'$. The elements u and $u_m = t_2^m \, u \, t_2^{-m}$ $(m \in \mathbb{Z})$ are parabolic. Property (4) implies that the groups $\langle u_m \rangle$ are maximal nonconjugate parabolic subgroups of F'. So, Property (a) holds. Property (b) also follows from (4). Moreover, according to (3), the manifold $M(K_n)$ has infinitely many cusp ends, giving an infinite system of independent cocycles in $H_{n-1}(M(K_n), \mathbb{Q})$. This consideration completes the proof of the theorem.

2.2. The proof of Corollary 2 proceeds as follows. We slightly enlarge the spheres Σ_1 and Σ_4 so that the isometric sphere $I(\beta_2)$ intersects L_2 at an angle $\pi/2q$, $q \in \mathbb{Z}\backslash\{0\}$. Next, repeating the previous construction, we obtain representations $\rho_q : K_3 \to \mathrm{Mob}(S^3)$, $\lim_{q\to\infty} \rho_q = \mathrm{id}$. Then the elements $\rho_q(u_m)$ become elliptic of order q. These elliptic elements are not conjugate in $\rho_q(K_3)$ which happens to be Kleinian (if q is sufficiently large).

§3. Proof of the theorem

3.0. Notation. Below, we shall denote by $P(K)$ an isometric fundamental domain for a Kleinian group K and by $\Lambda(K)$ its limit set. If S is a closed surface in $\mathbb{R}^3$, then $\mathrm{int}(S)$ will denote the interior of the compact component of $\mathbb{R}^3\backslash S$ and $\mathrm{ext}(S) = S^3 \backslash \mathrm{cl}(\mathrm{int}(S))$.

Recall that a subset S of S^n is called *precisely invariant* under a subgroup H of a discrete group $G \subset \mathrm{Mob}(S^n)$, if $h(S) = S$ for every $h \in H$ and $g(S) \cap S = \varnothing$ for any $g \in G\backslash H$.

3.1. First we recall several constructions of our previous paper **[K-P 1]**. Let us consider the unit sphere $\Sigma_1 \subset \mathbb{R}^3$ centered at zero; $p_1 = (0,1,0)$, $p_2 = (0,0,1)$. Let Π_i be the extended Euclidean planes tangent to Σ at the points p_i (see Figure 1) and let Π_i^- be the component of $\mathbb{R}^3\backslash\Pi_i$ such that $\Pi_i^- \cap \Sigma = \varnothing$ $(i = 1, 2)$.

In **[K-P 1,** Lemma 1 and 2] we constructed a Kleinian group Γ_1 with the following properties:

(1) Γ_1 contains maximal parabolic subgroups $\widetilde{H}_1$, $\widetilde{H}_2$ with limit points p_1, p_2, respectively.
(2) Γ_1 contains a free finitely generated normal subgroup F_1 such that: $\Gamma_1/F_1 \cong \mathbb{Z}$, $\langle F_1, \widetilde{t}_2 \rangle = \Gamma_1$ for some $\widetilde{t}_2 \in \widetilde{H}_2$; $\langle \widetilde{t}_2 \rangle \oplus \langle \beta_2 \rangle = \widetilde{H}_2$.
(3) The group Γ_1 has a fundamental set P such that $P \cap \mathrm{cl}(\Pi_i^-)$ is an isometric fundamental domain for the action of the group $\widetilde{H}_i$ on $\mathrm{cl}(\Pi_i^-)$.
(4) The group Γ_1 has the S-RF property for every geometrically finite subgroup $S \subset \Gamma_1$; i.e., for any element $g \in \Gamma_1\backslash S$ there is a finite-index subgroup $\Gamma_1^0 \subset \Gamma_1$ that contains S but not g.

REMARK. Without loss of generality, we can assume that the center of $I(\beta_2)$ has coordinates $(0, a_2, a_3)$.

Denote by L_2 the plane $\{(x_1, \lambda, x_3) : (x_1, x_3) \in \mathbb{R}^3\}$ such that $\lambda > 1$ and L_2 is tangent to the isometric spheres $I(\beta_2)$, $I(\beta_2^{-1})$ (see Figure 2). Let $\overline{L}_2 = L_2 \cup \{\infty\}$, $L_2^- = \{(x_1, x_2, x_3) : x_2 > \lambda_2\}$. Denote by τ_1 the reflection in the plane Π_1 and by τ_2' the reflection in the plane L_2. Put $L_2^+ = \tau_2'(L_2^-)$, $\Pi_i^+ = \tau_1(\Pi_1^-)$.

LEMMA 1. *These exists a finite-index subgroup $\Gamma_1' \subset \Gamma_1$ possessing the following properties*:

(a) $\beta_2 \in \Gamma_1'$;
(b) $\Gamma_1' = \langle F_1', t_2 = (\widetilde{t}_2)^n\rangle$ *for some finitely generated free normal subgroup* $F_1' \subset \Gamma_1'$ *and* $n \in \mathbb{Z}$;
(c) *if* $\gamma \in \Gamma_1' \backslash \{\beta_2, \beta_2^{-1}, 1\}$, *then* $I(\gamma) \cap L_2' = \varnothing$;
(d) *let* $H_i' = \widetilde{H}_i \cap \Gamma_1'$ $(i = 1, 2)$. *If* $\gamma \in \Gamma_1' \backslash H_1'$ *then* $I(\gamma) \cap (\Pi_1 \backslash P(H_1')) = \varnothing$.

PROOF. Note that $\overline{L}_2'$ and $\mathrm{cl}(\Pi_1 \backslash P(H_1'))$ are compact subsets of $H^3 = \mathrm{ext}\,\Sigma_1$. Then there exist only finitely many elements $\{\gamma_1, \dots, \gamma_p\} \subset \Gamma_1$ such that $I(\gamma_i) \cap \overline{L}_2' \neq \varnothing$. The group Γ_1 has the S-RF property for geometrically finite subgroups S. Hence we can find a finite-index subgroup $\Gamma_1'' = \langle F_1'', t_2 = (\widetilde{t}_2)^n\rangle < \Gamma_1$ such that $\{\gamma_1, \dots, \gamma_p\} \cap \Gamma_1'' \backslash \{\beta_2, \beta_2^{-1}, 1\} = \varnothing$. Let $H_2' = \Gamma_1'' \cap \widetilde{H}_2$. Then there are only finitely many elements $\{\gamma_1', \dots, \gamma_k'\} \subset \Gamma_1'' \backslash H_2'$ such that $I(\gamma_i') \cap \mathrm{cl}(\Pi_1 \backslash P(H_1')) \neq \varnothing$. Property (3) of the group Γ_1 implies that $(\widetilde{H}_2 * \widetilde{H}_1) \backslash H_2' \cap \{\gamma_1', \dots, \gamma_k'\} = \varnothing$. The Schottky-type group $H_2' * H_1'$ is geometrically finite. Hence Γ_1'' has the $H_2' * H_1'$–RF property and we can find a finite-index subgroup $\Gamma_1' \subset \Gamma_1''$ such that $H_2' \cup H_1' \subset \Gamma_1'$, $\Gamma_1' \cap \{\gamma_1', \dots, \gamma_q'\} = \varnothing$. The group Γ_1' has all the desired properties.
Lemma 1 is proved. □

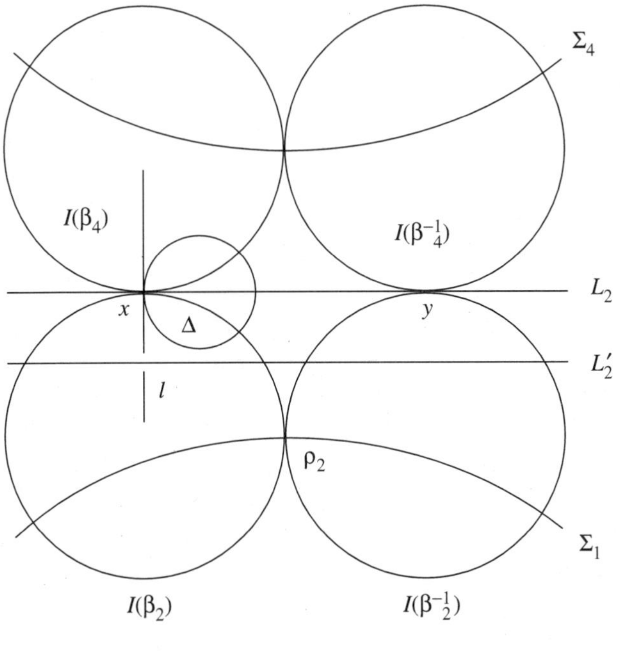

FIGURE 2

3.2. Let Θ_2 be a sphere that touches Σ_1 at the point p_1 and touches L_2 at some point $z \in \Pi_1^-$. Next, let $x = L_2 \cap I(\beta_2)$ and $y = L_2 \cap I(\beta_2^{-1})$. There exists a unique orientation-reversing Möbius transformation T that maps p_2 to z and commutes with every element of H_1'. It is easy to see that $T^{-1}(L_2^-) \subset \Pi_2^-$ and that $T^{-1}(\bar{L}_2^-)$ touches Σ_1 at the point p_2. Hence, it follows from property (3) above that L_2^- is precisely invariant under $H_3' = T H_2' T^{-1}$ in the group $\Gamma_2' = T\Gamma_1 T^{-1}$. Also, for the groups $\Gamma_1' = \Gamma_1$, Γ_2' and the amalgamated subgroup $\widetilde{H}_1$ the conditions of the first Maskit Combination Theorem are fulfilled.

3.3. LEMMA 2. *The group* $G_1' = \langle \Gamma_1', \Gamma_2' \rangle$ *is Kleinian and has a simply connected invariant component* $\Omega_1' \ni \infty$. *The manifold* Ω_1'/G_1' *is obtained by gluing together two hyperbolic components homeomorphic to* $(\mathrm{ext}\, \Sigma_1 \setminus \Gamma_1'(\Pi_1^-))/\Gamma_1'$. *The manifold* Ω_1'/G_1' *fibers over* S^1. *The group* $\Phi_1' = \langle F_1', T F_1' T^{-1} \rangle$ *is a finitely generated free normal subgroup of* G_1' *corresponding to the fiber* $G_1'/\Phi_1' \cong \mathbb{Z}$. *There is a fundamental set* D_1 *for the action of* G_1' *on* Ω_1' *such that:*

(i) $(D_1 \cap \mathrm{cl}\, L_2^-) \cup \{y\}$ *is a fundamental set for the action of* H_3' *on* $\mathrm{cl}(L_2^-)$;

(ii) *for some* $\lambda' < \lambda$ *and plane* $L_2' = \{(x_1, \lambda', x_3) : (x_1, x_3) \in \mathbb{R}^2\}$ *the intersection* $L_2' \cap D_1 \cap \tau_1(\Pi_1^-)$ *coincides with* $L_2' \cap P(\langle \beta_2 \rangle) \cap \tau_1(\Pi_1^-)$.

PROOF. All statements, except (i) and (ii), easily follow from the Maskit Combination Theorem (see the proof of Lemma 3 in **[K-P 1]**). Let $D_1 = (P \setminus \Pi_1^-) \cup T(P \setminus \Pi_1^-)$. Then statements (i), (ii) follow from the properties of the group Γ_1' listed above. □

3.4. Introduce the following notation: $J = H_3'$, $X_2 = L_2^- \cup \left(\overline{L}_2 \setminus J(\{x, y, z\})\right)$, $X_1 = \tau_2'(X_2)$, $G_2' = \tau_2' G_1' \tau_2'$, $\Phi_2' = \tau_2' \Phi_1' \tau_2'$, $D_2 = \tau_2'(D_1)$.

Direct considerations based on Lemma 1 imply that the triple (G_1', G_2', J) is proper interactive **[Mk,** Chapter VII] for the pair of sets (X_1, X_2). Moreover, $D_i \cap X_i \subset \overline{L}_2$ for $i = 1, 2$.

3.5. LEMMA 3.

(1) *The groups* G_1', G_2', J *satisfy the conditions of the weak Maskit Combination Theorem* **[Mk,** Chapter VII, Theorem A.15].

(2) *The group* $G' = \langle G_1', G_2' \rangle$ *is isomorphic to* $G_1' *_J G_2'$.

(3) *The set* $D = (D_1 \cap X_1) \cup (D_2 \cap X_1)$ *does not contain points equivalent under the action of* G'.

(4) $\mathrm{int}(D) \subset \Omega(G')$.

PROOF. The first statement follows from item 3.4; the remaining statements follow from the weak Maskit Combination Theorem. □

3.6. Denote by Ω' the component of $\Omega(G')$ containing the point ∞. It is easy to see that $G'(\Omega') = \Omega'$. Let $\beta_4 = \tau_2' \beta_2 \tau_2'$; then the element $u = (\beta_4)^{-1} \circ \beta_2$ is composition of the inversions in the spheres $I(\beta_2)$, $I(\beta_4)$. Hence u is a parabolic transformation conjugate to Euclidean translation; $u(x) = x$.

3.7. LEMMA 4. *Let* G^n *be the conformal extension of* G' *to the space* $\mathbb{R}^n$ $(n \geq 3)$. *Then the point* x *is a parabolic cusp point for the group* G^n. *If* $g \in G^n$ *stabilizes the point* x, *then* $g \in \langle u \rangle$.

PROOF. First we construct a cusp-neighborhood of the point x in $\mathbb{R}^3$. Let l be the straight line which passes through x and is orthogonal to Π_2. Let Δ be a closed disk with the following properties (see Figure 2):

(i) $\Delta \subset \{(0, x_2, x_3)\colon (x_2, x_3) \in \mathbb{R}^2\}$;

(ii) $\partial\Delta$ touches l at x;

(iii) the diameter of Δ is less than the radius of $I(\beta_2)$;

(iv) $\Delta \cap L_2' = \varnothing$.

Then we define $\mathscr{O}$ to be the open body in $\mathbb{R}^3$ obtained by rotating $\operatorname{int}(\Delta)$ around l. Obviously we have $u(\mathscr{O}) = \mathscr{O}$. Next we shall prove $\mathscr{O}$ to be precisely invariant under $\langle u\rangle$ in G'.

By Lemmas 2 and 3, the intersection $\mathscr{O}_- = \mathscr{O} \cap \operatorname{cl}\operatorname{ext} I(\beta_2) \cap \operatorname{cl}\operatorname{ext} I(\beta_4)$ lies in D and contains no G'-equivalent points.

Let $w\colon \mathbb{R}^3 \to \mathbb{R}^3$ be the translation $w\colon A \mapsto A + (y - x)$. Then the set $w(\mathscr{O}_-)\backslash\big(I(\beta_2^{-1}) \cup I(\beta_4^{-1})\big) = \mathscr{O}_-'$ is also contained in D.

Hence, $\mathscr{O}_- \cup \beta_2^{-1}(\mathscr{O}_-') = \mathscr{O}_+$ contains no G'-equivalent point; this set is a fundamental domain for the action of $\langle u\rangle$ in $\mathscr{O}$. Thus $\mathscr{O}$ is precisely invariant under $\langle u\rangle$, x is a cusp point for G', and $\langle u\rangle$ is a maximal elementary subgroup of G'.

It remains to verify the statement concerning the conformal extension G^n $(n \geq 4)$ of the group $G^3 = G'$. The parabolic transformation $u \in G^n$ is conjugate to a Euclidean translation. Then the existence of a precisely invariant cusp-neighborhood $\mathscr{O}_n$ of the point x (with respect to G^n, $n \geq 4$) easily follows from the properties of $\mathscr{O}$ and **[W]**. Lemma 4 is proved. □

3.8. Let $F' = \langle\Phi_1', \Phi_2'\rangle$ (see 3.4). Then Lemma 3 implies that F' is normal in G' and $G'/F' \cong \mathbb{Z}$. Also, we have that F' is a finitely generated free group; $G' = \langle F', t_2\rangle$, $\{\beta_2, \beta_4\} \subset F'$. Hence, the element $u = \beta_2 \circ (\beta_4)^{-1}$ is contained in F'.

Next, we put $u_m = t_2^m\, u\, t_2^{-m}$, $m \in \mathbb{Z}$. Every u_m belongs to F'. Also, if $g'(x) = x$, $g' \in G'$, then $g' \in \langle u\rangle$. Suppose for a moment that $g\, u_m\, g^{-1} = u_k$, $g \in F'$. Then $(t_2^{-k} g\, t_2^m) u (t_2^{-m} g^{-1}\, t_2^k) = u$ and $t_2^{-k} g\, t_2^m(x) = x$. Hence $t_2^{-k} g\, t_2^m = u^r$ and $t_2^{m-k} \in F'$. However, it now follows that $t_2^{m-k} = 1$, $m = k$, and $u_m = u_k$.

Thus, the parabolic groups $\langle u_k\rangle$, $k \in \mathbb{Z}$, are maximal nonconjugate parabolic subgroups of $F' = K_3$ and we obtain property (a) (see the main theorem).

3.9. The point x admits a precisely invariant cusp-neighborhood $\mathscr{O} \subset \mathbb{R}^3$ with respect to the group G'. Hence **[W]** implies that x has a precisely invariant horoball neighborhood U_0 in $\mathbb{H}^4$. The desired horoball neighborhoods $U_i \subset \mathbb{H}^4$ of the points $t_2^i(x)$ are equal to $t_2^i(U_0)$. Thus, we have proved property (b) of the group $K_3 = F'$.

REMARK. In fact, the configuration in Figure 1 is familiar from the theory of planar b-groups; the element u can be considered to be an accidental parabolic element. The only unusual thing is that the action $\operatorname{Ad}(t)$ does not preserve the F'-conjugacy class of $\langle u\rangle$.

3.10. Proof of assertion (c) of the theorem. Every point $x_i = t_2^i(x)$ $(i \in \mathbb{Z})$ has a precisely invariant cusp–neighborhood $O_{n,i} = t_2^i(O_n) \subset \mathbb{R}^n$ such that $O_{n,i} \cap O_{m,i} = \varnothing$ if $i \neq j$. Each neighborhood $O_{n,i}$ is conformally equivalent to $\big(\mathbb{R}^n\backslash\text{solid cylinder}\big) = \big(\mathbb{R}^{n-1}\backslash\text{unit ball}\big) \times \mathbb{R} = [0, \infty) \times S^{n-2} \times \mathbb{R}$. The projection $E(n, i)$ of $O_{n,i}$ to the manifold $M(K_n) = \Omega(K_n)/K_n$ is homeomorphic to $[0, \infty) \times S^{n-2} \times S^1$. Hence, $\partial E(n, i) = S^{n-2} \times S^1$, $E(n, i)$ represents one end $\big[E(n, i)\big]$ of the manifold $M(K_n)$,

$i \in \mathbb{Z}$; $[E(n,i)] \neq [E(n,j)]$ if $i \neq j$. Hence, the cycles $[\partial E(n,i)] \in H_{n-1}(M(K_n), \mathbb{Q})$ are independent and $\operatorname{rank}\left(\mathbb{H}_{n-1}(\Omega(K_n)/K_n, \mathbb{Q})\right) = \infty$. So the theorem is completely proved.

§4. Proof of Corollary 1

Corollary 1 directly follows from part (c) of the theorem.

§5. Proof of Corollary 2

5.1. First we construct a sequence of representations $\rho_q : K_3 \to \operatorname{Mob}(S^3)$ such that $\lim_{q\to\infty} \rho_q = \rho_\infty = \mathrm{id}$ and $\operatorname{order}(\rho_q(u_m)) = q$.

Let $\Sigma_1(s)$ be the family of Euclidean spheres tangent to each other at the point p_1, $\operatorname{radius}(\Sigma_1(s)) = (\lambda - 1)s + 1 = r(s)$, $-1 \le s \le 1$. Define $p_2(s)$ to be the point of $\Sigma_1(s)$ with coordinates $(0, r(s), *)$. Choose a parabolic transformation ξ_s that commutes with H_1' and maps p_2 to $p_2(s)$. Let $\beta_2(s) = \xi_s \beta_2 \xi_s^{-1}$. It is easy to see that the isometric spheres $I(\beta_2(s))$, $I(\beta_2^{-1}(s))$ intersect L_2 at equal angles $\varphi(s)$; $\varphi(0) = 0$, $\varphi(1) = \pi/2$, φ is continuous function. Let $0 \le s(q) \le 1$ be a sequence of numbers such that $\varphi(s(q)) = \pi/2q$.

Let $\rho_q : \Gamma_1' \to \operatorname{Mob}(S^3)$ be the representation given by $\rho_q(\gamma) = \xi_{s(q)}\, \gamma\, \xi_{s(q)}^{-1}$. Hence, $\rho_q|_{H_1'} = \mathrm{id}$. Define $\rho_q : G_1' \to \operatorname{Mob}(S^3)$ to be a homomorphism such that $\rho_q|_{\Gamma_2'} = \mathrm{id}$. Next, $\rho_q : G_2' \to \operatorname{Mob}(S^3)$ is given by the formula $\tau_2' \rho_q(\tau_2' g \tau_2')\tau_2' = \rho_q(g)$, $g \in G_2'$.

Clearly, ρ_q is a homomorphism and $\lim_{q\to\infty} \rho_q = \rho_\infty = \mathrm{id}$.

The element $\rho_q(u)$ is the composition of inversions in the two spheres $I(\rho_q(\beta_2))$, $I(\rho_q(\beta_4))$, which intersect at the angle π/q. Hence $\rho_q(u)$ is a rotation of order q around the circle $\mathscr{L}_q = I(\rho_q(\beta_2)) \cap I(\rho_q(\beta_4))$.

5.2. In this subsection we consider the geometric and algebraic properties of the group $\langle \rho_q(\Gamma_1'), \rho_q(\Gamma_4')\rangle$ for large q.

Define Γ_{14} to be the group $\langle \Gamma_1', \Gamma_4' \rangle$, $\Gamma_i(q) = \rho_q \Gamma_i'$, $\Gamma_{14}(q) = \langle \Gamma_1(q), \Gamma_4(q)\rangle$.

There exists a number $q_0 \in \mathbb{N}$ such that for every $q \ge q_0$ $\partial P(\Gamma_1(q)) \cap \partial P(\Gamma_4(q)) = \partial P\left(\langle \rho_q(\beta_2)\rangle\right) \cap \partial P\left(\langle \rho_q(\beta_4)\rangle\right) \subset L_2$ (this follows directly from Lemma 1). Define $Q_{14}(q) = P\left(\langle \rho_q(\Gamma_1')\rangle\right) \cap P\left(\langle \rho_q(\Gamma_4')\rangle\right)$, $q_0 \le q \le \infty$.

LEMMA 5. *For every $q \ge q_0$ we have*

(i) *The polyhedron $Q_{14}(q)$ is fundamental for the group $\Gamma_{14}(q)$;*

(ii) *The circle $\mathscr{L}_q$ has a regular neighborhood $\mathscr{N}(\mathscr{L}_q)$ which is precisely invariant under $\langle \rho_q(u)\rangle$ in $\Gamma_{14}(q)$ $(q < \infty)$;*

(iii) *The group $\Gamma_{14}(q)$ is isomorphic to*

$$(\Gamma_{14}(\infty) = \Gamma_1' * \Gamma_4')/\langle\langle u^q \rangle\rangle.$$

PROOF. Statement (i) follows from the Poincaré theorem on fundamental polyhedra [**Mk**]. Statement (i) implies (ii) in the same way as the properties of G' imply Lemma 4.

Consider (iii). Every relation in $\Gamma_{14}(q)$ follows from relations corresponding to edge cycles on $\partial Q_{14}(q)$. Let $c = \{e_1, \dots, e_p\}$ be any edge cycle on $\partial Q_{14}(q)$. Then we have three possibilities:

(a) $c \subset \partial P(\Gamma_1(q))$,
(b) $c \subset \partial P(\Gamma_4(q))$,
(c) $c = \mathscr{L}_q \cup \rho(\beta_4)(\mathscr{L}_q)$ (if $q < \infty$).

In the cases (a),(b) the relation $R(c)$ follows from the relations of the groups Γ_1', Γ_4', respectively. In the case (c) we have the relation $(\rho_q(u)^q)$.

The lemma is proved. □

5.3. Lemma 6. *The half-space $\Pi_1 \cup \Pi_1^- = V_1$ is precisely invariant in the group $\Gamma_{14}(q)$ under the subgroup*

$$H(q) = H_1' * \rho_q(\tau_2' H_1' \tau_2').$$

Proof. First consider the plane domain

$$\mathscr{R}_q = \Big(P(H(q)) = P(H') \cap P\big(\rho_q(\tau_2' H_1' \tau_2')\big)\Big) \cap \Pi_1.$$

The Klein Combination Theorem implies that $\mathscr{R}_q$ is an isometric fundamental domain for the action of $H(q)$ in Π_1. It follows from Lemma 1 and Lemma 2 that the domain $\mathscr{R}_q$ is contained in $Q_{14}(q)$. Hence, the sphere Π_1 is precisely invariant under $H(q)$ (cf. **[K-P 1]**). The projection of $\mathscr{R}_q$ to $M(\Gamma_{14}(q)) = \Omega(\Gamma_{14}(q))/\Gamma_{14}(q)$ is a regular compact surface $\widehat{\mathscr{R}}_q$. There are two cases:

(a) $\widehat{\mathscr{R}}_q$ divides $M(\Gamma_{14}(q))$,
(b) $\widehat{\mathscr{R}}_q$ does not divide $M(\Gamma_{14}(q))$.

Consider (a). Then there exists an element $\gamma \in \Gamma_{14}(q)$ such that $\gamma(\Pi_1^+) \subset V_1$, where $\Pi_1^+ = \tau_1(\Pi_1^-)$. Note that $\Pi_1^+ \supset \operatorname{int}\Lambda(\Gamma_1(q)) \cup \operatorname{int}\Lambda(\Gamma_4(q))$. It is easy to see that $g(\Pi_1^-) \cap \Big(\operatorname{int}\Lambda(\Gamma_1(q)) \cup \operatorname{int}\Lambda(\Gamma_4(q))\Big) = \varnothing$ for every $g \in \Gamma_{14}(q)$. Hence, for the element γ we have $\gamma(\Pi_1^+) \supset \operatorname{int}\Lambda(\Gamma_1(q)) \cup \operatorname{int}\Lambda(\Gamma_4(q))$, so that $\gamma(\Pi_1^+) \cap \Pi_1^+ \neq \varnothing$. This contradiction shows that the case (a) does not hold.

Consider (b). Then there is an element $\gamma \in \Gamma_{14}(q)$ such that $\gamma(\Pi_1^-)$ is the complement to an open ball in ext $\Lambda(\Gamma_1(q)) \cap$ ext $\Lambda(\Gamma_4(q))$. Then $\gamma(\Pi_1^-) \supset \operatorname{int}\Lambda(\Gamma_1(q)) \cap \operatorname{int}\Lambda(\Gamma_4(q))$ and we get a contradiction as above.

The lemma is proved. □

5.4. Lemma 7.

(i) *The pair of groups $(\Gamma_{14}(q), \langle\Gamma_2', \Gamma_3'\rangle)$ with amalgamated subgroup $H(q)$ satisfies the conditions of the first Maskit Combination Theorem.*
(ii) *The group $G(q) = \rho_q(G')$ is isomorphic to $G'/\langle\langle u^q\rangle\rangle$.*
(iii) *The regular neighborhood $\mathscr{N}(\mathscr{L}_q)$ of $\mathscr{L}_q$ is precisely invariant in $G(q)$ under $\langle\rho_q(u)^q\rangle$.*

Proof. As we have proved in Lemma 6, the half-space V_1 is precisely invariant in $\Gamma_{14}(q)$ under $H(q)$. Consider the group $\Gamma_{23}(q) = \Gamma_{23} = \langle\Gamma_2', \Gamma_3'\rangle$. The sets $P_2 = T(P)$ and $\tau' T(P) = P_3$ are fundamental for the groups Γ_2', Γ_3', respectively (see 3.2). We have $P_2 \cap L_2' = P_3 \cap L_2' = P(J) \cap L_2'$. Hence, Γ_{23} results from the Maskit combination of the groups Γ_2', Γ_3' along the subgroup J. The domain $Q_{23} = P_2 \cap P_3$ is fundamental for the group Γ_{23}. The property (d) of the group Γ_1' implies that

$$\Pi_1 \cap Q_{23} = \Pi_1 \cap P(H(q)).$$

Similarly to Lemma 6, we see that $W_1 = \Pi_1 \cup \Pi_1^+$ is precisely invariant under $H(q)$ in Γ_{23}. Then assertion (1) is true. As a fundamental set for the group

$$G(q) = \Gamma_{14}(q) *_{H(q)} \Gamma_{23}$$

we shall use $Q_{23} \cap Q_{14}(q) = Q(q)$. Then Lemma 5 implies assertion (iii).

Next we note that

$$\Gamma(q) \cong (\Gamma_1 * \Gamma_4/\langle\langle u^q \rangle\rangle) *_{H(q)} \Gamma_{23} \quad \text{and} \quad G' \cong \Gamma_1 * \Gamma_4 *_{H(q)} \Gamma_{23}.$$

Hence $G(q) \cong G'/\langle\langle u^q \rangle\rangle$.

The lemma is proved. □

5.5. Now, using assertion (ii) of the previous lemma, arguing analogously as in subsection 3.8 we deduce that the elliptic elements $\rho_q(u_k)$, $\rho_q(u_m)$ are not conjugate for any $q_0 \le q < \infty$, $m, k \in \mathbb{Z}$, $m \neq k$. So the finitely generated Kleinian group $\rho_q(F')$ contains infinitely many conjugacy classes of finite order elements.

Consider the representation $\psi_q = \rho_q|_{F'}$. We know that $\mathrm{Ker}(\rho_q) : G' \to G(q)$ is the normal closure of $\langle u_q \rangle$ in G'. Since $\langle u \rangle \subset F'$ and F' is normal in G', we obtain that $\mathrm{Ker}(\psi_q)$ is the normal closure of $\bigcup_{m\in\mathbb{Z}} t_2^m \langle u^q \rangle t_2^{-m}$ in F'. So the group $\psi_q(F')$ has the presentation

$$\langle x_1, \dots, x_r : u_m^q,\ m \in \mathbb{Z} \rangle .$$

It is easy to see that the elements β_2, β_4 can be included in a system of free generators of the free group F'. Hence, the same is true for the element $u = \beta_4^{-1}\beta_2$. Then the group $\psi_q(F')$ has the desired representation

$$\left\langle x_1, \dots, x_r : \theta^m(x_1^q) = 1,\ m \in \mathbb{Z} \right\rangle \cong F_{r,q},$$

where θ is the automorphism of free group F_r induced by $\mathrm{Ad}(t_2)$. Corollary 2 is proved. □

§6. Description of the quotient manifolds

In this section we describe briefly the topology of the manifolds $M(K_3)$ and $O(F_{r,q}) = \Omega(F_{r,q})/F_{r,q}$, $q \ge q_0$.

6.1. We start with the sphere $\Sigma_i(s)$ for some $s > 0$, $s < 1/(\lambda - 1)$. Then we construct a representation $\rho_s : G' \to G(s) \subset \mathrm{Mob}(S^3)$ in the same manner as in subsection 5.1. A fundamental polyhedron $Q(s)$ for the group $G(s)$ is equal to $P_2 \cap P_3 \cap P(\xi_s \Gamma_1' \xi_s^{-1}) \cap P(\tau_2' \xi_s \Gamma_4' \xi_s^{-1} \tau_2')$. As in Lemma 7, we conclude that $G(s)$ is a Kleinian group. Let $\Omega(s)$ be the infinite component of $\Omega(G(s))$. Then, according to the Maskit Combination Theorem, the manifold $M(s) = \Omega(G(s))/G(s)$ is homeomorphic to a fiber bundle over S^1 (cf. Lemma 3 in **[K-P 1]**). Let $x_2(s)$, $y_2(s) = \rho_s(\beta_2)(x_2(s))$ be distinct points of $\partial P(\rho_s(\Gamma_1')) \cap \partial P(\langle \rho_s(\beta_2) \rangle)$; $x_4(s) = \tau_2'(x_2(s))$ and $y_2(s) = \tau_2'(y_2(s))$. Join the pair of points $x_2(s)$ and $x_4(s)$ by a Euclidean segment I_s. Join the pair of points $y_2(s)$ and $y_4(s)$ by a Euclidean segment I_s'.

The union $I_s \cup I_s'$ projects to a circle C in $M(s)$.

6.2. Proposition 1. *The manifold Ω'/G' is homeomorphic to $M(s)\backslash C$; $\pi_1(\Omega')$ $= 1$.*

Proof. Note that the removal of $I_s \cup I_s'$ from $Q(s)$ is equivalent to deleting $\{x, y\}$ from $\mathrm{cl}(Q(\infty) = P(G'))$. Then the proof is concluded as in **[K-P 1,** Lemma 3**]**.

6.3. Let $\mathfrak{F}$ be a fiber of the manifold $M(s)$. The cyclic covering $\eta_s : \Omega(s)/\rho_s F' \to \Omega(s)/G'(s)$ is induced by the embedding $\pi_1(\mathfrak{F}) \to \pi_1(M(s))$. The manifold $X_r = \Omega(s)/\rho_s F'$ is homeomorphic to the handlebody $\mathfrak{F} \times \mathbb{R}$ and $\pi_1(\mathfrak{F})$ is a free group of rank r. Then the manifold Ω'/F' is homeomorphic to $X_r \backslash (\eta_s^{-1}(C) = L_\infty)$, which is the complement to the infinite-component link in the handlebody.

Let $\mathcal{N}(C)$ be an open regular neighborhood of the knot $C, M^-(s) = M(s) \backslash \mathcal{N}(C)$.

6.4. PROPOSITION 2. *Let Ω_q be the infinite component of $\Omega(G(q))$. Then the orbifold $\Omega_q/G(q)$ is homeomorphic to $M^-(s) \bigcup_{\partial \mathcal{N}(C)} D(q) \times S^1$, where $D(q)$ is 2-disc with one singular point of order q.*

PROOF. Evident (see Lemmas 5–7 and the considerations above). □

6.5. Thus, the quotient orbifold $\Omega_q/F(q)$ is supported by an open handlebody X_r and the singular set is the infinite-component link L_∞ of order q.

REMARK. Every quotient manifold $\Omega(F(q))/F(q)$ $(q_0 \leq q \leq \infty)$ has also four components besides $\Omega_q/F(q)$. These components are homeomorphic to the handlebody $\operatorname{int}\Big(\Lambda(F(q))\Big)/(F(q) \cap \Gamma_i(q))$, $i = 1, \dots, 4$.

§7. Finite order elements in discrete subgroups of $PSL(2, \mathbb{C})$

PROPOSITION 3. *Let Γ be any discrete finitely generated subgroup of $PSL(2, \mathbb{C})$. Then Γ contains only a finite number of conjugacy classes of finite order elements.*

PROOF. Denote by $M(G)$ the factor orbifold for a discrete group G in $PSL(2, \mathbb{C})$. Let $\Gamma_0 < \Gamma$ be a torsion-free finite index normal subgroup in Γ (which exists according to Selberg's lemma). Then the finite group $F = \Gamma/\Gamma_0$ acts on the manifold $M(\Gamma_0)$, preserving the peripheral structure of the fundamental group $\pi_1(M(\Gamma_0)) \cong \Gamma_0$. Then it is easy to see that the manifold $M(\Gamma_0)$ has a compact Scott core $C(\Gamma_0)$ **[Sc]** invariant under F. The compact orbifold $O(\Gamma) = C(\Gamma_0)/F$ has the singular set $\Sigma(\Gamma)$, which is a finite graph. Vertices of $\Sigma(\Gamma)$ and vertex–free components of it are in one-to-one correspondence to the Γ-conjugacy classes of maximal finite order subgroups of Γ. Hence, Γ contains only finitely many conjugacy classes of finite order subgroups. The proposition is proved. □

§8. On the Abikoff conjecture about noncone limit points for finitely generated discrete subgroups of $PSL(2, \mathbb{C})$

8.1. Let $\Gamma \subset \mathrm{Mob}(S^n)$ be a discrete group. Then there is a hierarchy of limit points $x \in \Lambda(\Gamma)$ according to the way in which they can be approximated by an orbit $\Gamma(z)$, $z \in H^{n+1}$. A point x is said to be an approximation or cone limit point if up to an infinite subsequence the family $\Gamma(z)$ lies in some Euclidean cone $K \subset H^{n+1}$. This may be considered as the best (fastest) approximation. The set of cone limit points is denoted by $\Lambda_c(\Gamma)$. The dynamics of Γ near an approximation point is similar to the dynamics near a fixed point of a loxodromic element. In contrast, parabolic fixed points cannot be cone limit points. Unlike cone limit points, nonapproximation limit points are a much more intriguing matter. In a very instructive survey **[Ab 2]** Abikoff proposed the following

CONJECTURE. *Let $\Gamma \subset PSL(2, \mathbb{C})$ be any finitely generated discrete group. Then $(\Lambda(\Gamma) \backslash \Lambda_c(\Gamma))/\Gamma$ is a finite set.*

REMARK. This conjecture was motivated by Sullivan's finiteness theorem (parabolic fixed points modulo Γ form a finite set).

8.2. The main aim of this section is to disprove the above conjecture. There are two types of counterexamples. In fact, both are from Abikoff's paper.

8.3. Next, we recall some constructions from **[Ab 2]**. Consider a Fuchsian torsion-free group $F \subset PSL(2, \mathbb{R})$ such that $S = H^2/F$ is a compact surface. Let $h : F \to F$ be a pseudo-Anosov homomorphism with attractive and repulsive foliations $\mathscr{L}$ and $\mathscr{L}^*$ respectively. We realize H^2 as the unit disk $\Delta \subset \overline{\mathbb{C}}$, $\Delta^* = \mathrm{ext}(\Delta)$. Lift the foliations $\mathscr{L}$, $\mathscr{L}^*$ to foliations $\mathscr{N} \subset \Delta$, $\mathscr{N}^* \subset \Delta^*$ invariant under F. Add to any geodesic in $\mathscr{N}$, $\mathscr{N}^*$ its end-points; the resulting "foliations" will be denoted by $\mathscr{J}$, $\mathscr{J}^*$. Consider two equivalence relations on $\overline{\mathbb{C}}$:

(1) $x \sim y$, if x and y belong to a path-connected component of $\mathscr{J}$ or to the closure in $\mathbb{C}$ of some component of $\Delta \backslash \mathscr{J}$;

(2) $x \approx y$, if x and y belong to a path-connected component of $\mathscr{J} \cup \mathscr{J}^*$ or to the closure in $\overline{\mathbb{C}}$ of some component of $\Delta \backslash (\mathscr{J} \cup \mathscr{J}^*)$.

These equivalence relations are invariant under the action of F. Hence, the action of F descends to $\overline{\mathbb{C}}/\sim$ and $\overline{\mathbb{C}}/\approx$. Thus, we obtain topological models for the action on $\overline{\mathbb{C}}$ of singly (case 1) and doubly (case 2) degenerate groups **[Ab 2]**. Let us denote the corresponding discrete subgroups of $PSL(2, \mathbb{C})$ by F_1 and F_2; let ζ_1 be the projection: $\overline{\mathbb{C}} \to \overline{\mathbb{C}}/\sim$, $\quad \zeta_2 : \overline{\mathbb{C}} \to \overline{\mathbb{C}}/\approx$.

PROPOSITION 4. *The sets $\Lambda_0(F_i) = \Lambda(F_i) \backslash \Lambda_c(F_i)$ contain continua $\zeta_i(\overline{\mathbb{C}} \backslash \partial\Delta) \cap \Lambda(F_i)$.*

REMARK. Indeed, $\zeta_1(\Delta) = \Lambda(F_1) \backslash \Lambda_c(F_1)$ consists of the end-points of the tree $\Lambda(F_1)$.

PROOF OF PROPOSITION 4. We shall discuss only case (1); the second case is essentially similar. □

8.4. First we recall another definition of approximation points **[B-M]**.

Let $\Gamma \subset \mathrm{Mob}(S^n)$ be a discrete group, $x \in \Lambda(\Gamma)$. Then $x \in \Lambda_c(\Gamma)$ if and only if there exists an infinite sequence $\{\gamma_m\} \subset \Gamma$ such that for every point $y \in S^n$

(a) the limit $\lim_{m \to \infty} \gamma_m(y) = y^*$ exists,

(b) $y^* \neq x^*$ for all $y \neq x$, and

(c) the point $z = y^*$ is one and the same for all $y \neq x$.

8.5. Suppose that $x \in \Lambda_c(F_1) \cap \zeta_1(\Delta)$. Hence $x = \zeta_1(l)$, where l is some geodesic in $\mathscr{J}$ and x admits a sequence $\{f_m\}$ with the properties (a)–(c) above. We shall denote by $\{f_m\}$ the corresponding sequence in F. Let $\{\alpha, \beta\}$ be the set of end-points of l. The sequence $\{f_m\}$ is not relatively compact in $PSL(2, \mathbb{C})$; hence there exists a subsequence $\{f_{m_s}\} \subset \{f_m\}$ and points $v, w \in \Lambda(F)$ such that:

$$\lim_{s \to \infty} \{f_{m_s}(z)\} = w \in \Lambda(F) \quad \text{for every } z \in \overline{\mathbb{C}} \backslash \{v\}.$$

Then for point α or β (say α) we have that $w = \lim_{s \to \infty} \{f_{m_s}(z)\} = \lim_{s \to \infty} \{f_{m_s}(\alpha)\}$ for all but one point $z \in \overline{\mathbb{C}}$. Consequently $\lim_{s \to \infty} \{f_{m_s} \zeta_1(x)\} = \lim_{s \to \infty} \{f_{m_s} \zeta_1(\alpha)\} = \lim_{s \to \infty} \{f_{m_s} \zeta_1(z)\}$. This contradiction proves that x is not an approximation limit point.

8.6. Now we prove that every $\widehat{x} \in \Lambda(F_1)\backslash\zeta_1(\Delta)$ is an approximation point. Let $x = \zeta_1^{-1}(\widehat{x})$; clearly, this is not an end-point of any geodesic from $\mathcal{J}$. The group F is geometrically finite and it is easy to construct a sequence $\{f_m\} \subset F$ possessing the properties (a)–(b) with respect to x, such that

(d) y^* is not an end-point of geodesics from the foliation $\mathcal{J}$.

Then x^* is not equivalent to the point y^*. So, the sequence $\{f_m\} \subset F_1$ has the properties (a)–(c) with respect to $\widehat{x}$.

The proposition is proved. □

REMARK. The results of 8.5 can be proved more geometrically without passing to a Fuchsian group, as in **[Ab 2]**. However that proof cannot be generalized to the case of doubly degenerate groups.

References

[Ab 1] W. Abikoff, *The Euler characteristic and inequalities for Kleinian groups*, Proc. Amer. Math. Soc. **97** (1986), 593–601.

[Ab 2] ———, *Kleinian groups—geometrically finite and geometrically perverse*, Geometry of Group Representations, Contemp. Math., vol. 74, Amer. Math. Soc., Providence, RI, 1988, pp. 1–50.

[Ah] L. V. Ahlfors, *Finitely generated Kleinian groups*, Amer. J. Math. **86** (1964), 413–429; Correction **87** (1965), 759.

[B-M] A. F. Beardon and B. Maskit, *Limit points of Kleinian groups and finite sided fundamental polyhedra*, Acta Math. **132** (1974), 1–12.

[Gr] M. L. Gromov, *Hyperbolic groups*, Math. Sci. Res. Inst. Publ., vol. 8, Springer-Verlag, Berlin and New York, 1987, pp. 75–263.

[K1] M. E. Kapovich, *On absence of Sullivan's cusp finiteness theorem in higher dimensions*, Abstract, International Conference on Algebra (Novosibirsk, 1989), Novosibirsk.

[K2] ———, *Flat conformal structures on 3-manifolds (survey)*, Proc. Internat. Conf. on Algebra, Part 1 (Novosibirsk, 1989), Contemp. Math., vol. 131, Amer. Math. Soc., Providence, RI, 1992, pp. 551–570.

[K-P 1] M. E. Kapovich and L. D. Potyagailo, *On absence of Ahlfors' finiteness theorem for Kleinian groups in dimension three*, Topology Appl. **40** (1991), 83–91.

[K-P 2] ———, *On the absence of finiteness theorems of Ahlfors and Sullivan for Kleinian groups in higher dimension*, Sibirsk. Mat. Zh. **32** (1991), no. 2, 61–73; English transl. in Siberian Math. J. **32** (1991).

[Kr 1] I. Kra, *Automorphic forms and Kleinian groups*, Benjamin, Reading, MA, 1972.

[Kr 2] ———, *On cohomology of Kleinian groups. IV. The Ahlfors-Sullivan construction of holomorphic Eichler integrals*, J. Analyse Math. **43** (1983), 51–87.

[Ku] R. S. Kulkarni, *A finiteness theorem for planar discontinuous groups*, J. Indian Math. Soc. **55** (1990), 37–43.

[K-S] R. S. Kulkarni and P. B. Shalen, *On Ahlfors' finiteness theorem*, Adv. Math. **76** (1989), 155–169.

[Mk] B. Maskit, *Kleinian groups*, Springer-Verlag, Berlin and New York, 1988.

[Sc] G. P. Scott, *Finitely generated 3-manifold groups are finitely presented*, J. London Math. Soc. (2) **6** (1973), 437–448; *Compact submanifolds of 3-manifolds*, J. London Math. Soc. (2) **7** (1973), 246–250.

[Se] A. Selberg, *On discontinuous groups in higher-dimensional symmetric spaces*, Contributions to Function Theory, Tata Inst. Fund. Res., Bombay, 1960, pp. 147–164.

[Su] D. Sullivan, *A finiteness theorem for cusps*, Acta Math. **147** (1981), 289–299.

[W] N. J. Wielenberg, *Discrete Moebius groups: fundamental polyhedra and convergence*, Amer. J. Math. **99** (1977), 861–877.

COMPUTER CENTER INSTITUTE FOR APPLIED MATHEMATICS

Amer. Math. Soc. Transl.
(2) Vol. **163**, 1995

The Variety of All Rings Has Higman's Property

G. Kukin

G. Higman [**1**] proved that a recursively presented (r.p.) group may be imbedded into some finitely presented (f.p.) group. Therefore, a finitely generated (f.g.) group is an r.p. group if and only if it is imbeddable into some f.p. group.

A variety (or quasivariety) of algebraic systems with an analogous characterization of its r.p. objects is named a Higman variety (or quasivariety). The problem of describing such (quasi)varieties was raised by L. Bokut′ and the author [**2**]. It is far from settled.

The following varieties have Higman's property:

- the variety of all semigroups (V. Murskiĭ [**3**]);
- the variety of all inverse semigroups (V. Belyaev [**4**]);
- the variety of all associative rings (or algebras over a field that is f.g. over its prime subfield) (V. Belyaev [**5**]);
- the variety of all Lie rings or algebras with the same condition on the ground field (G.Kukin [**6**]).

O. Belegradek and N. Koveshnikova [**7**] remarked that it is more natural to raise the question about Higman classes for quasivarieties. Sufficient conditions for a (quasi)variety of algebras to be Higman were found in [**8**]. O. Gatelŭk extended this analysis and found a series of Higman varieties of algebras.

It is clear that there are f.p. objects with unsolvable word problem in a Higman (quasi)variety. Consequently, the variety of all algebras over fields (or all commutative, anticommutative algebras) does not have the Higman property. A. Zhukov and A. Shirshov [**9, 10**] solved (positively) the word problem in such varieties.

Recently, Yu. Vazhenin constructed a f.p. ring (without identity) where the word problem is unsolvable [**11**]. V. Belyaev conjectured that the variety of all rings has Higman's property. We prove this conjecture in this article.

The author considers it a pleasant duty to thank Yu. Vazhenin for the opportunity to use his article before publication and V. Belyaev for formulating the conjecture.

THEOREM. *The variety of all rings has Higman's property.*

The analogous assertion is true for the variety of all (anti)commutative rings.

PROOF. The technique of the proof is close to that of [**6, 8**]. But the specifics of our case do not let us to use references instead of new calculations.

We consider rings with a finite set of generators b_α, x_*, x_β, y.

1991 *Mathematics Subject Classification*. Primary 16R10; Secondary 20E10.

We pay attention to the words (or monomials)

$$\underbrace{b_\alpha\, x_\beta \dots x_\beta\, y}_{k} \qquad (\stackrel{\text{def}}{=} b_\alpha x_\beta^k y)$$

with "right ordered" parentheses

$$abc \dots d = (\dots((ab)c)\dots)d.$$

The natural number n is coded by the word

$$\underbrace{b_\alpha\, x \dots x\, y}_{1+n} = b_\alpha x^{1+n} y$$

$(x \in \{x_*, x_\beta\})$ or analogous ones; the pair of numbers (m, n) is coded by the word

$$b_\alpha x_*^{1+m} x_\beta^{1+n} y,$$

the vector $(n_1, \dots, n_k)$ by the sum of k words

$$b_\alpha x_\beta^{1+n_1} y + \dots + b_\tau x_\gamma^{1+n_k} y.$$

Further we interpret the calculation of every recursive function $y = f(n)$, $n = 0, 1, 2, \dots$. It is well known **[12]** that it is necessary for this to interpret (in f.p.rings) the calculation of the functions $n \to n+1$, $n \to n \dot{-} 1$, $(n_1, \dots, n_k) \to n_i$, the scheme of primitive recursion, the operator of minimization, and the composition of functions.

We prove a series of lemmas to describe the steps of "assembly" for a recursive function using these "details".

LEMMA 1. *Let L be a free ring with free generators b_α, x_*, x_β, y (a finite set), let B be its subring generated by all words of the type*

$$h_k = b_\alpha x_*^\tau x_\beta^n y,$$

and let B_0 be the subring generated by the elements of type h_k or their sums.

If I is a two-sided ideal in B_0, and I_L is the ideal of L generated by I, then

$$I_L \cap B_0 = I.$$

PROOF. The inclusion $I \subseteq I_L \cap B_0$ is obvious. Let $f_l(h_1, \dots, h_t)$ be the generators of the ideal I in B_0. Then the additive group I is generated by elements of the type

$$f_l U_{s_1} \dots U_{s_q}, \tag{1}$$

where $U = \mathscr{R}$ or $U = \mathscr{L}$ is the operator of right or left multiplication by a monomial in b_τ, x_*, x_β, y. Let us suppose that a $\mathbb{Z}$-linear combination C of elements $s_\varkappa$ the form (1) belongs to subring B_0. The summand c in C (which has one or more monomial $s_\varkappa$, cf. (1), that have a type different from $b_\tau x_*^\rho x_\beta^\sigma y$) must cancel in the sum; otherwise the inclusion

$$c = \sum z_l f_\gamma U_{s_1} \dots U_{s_q} \in B_0 \qquad (z_l \in \mathbb{Z})$$

is impossible. Thus, we can assume that such summands c are absent.

Consequently, the calculations involve only the free subring B (and the h_k are in the set of its free generators), and B_0 is a free factor in B:

$$B = B_0 \times B_1$$

for some subring B_1. The assertion of the lemma is obvious now. □

LEMMA 2 (about the graph of the function $s(n) = n + 1$). *Let the ring $\bar{L}$ have generators b_1, b_2, b_3, x_1, y and the defining relation*

$$b_1 x_1 = b_2 x_1 + b_3 x_1^2.$$

The equality

$$b_1 x_1^{1+n} y = b_2 x_1^{1+u} y + b_3 x_1^{1+v} y \tag{2}$$

is valid in I if and only if $u = n$, $v = n + 1$.

PROOF. Let us multiply the relation from the right by x_1 (n times); we obtain the equality

$$b_1 x_1^{1+n} = b_2 x_1^{1+n} + b_3 x_1^{1+(n+1)}.$$

Now we multiply it by y; this gives (2).

It is clear that an element f of the ideal I generated by

$$g = b_1 x_1 - b_2 x_1 - b_3 x_1^2$$

(in the free ring with generators b_α, x_1, y) is an element of the subring B generated by $b_\alpha x_1^\tau y$ iff

$$f = g \underbrace{x_1 \ldots x_1}_{m} y.$$

This proves the lemma. □

The next lemmas (numbered 3–7) can be proved analogously. Here we give statements only. The reader will note that it is easier to recall the general principles and to write the proof by himself than to check a written text (as in the case when we work with Turing machines).

LEMMA 3 (about the graph of the function $n \to n \dot{-} 1$). *Let the ring $\bar{L}$ have generators b_1, b_2, b_3, x_1, y and defining relations*

$$b_1 x_1^2 = b_2 x_1^2 + b_3 x_1, \qquad b_1 x_1 y = b_2 x_1 y + b_3 x_1 y.$$

The equality

$$b_1 x_1^{1+n} y = b_2 x_1^{1+u} y + b_3 x_1^{1+v} y$$

is valid in $\bar{L}$ if and only if $u = n$, $v = n \dot{-} 1$.

LEMMA 4 (about the graph of function $I^q_m(n_1, \dots, n_q) = n_m$, where $1 \le m \le q$). *We code the vector $(n_1, \dots, n_q)$, $n_m \in N$, by the sum of q monomials*

$$\sum_{i=1}^{q} b_i x_1^{1+n_i} y.$$

Let us consider the ring $\bar{L}$ with generators b_i, b'_j, b''_m, x_1, y $(j \neq m)$ and relations (p is a fixed prime number)

$$pb_i x_1 = b'_i x_1 \quad (i \neq m), \qquad pb_m x_1 = b'_m x_1 + b''_m x_1.$$

The equality

$$t \sum_{i=1}^{q} b_i x_1^{1+n_i} y = \sum_{i \neq m} b'_i x_1^{1+u_i} y + b''_m x_1^{1+v} y$$

is valid in $\bar{L}$ (for some $t \in \mathbb{Z}$) if and only if $u_i = n_i$, $v = n_m$, and then $t = p$.

These (and the next) lemmas have analogs concerning the interpretation of functions rather than graphs. For this we use the operators of multiplication by prime numbers p_λ, which remove unnecessary information. For example, we use in the ring $\bar{L}$ (Lemma 4), in addition, the defining relations

$$pp' b_j x_1 = \hat{b}_j x_1, \qquad p' b'_i x_1 = 0 \quad (i \neq m), \qquad p' b''_m x_1 = \tilde{b}_m x_1.$$

Then an equality of the type

$$\sum \hat{b}_i x_1^{1+n_i} y = \sum_{i \neq m} b'_i x_1^{\alpha_i} y + \tilde{b}_m x_1^{1+v} y$$

is valid if and only if the summands $b'_i x_1^{\alpha_i} y$ are absent, $v = n_m$.

LEMMA 5 (about the graph of the composition of functions). *Let f, h be recursive functions with their calculation interpreted in the f.p. rings L_f, L_h. We suppose that the equalities*

$$t b_1 x_1^{1+a} y = b_2 x_1^{1+a} y + b_3 x_1^{1+f(a)} y$$

are interpreted in L_f and

$$t' b_3 x_1^{1+c} y = b_4 x_1^{1+c} y + b_5 x_1^{1+h(c)} y$$

in L_h (for suitable t, $t' \in N$). We shall consider that the set of generators $\{b_\alpha, \dots\}$ for L_f contains no generators of L_h and conversely, except for the letter b_3.

Let us consider the ring L_{fh} with generators the union of those for L_f, L_h and with defining relations both those of L_f, L_h as well as

$$pb_1 x_1 = b_1 x_1, \quad pb_3 x_1 = b_3 x_1, \quad pb_4 x_1 = 0, \quad pb_5 x_1 = b_5 x_1, \quad pb_2 x_1 = b_2 x_1.$$

(The operator p is a prime number that did not appear in the defining relations for L_f, L_h.)

The equality

$$rb_1x_1^{1+c}y = b_2x_1^{1+u}y + b_5x_1^{1+v}y$$

is valid in L_{fh} *(for some* $r \in N$*) iff*

$$u = c, \qquad v = h\big(f(c)\big).$$

We note that coefficients t, t' and others in Lemmas 4, 5, and their analogs may be removed, i.e., changed to 1. We show this fact for Lemma 5.

Let $p_1, \dots, p_\tau$ be all prime numbers involved in the defining relations for L_{fh} (they can only be prime factors of r). We consider new operators which are prime numbers $q, q_1, \dots, q_\tau$ and have not appeared earlier. Also we consider new relations

$$qb_1x_i = \tilde{b}_1x_i, \qquad qb_2x_i = \tilde{b}_2x_i, \qquad qb_5x_i = \tilde{b}_5x_i,$$
$$p_iq_i\tilde{b}_1x_i = \tilde{b}_1x_i, \qquad q_i\tilde{b}_2x_i = \tilde{b}_2x_i, \qquad q_i\tilde{b}_5x_l = \tilde{b}_5x_l.$$

Then it is obvious that the equality

$$\tilde{b}_1x_1^{1+c}y = \tilde{b}_2x_1^{1+u}y + \tilde{b}_5x_1^{1+v}y$$

is valid if and only if $u = c$, $v = h\big(f(c)\big)$.

Now we show that the pair of numbers (m, n) coded by the sum

$$b_1x_1^{1+m}y + b_1x_2^{1+n}y$$

can be coded by the monomial $b_2x_1^{1+m}x_2^{1+n}y$. Let us consider the relations (3)–(5)

(3) $$b_1x_1 + b_1x_2 = b_2x_1x_2$$

(we can obtain the equality $b_1x_1x_2^l + b_1x_2^{1+l} = b_2x_1x_2^{1+l}$ using multiplication by x_2).

Let

(4) $$b_1x_1x_2 = b_1x_1;$$

then $b_1x_1 + b_1x_2^{1+l} = b_2x_1x_2^{1+l}$. The relations

(5) $$pb_1x_1 = b_1x_1^2, \qquad pb_1x_2 = b_1x_2, \qquad pb_2x_1 = b_2x_1^2$$

give the equalities

$$b_1x_1^{1+m}y + b_1x_2^{1+n}y = b_2x_1^{1+m}x_2^{1+n}y.$$

Analogously, the number k coded by the word $b_1x_1^{1+k}y$ can be coded by $b_2x_2^{1+k}y$, precisely because the relations

(6) $$b_1x_1 = b_2x_2, \qquad pb_1x_1 = b_1x_1^2, \qquad pb_2x_2 = b_2x_2^2$$

imply the equalities $b_1x_1^{1+k}y = b_2x_2^{1+k}y$.

Let $g\colon N \to N$, $h\colon N^2 \to N$ be given recursive functions. Then the function $\varphi\colon N^2 \to N$ is said to be constructed from the functions h by the scheme of primitive recursion if it is obtained as shown in Figure 1 (next page).

We will refer to this recursion scheme as Scheme 1.

Let us suppose that the calculation of the graphs for g, h is interpreted in the f.p. rings L_g and L_h as earlier for concrete functions and their compositions.

The defining relations of L_g, L_h are $\{\rho\}$ and $\{\sigma\}$ respectively.

$$\varphi(n,0) = g(n) \quad \text{for all } n \in N,$$
$$\varphi(n,m+1) = h(\varphi(n,m),n) \quad \text{for all } n,m \in N.$$

$$b_2 x_1^{1+n} y + b_2 x_1^{1+m} y \xrightarrow{\{\rho\}} b_1 x_2^{1+m} y + b_4 x_2^{1+n} y + b_4 x_3^{1+g(n)} y$$

$$(b_1 x_2 = b_3 x_2 + b_3' x_2) \Big\downarrow$$

$$b_3 x_2^{1+m} y + b_3' x_2^{1+m} y + b_4 x_2^{1+n} y + b_4 x_3^{1+g(n)} y$$

$(t=0) \;\|\|\|$

$\longrightarrow b_3 x_2^{1+m} y + b_3' x_2^{1+m-t} y + b_4 x_2^{1+n} y + b_4 x_3^{1+\varphi(n,t)} y$

$\Big\downarrow \quad$ (type (3)–(5))

$b_3 x_*^{1+m-t} x_2^{1+m} y + b_4 x_*^{1+m-t} x_2^{1+n} y + b_4 x_*^{1+m-t} x_3^{1+\varphi(n,t)} y$

$\Big\downarrow \quad (p_2 b_i x_*^2 = b_5 x_*) \qquad (p_1 b_j x_* x_i = b_9 x_i)$

$b_5 x_*^{1+m-(t+1)} x_2^{1+m} y +$

$+ b_5 x_*^{1+m-t-1} x_2^{1+n} y +$

$+ b_5 x_*^{1+m-t-1} x_3^{1+\varphi(n,t)} y \qquad \boxed{b_9 x_2^{1+m} y + b_9 x_2^{1+n} y + b_4 x_3^{1+\varphi(n,t)} y}$

$i = 2, j = 3$ or $i = 2, 3, j = 4$

$\Big\downarrow \quad$ (type (3)–(5))

$b_6 x_2^{1+m} y + b_6' x_2^{1+n} y + b_6 x_3^{1+\varphi(n,t)} y + b_6' x_3^{1+m-(t+1)} y$

$\Big\downarrow \quad \{\sigma\}$

$b_6 x_2^{1+m} y + b_7 x_2^{1+n} y + b_7 x_3^{1+\varphi(n,t)} y + b_7 x_6^{1+h(\varphi(n,t),n)} y + b_6' x_3^{1+m-(t+1)} y$

$\Big\downarrow \quad (p_3 b_7 x_3 = 0, \quad b_3 b_7 x_2 = b_8' x_2, \quad p_3 b_6 x_2 = b_8 x_2, \quad p_3 b_6' x = b_8'' x_3)$

$b_8 x_2^{1+m} y + b_8' x_2^{1+n} y + b_8 x_6^{1+\varphi(n,t+1)} y + b_8'' x_3^{1+m-(t+1)} y$

$\Big\downarrow \quad$ (type (6))

$b_8 x_2^{1+m} y + b_8' x_2^{1+n} y + b_8' x_3^{1+\varphi(n,t+1)} y + b_8'' x_2^{1+m-(t+1)} y$

$(p_4 b_8 x_2 = b_3 x_2, \quad p_4 b_8'' x = b_3' x_2, \quad p_4 b_8' x_2 = b_4 x_3)$

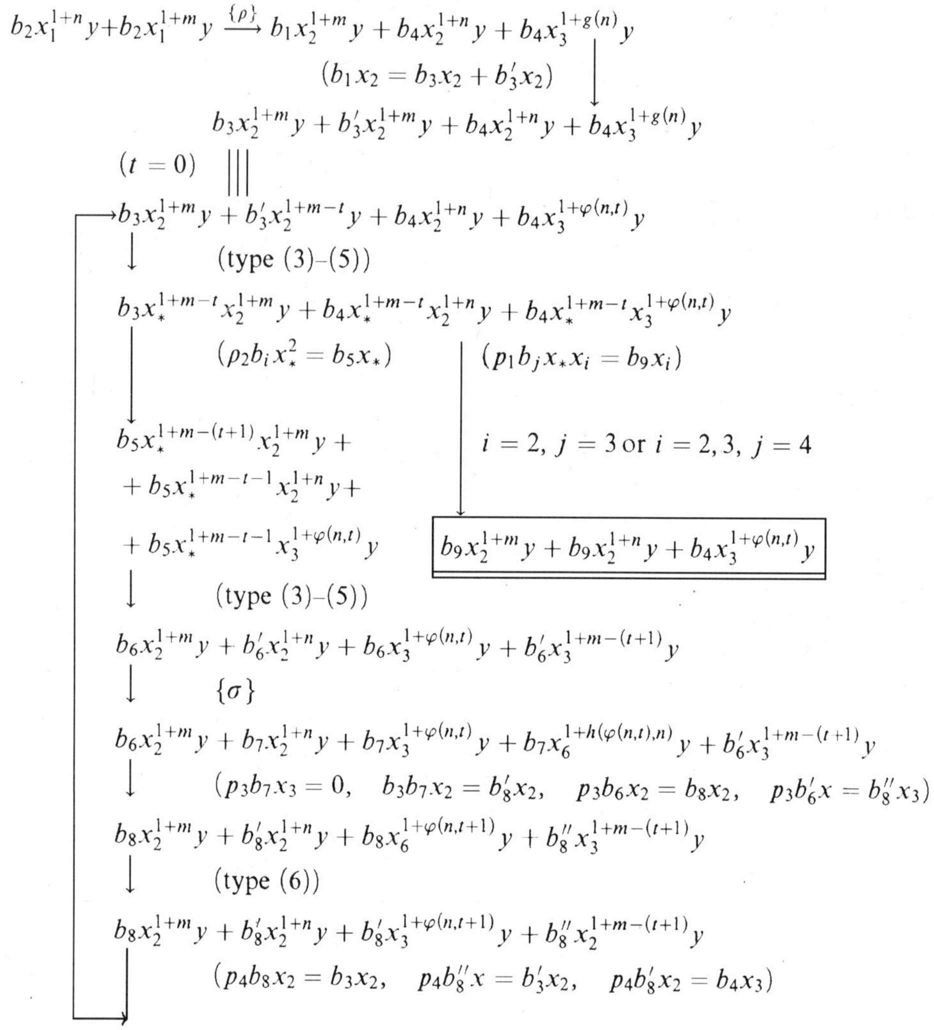

FIGURE 1. SCHEME 1

We assume that the equalities

$$b_2 x_1^{1+n} y = b_4 x_2^{1+n} y + b_4 x_3^{1+g(n)} y$$

are interpreted in L_g.

The letters (prime numbers) and letters-generators used in L_g or in L_h are supplemented by new letters p_j, b_1, b_3, b_5, b_8, b_9 (and others).

LEMMA 6 (about the graph of φ constructed by g, h using the scheme of primitive recursion). *The construction of the graph for* $\varphi : N^2 \to N$ *is interpreted in the following*

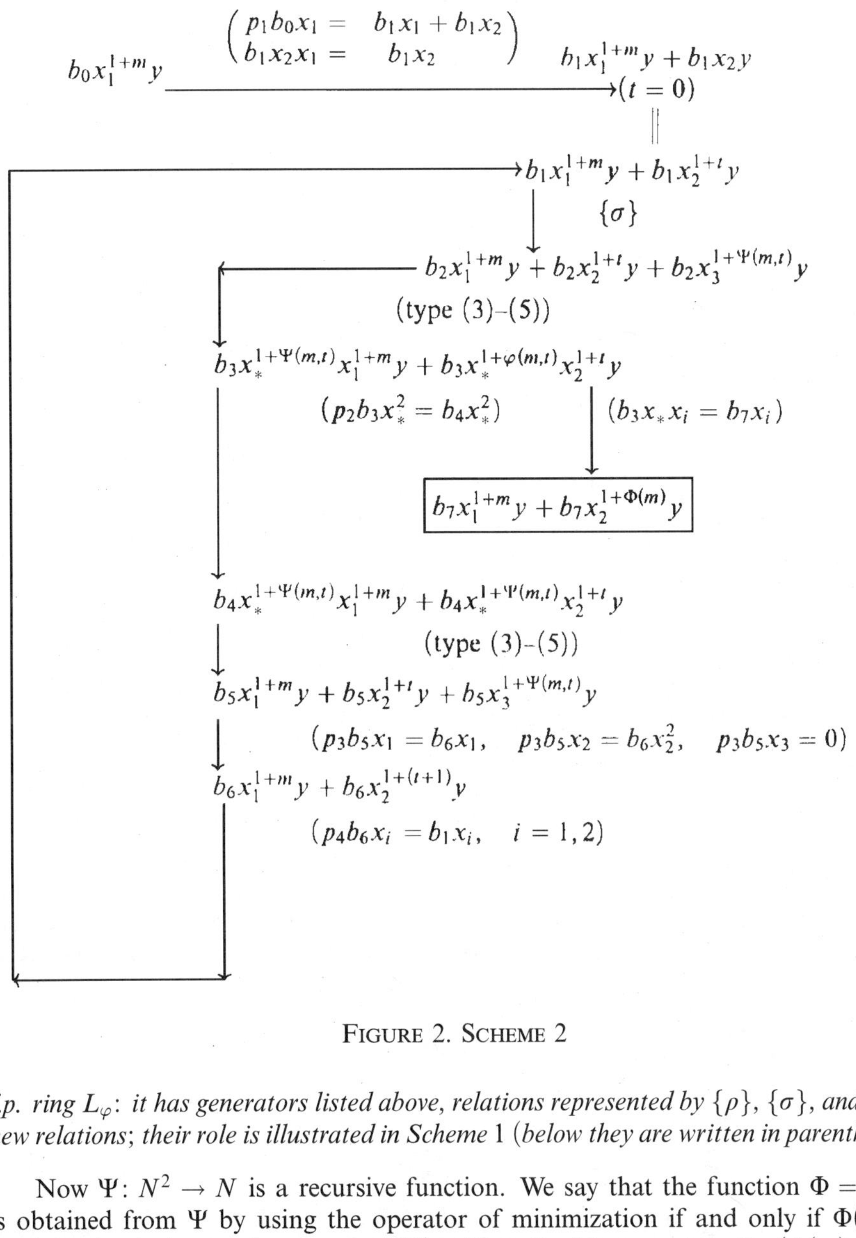

FIGURE 2. SCHEME 2

f.p. ring L_φ: it has generators listed above, relations represented by $\{\rho\}$, $\{\sigma\}$, and some new relations; their role is illustrated in Scheme 1 (below they are written in parenthesis).

Now $\Psi\colon N^2 \to N$ is a recursive function. We say that the function $\Phi = \mu_Y\Psi$ is obtained from Ψ by using the operator of minimization if and only if $\Phi(m)$ is a minimal solution of the equation $\Psi(m, Y) = 0$ with respect to Y ($\Phi(m)$ is not defined if the equation $\Psi(m, y) = 0$ has no solution for the given m).

We assume that the construction of the graph for $\Psi\colon N^2 \to N$ is interpreted in L_Ψ as above; $\{\sigma\}$ are defining relations for L_Ψ. The equalities

$$b_1x_1^{1+m}y + b_1x_1^{1+n}y = b_2x_1^{1+m}y + b_2x_2^{1+n}y + b_3x_3^{1+\Psi(m,n)}y$$

are valid in L_Ψ.

We supplement the list of letters-prime numbers and letters-generators for L_Ψ with new letters p_j, b_0, b_i ($i > 2$) and others (see Figure 2).

LEMMA 7 (about the graph of the operator of minimization). *The construction of the graph of* $\Phi = \mu_Y \Psi$ *is interpreted in the f.p. ring* L_Φ *with the generators listed above, relations*

$$p_1 b_0 x_1 = b_1 x_1 + b_1 x_2, \qquad p_2 b_3 x_*^2 = b_4 x_*^2, \qquad b_3 x_* x_i = b_7 x_i,$$

$$p_3 b_5 x_j = b_6 x_j, \qquad p_4 b_6 x_j = b_1 x_j \quad (j = 1, 2),$$

and the additional relations illustrated in Scheme 2.

So, we can construct an f.p. ring L_φ with generators and relations as in Lemmas 2–7 for every recursive function $\varphi\colon N \to N$ such that the element $b_1 x_1^{1+m} y - b_2 x_1^{1+u} y$ is equal to 0 in L_φ iff $u = \varphi(m)$.

Below we will need a similar interpretation of functions $\varphi\colon N^2 \to N$ in two variables. We show this construction now.

First we interpret functions $\varphi_1\colon m \mapsto 2^m$, $\varphi_2\colon n \mapsto 3^n$, such that the element

$$b_1 x_1^{1+m} y - b_2 x_1^{1+u} y$$

(or, respectively, $b_3 x_1^{1+n} y - b_4 x_1^{1+v} y$) is equal to 0 if and only if $u = 2^m$, $v = 3^n$. The operator of primitive recursion is interpreted in Lemma 6. Consequently, the function $\theta(m, n) \mapsto mn$ can be interpreted. As a result we obtain that the element

$$b_1 x_1^{1+m} y + b_3 x_1^{1+n} y = b_5 x_1^{1+u} y$$

is equal to 0 in the f.p. ring L_1 if and only if $u = 2^m \cdot 3^n$.

The map $\tau\colon 2^m \cdot 3^n \mapsto \Psi(m, n)$ and undefined on $k \notin \{2^m \cdot 3^n\}$ is a recursive function of one variable. We can interpret such a function. That is, we can construct an f.p. ring $\bar{L}_\Psi$ for the function $\Psi\colon N^2 \to N$ such that the element $b_1 x_1^{1+m} y + b_3 x_1^{1+n} y - b_6 x_1^{1+u} y$ is equal to 0 in $\bar{L}_\Psi$ iff $u = \Psi(m, n)$.

Let C be an arbitrary r.p. ring, and let $\nu\colon N \to C$ be its enumeration. Then the ring C is defined by a system of defining relations, which can be devided into two groups.

The table of multiplication is

$$c_i c_j = c_{\varphi(i,j)},$$

where $c_i \in C$, $\varphi\colon N^2 \to N$ is a recursive function.

The table of addition is given by the equality

$$c_i + c_j = c_{\Psi(i,j)},$$

with $c_i = c_j$ if and only if $\langle i, j \rangle \in D \subset N^2$, where D is a recursively enumerable set with characteristic function $\theta\colon N^2 \to \{0, 1\}$ and Ψ denotes a recursive function.

We denote by $C^{(\cdot)}$ the ring with the multiplication table as the list of defining relations, and use the notation $C^{(+)}$ for the ring defined by the table of addition and equality.

LEMMA 8 (about interpretation of the multiplication table in an f.p. ring). *Let $\bar{L}$ be an f.p. ring, let b_τ, x_*, x_i, y be its generators, and let*

$$(b_1x_1y)\cdot(b_2x_1y) = b_3x_1y + b_4x_1y,$$
$$p_1b_1x_1 = b_1x_1^2, \qquad p_1b_3x_1 = b_3x_1^2,$$
$$p_2b_2x_1 = b_2x_1^2, \qquad p_2b_4x_1 = b_4x_1^2$$

be its defining relations. Then

$$(b_1x_1^{1+m}y)\cdot(b_2x_1^{1+n}y) = b_3x_1^{1+m}y + b_4x_1^{1+n}y$$

are the defining relations for the subring B_0 generated by $\{b_\tau x_1^{1+i}y\}$.

In addition, if we consider defining relations for $\bar{L}$ similar to those in Lemmas 2–7 to interpret the calculation of the function $\varphi\colon N^2 \to N$, then the element

$$b_3x_1^{1+m}y + b_4x_1^{1+n}y - b_5x_1^{1+u}y$$

is equal to 0 if and only if $u = \varphi(m,n)$. Hence, the element

$$(b_1x_1^{1+m}y)\cdot(b_2x_1^{1+n}y) - b_5x_1^{1+u}y$$

is equal to 0 if and only if $u = \varphi(m,n)$.

This lemma is a direct corollary of Lemmas 1–7.

Let us consider the ring $\bar{L}$ of Lemma 8. We will add a new generator b_6 and new relations (7) (where p, p' denote prime numbers that do not appear in the defining relations for $\bar{L}$):

$$pb_1x_1 = b_6x_1, \qquad pb_5x_1 = b_6x_1,$$
$$p'b_2x_1 = b_6x_1, \qquad p'b_6x_1 = b_6x_1. \tag{7}$$

COROLLARY. *Let $L^{(\cdot)}$ be the subring generated by $\{\tilde{c}_l = b_6x_1^{1+l}y,\ l \geq 0\}$ in the ring $\bar{L}$ obeying the relations of Lemma 8 and, in addition, formula (7). Then $L^{(\cdot)}$ has*

$$\tilde{c}_i\tilde{c}_j = \tilde{c}_{\varphi(i,j)}$$

as defining relations. So we have imbedded the ring $C^{(\cdot)} \cong L^{(\cdot)}$ into an f.p. ring.

LEMMA 9 (about the interpretation of the table of addition and equality in an f.p. ring). *Let the ring $\bar{L}$ have generators $\hat{b}_\tau$, x_*, x_i, y and three groups of defining relations:*

1) *The first group of relations (as in Lemmas 2–7) which give the following property:*

$$\hat{b}_1x_1^{1+i}y + \hat{b}_2x_1^{1+j}y - \hat{b}_3x_1^{1+u}y$$

is equal to 0 if and only if $u = \Psi(i,j)$, Ψ as in the table of addition.

2) *The analogous relations: the element*

$$\hat{b}_4x_1^{1+i}y - \hat{b}_5x_1^{1+j}y - \hat{b}_6x_1^{1+v}y$$

is equal to 0 if and only if $v = \theta(i,j)$, where θ is the characteristic function of D in the table of addition.

3) *In addition to items* 1) and 2),

$$\hat{b}_6 x_1^2 y = 0, \qquad p\hat{b}_1 x_1 = \hat{b}_7 x_1 \quad (1 \le i \le 6).$$

(*Here* p *is a prime number that does not appear in defining relations of this lemma*). *Then the subring* $L^{(+)}$ *generated by*

$$\hat{c}_i = \hat{b}_7 x_1^{1+i} y$$

is isomorphic to the ring $C^{(+)}$ *considered above.*

PROOF. It is analogous to proofs of the previous lemmas. □

To prove the main result formulated in the introduction we consider a finitely defined ring $\bar{L}$. Its set of generators unites those in Lemmas 8, 9 (i.e., b_τ, $\tilde{b}_\tau$, $\hat{b}_\tau$, etc.); the set of defining relations is obtained analogously.

Let p be a prime number that does not appear above. Now let L_1 be the ring with the same generators as $\bar{L}$, the same relations, and, in addition,

$$p\hat{b}_7 x_1 = b_6 x_1.$$

It is obvious that the defining relations of the subring C_0 generated by the elements $c_i = b_6 x_1^{1+i} y$ are those for the given recursively defined ring C. This proves the theorem. □

Now we can give some applications of our techniques to the word problem in ring varieties. Let $\mathfrak{M}$ be the variety of rings, and L the free ring generated by b_τ, x_*, x_i, y.

We assume that all words $b_\tau x_*^\alpha x_i^\beta y x_*^\gamma x_i^\delta$ are $\mathbb{Z}$-linearly independent in L. Then our techniques to construct a ring (f.p. in $\mathfrak{M}$) with unsolvable word problem are suitable for $\mathfrak{M}$ or for any greater variety $\mathfrak{N}$ ($\mathfrak{N} \supseteq \mathfrak{M}$) (see **[8]**).

For example, in just the same way as for algebras we can consider one of the objects in the following list in the role of $\mathfrak{M}$:

- $\mathfrak{M}_1$ (all associative rings with the identity $[y_1, y_2][y_3, y_4][y_5, y_6][y_7, y_8] = 0$);
- $\mathfrak{M}_2$ (all Lie rings whose commutant is nilpotent of index 3);
- $\mathfrak{M}_3(\varepsilon)$ (the identity $(xy)z = \varepsilon(zy)x$, $\varepsilon = \pm 1$);
- $\mathfrak{M}_4$ (the identity $(xy)z = (yz)x$).

The new varieties with unsolvable word problem that are constructed using our techniques are $\mathfrak{M}_5(\varepsilon)$ (commutative rings for $\varepsilon = 1$, anticommutative rings for $\varepsilon = -1$), $\mathfrak{M}_6$ (the left nilpotency $x(yz) = 0$), and certainly $\mathfrak{M}_6^*$ (the identity $(xy)z = 0$).

For a ring variety to have Higman's property it is sufficient that it be an $\mathfrak{M}$-free ring (as in Lemma 1). For example, $\mathfrak{M}$ can be the variety of all associative (or Lie) rings (see the introduction), or $\mathfrak{M} = \mathfrak{M}_s(\varepsilon)$.

References

1. G. Higman, *Subgroups of finitely presented groups*, Proc. Roy. Soc. London Ser. A **262** (1961), 455–475.
2. V. A. Andrunakievich (ed.), *Dniester notebook*, 3rd ed., Akad. Nauk SSSR Sibirsk. Otdel, Inst. Mat., Novosibirsk, 1982. (Russian)
3. V. Murskiĭ, *Isomorphic imbeddability of a semigroup with an enumerable set of defining relations into a finitely presented semigroup*, Mat. Zametki **1** (1967), no. 2, 217–224; English transl. in Math. Notes **1** (1967).
4. V. Ya. Belyaev, *Imbeddability of recursively defined inverse semigroups in finitely presented semigroups*, Sibirsk. Mat. Zh. **25** (1984), no. 2, 50–54; English transl. in Siberian Math. J. **25** (1984).

5. ______, *Subrings of finitely presented associative rings*, Algebra i Logika **17** (1978), no. 6, 627–638; English transl. in Algebra and Logic **17** (1978).
6. G. P. Kukin, *Subalgebras of finitely presented Lie algebras*, Algebra i Logika **18** (1979), no. 3, 311–327; English transl. in Algebra and Logic **18** (1979).
7. O. V. Belegradek and N. Koveshnikova, *On subalgebras of relatively finite defined algebras*, Preprint No. 961-80, Kemerovsk. Gos. Univ., Kemerovsk, 1980. (Russian)
8. V. I. Epanchintsev and G. P. Kukin, *Higman quasivarieties*, Computable Invariants in the Theory of Algebraic Systems, Akad. Nauk SSSR Sibirsk. Otdel., Vychisl. Tsentr, Novosibirsk, 1987, pp. 90–109. (Russian)
9. A. Zhukov, *Reduced systems of defining relations in non-associative algebras*, Mat. Sb. **27** (1950), no. 2, 267–280. (Russian)
10. A. Shirshov, *Some algorithmic problems for ε-algebras*, Sibirsk. Mat. Zh. **3** (1962), no. 2, 132–137. (Russian)
11. Yu. M. Vazhenin, *Nonassociative rings with one defining relation whose elementary theories are decidable*, Algebra i Logika **29** (1990), no. 5, 509–522; English transl. in Algebra and Logic **29** (1990).
12. A. Mal′cev, *Algorithms and recursive functions*, "Nauka", Moscow, 1965; English transl., Noordhoff, Groningen, 1970.
13. G. P. Kukin, *Variety of all rings has Higman's property*, School on Algebraic Systems Varieties, Magnitogorsk, 1990. (Russian)

OMSK STATE UNIVERSITY

Amer. Math. Soc. Transl.
(2) Vol. **163**, 1995

Boolean-Valued Introduction to the Theory of Vector Lattices

A. G. Kusraev

The theory of vector lattices appeared in early thirties of this century and is connected with the names of L. V. Kantorovich, F. Riesz, and H. Freudenthal. The study of vector spaces equipped with an order relation compatible with a given norm structure was evidently motivated by the general circumstances that brought to life functional analysis in those years. Here the general inclination to abstraction and uniform approach to studying functions, operations on functions, and equations related to them should be noted. A remarkable circumstance was that the comparison of the elements could be added to the properties of functional objects under consideration. At the same time, the general concept of a Banach space ignored a specific aspect of the functional spaces—the existence of a natural order structure in them, which makes these spaces vector lattices.

Along with the theory of ordered spaces, the theory of Banach algebras was being developed almost at the same time. Although at the beginning these two theories advanced in parallel, soon their paths parted. Banach algebras were found to be effective in function theory, in the spectral theory of operators, and in other related fields. The theory of vector lattices was developing more slowly and its achievements related to the characterization of various types of ordered spaces and to the description of operators acting in them was rather unpretentious and specialized.

In the middle of the seventies the renewed interest in the theory of vector lattices led to its fast development which was related to the general explosive developments in functional analysis; there were also some specific reasons, the main one being the use of ordered vector space in the mathematical approach to social phenomena, economics in particular. The scientific work and the unique personality of L. V. Kantorovich also played important role in the development of the theory of ordered spaces and in relating this theory to economics and optimization. Another, though less evident, reason for the interest in vector lattices was their rather unexpected role in the theory of nonstandard—Boolean-valued—models of set theory. Constructed by D. Scott, R. Solovay, and P. Vopenka in connection with the well-known results by P. G. Cohen about the continuum hypothesis, these models proved to be inseparably linked with the theory of vector lattices. Indeed, it was discovered that the elements of such lattices serve as images of real numbers in a suitably selected Boolean model. This fact not only gives a precise meaning to the initial idea that abstract ordered spaces are derived from real numbers, but also provides a new possibility to infer common

1991 *Mathematics Subject Classification*. Primary 46A40; Secondary 03C, 06E.

properties of vector lattices by using the fact that they, in a precise sense, depict the sublattices of the field $\mathbb{R}$. Indeed, this possibility was taken as a basis for the present minicourse of lectures.

The main attention in these lectures is paid to the fundamental concepts. For brevity, we usually skip the proofs of the formulated theorems.

The bibliography, both in the field of vector lattices and in nonstandard analysis, is by no means complete. With few exceptions, the list of references consists of monographs and survey articles containing extensive bibliographies. Other original works are cited for specific reasons.

LECTURE I. Vector lattices

We start with a brief description of basic concepts of the theory of vector lattices.[1] Details can be found in [**7, 13, 14, 38, 41, 51**].

1.1. Let F be a linearly ordered field. An ordered vector space over F is a pair $(E, \leq)$, where E is a vector space over the field F and $\leq$ is an order relation on E such that, in addition, the following conditions are fulfilled:

(1) if $x \leq y$ and $u \leq v$, then $x + u \leq y + v$ for any x, y, u, $v \in E$;

(2) if $x \leq y$, then $\lambda x \leq \lambda y$ for any x, $y \in E$ and $0 \leq \lambda \in F$.

Thus, in an ordered vector space inequalities can be added together and multiplied by positive elements from F. This can be expressed as follows: $\leq$ is an order relation compatible with the vector space structure or, in short, $\leq$ is a vector order.

The definition of a vector order on a vector space E over the field F is equivalent to specifying a certain set (called the positive cone) $E_+ \subset E$ with the following properties:

$$E_+ + E_+ \subset E_+, \quad \lambda E_+ \subset E_+ \quad (0 \leq \lambda \in F), \quad E_+ \cap (-E_+) = \{0\}.$$

Moreover, the order $\leq$ and the cone E_+ are connected by the relation

$$x \leq y \leftrightarrow y - x \in E_+ \quad (x, y \in E).$$

The elements of the cone E_+ are called positive.

1.2. An ordered vector space that is also a lattice is called a *vector lattice*. Hence, for any finite set $\{x_1, \dots, x_n\}$ in a vector lattice E, there exist the least upper bound $\sup\{x_1, \dots, x_n\} =: x_1 \vee \dots \vee x_n$ and the greatest lower bound $\inf\{x_1, \dots, x_n\} =: x_1 \wedge \dots \wedge x_n$ (these elements are, of course, unique). In particular, any element x of a vector lattice has a positive part $x^+ := x \vee 0$, a negative part $x^- := (-x)^+ = -x \wedge 0$, and a modulus $|x| := x \vee (-x)$.

The disjunction (disjointness relation) $\perp$ in a vector lattice E is defined by the following formula:

$$\perp := \big\{(x, y) \in E \times E : \quad |x| \wedge |y| = 0\big\}.$$

A set of the form

$$M^\perp := \big\{x \in E : (\forall y \in M)\, x \perp y\big\},$$

where M is an arbitrary nonempty set in E, is called a *component* or a *band* of the

[1] *Translator's note.* Vector lattices are also called Riesz spaces.

vector lattice E. The set of all bands of a vector lattice, ordered by inclusion, forms a complete Boolean algebra $\mathfrak{B}(E)$ under the following Boolean operations:

$$L \wedge K := L \cap K, \quad L \vee K := (L \cup K)^{\perp\perp}, \quad L^* := L^{\perp} \quad (L, K \in \mathfrak{B}(E)).$$

The algebra $\mathfrak{B}(E)$ is called the *base* of E.

An element $1 \in E$ is called an (*order*) *unit*, if $\{1\}^{\perp\perp} = E$, i.e., if E has no nonzero elements that are disjoint with 1. Let $e \wedge (1 - e) = 0$ for some $0 \leq e \in E$. Then e is said to be a unit element (with respect to 1). The set $\mathfrak{C}(1) := \mathfrak{C}(E)$ of all unit elements with the order induced from E is a Boolean algebra. The lattice operations in $\mathfrak{C}(1)$ are inherited from E, and the Boolean complement has the form $e^* := 1 - e$ $(e \in \mathfrak{C}(E))$.

Let K be a band of a vector lattice E. If there exists an element $\sup\{u \in K : 0 \leq u \leq x\}$ in E, then this element is called the *projection* of x to the band K and is denoted by $[K]x$ (or $\mathrm{Pr}_K\, x$). For an arbitrary $x \in E$ one defines $[K]x := [K]x^+ - [K]x^-$. The projection of an element $x \in E$ on K exists if and only if there is a decomposition $x = y + z$, where $y \in K$, $z \in K^{\perp}$. Moreover, in that case, $y = [K]x$ and $z = [K^{\perp}]x$. We shall assume that any element $x \in E$ has a projection on K. Then the operator $x \mapsto [K]x$ $(x \in E)$ is linear, idempotent, and $0 \leq [K]x \leq x$, for all $0 \leq x \in E$. One says that E is a vector lattice with the projection (principal projection) property if for any band (principal band) $K \in \mathfrak{B}(E)$ the projection operator $[K]$ is defined.[2]

1.3. A linear subspace I of a vector lattice is called an *order ideal* or an *o-ideal* (or just an ideal, if the rest is clear from the context), whenever the inequality $|x| \leq |y|$ implies that $x \in I$, for any $x \in E$ and $y \in I$.[3]

If an ideal I has the aditional property $I^{\perp\perp} = E$ (or $I^{\perp} = \{0\}$, which is the same), then it is called a *foundation* of E.[4]

A subspace $E_0 \subset E$ is called a *sublattice* of E if $x \wedge y, x \vee y \in E_0$ for any x, $y \in E_0$. It is then said that the sublattice E_0 is minorizing (or that it is a minorant) if for any $0 \neq x \in E_+$ there exists an element $x_0 \in E_0$ satisfying the inequalities $0 < x_0 \leq x$. We say that E_0 is a majorizing (or massive) sublattice if for any $x \in E$ there exists $x_0 \in E_0$ such that $x \leq x_0$. Thus, E_0 is a minorizing (majorizing) sublattice if and only if $E_+ \setminus \{0\} = E_+ + (E_{0+} \setminus \{0\})$ (respectively, $E = E_+ + E_0$).

Everywhere below, whenever the field F is not indicated explicitly, a vector lattice over the linear ordered field $\mathbb{R}$ of real numbers is implied. An order interval in E is a set of the form $[a, b] := \{x \in E : a \leq x \leq b\}$, where $a, b \in E$. A set in E is called (*order*) *bounded* (or *o-bounded*) if it is contained in some order interval. It is possible to introduce a seminorm on the ideal $I(u) := \bigcup_{n=1}^{\infty}[-nu, nu]$ generated by the element $0 \leq u \in E$:

$$\|x\|_u := \inf\{\lambda \in \mathbb{R}_+ : |x| \leq \lambda u\} \quad (x \in I(u)).$$

If $I(u) = E$, then u is called a strong unit and E is a vector lattice of bounded elements. The seminorm $\|\cdot\|_u$ is a norm if and only if the lattice $I(u)$ is Archimedean; i.e., for any $x \in I(u)$ the order boundedness of the set $\{n|x| : n \in \mathbb{N}\}$ implies that $x = 0$.

[2] *Translator's note.* The principal band generated by an element f is $\{f\}^{\perp\perp}$.

[3] *Translator's note.* A set D is called solid if $f \in D$, $|h| \leq |f| \Rightarrow h \in D$.

[4] *Translator's note.* This property is often called "quasi order dense" in the literature (cf. [**34**, p. 110]); "foundation" is the literal translation of the Russian term and is more evocative.

An element $x \geq 0$ of a vector lattice is said to be discrete if $[0, x] = [0, 1]x$, i.e., if $0 \leq y \leq x$ implies $y = \lambda x$ for some $0 \leq \lambda \leq 1$. A vector lattice E is called discrete if for every $0 \neq y \in E_+$ there exists a discrete element $x \in E$ such that $0 < x \leq y$. If E has no nonzero discrete elements we say that E is continuous.

1.4. A vector lattice over the field of real numbers in which every nonempty order bounded set has an infimum and a supremum is called a *Kantorovich space*, or, in short, a *K-space*.[5] Sometimes instead of a K-space a more descriptive term is applied, namely (relatively) order complete vector lattice. If infima and suprema exist only for countable bounded sets, then the corresponding vector lattice is called a *K_σ-space*. Any K_σ-space, hence any K-space, is Archimedean. It is said that a K-space (K_σ-space) is extended[6] if any set (any countable set) in it consisting of pairwise disjoint elements is bounded.

A K-space has a projection onto every band. The set of all projections onto the bands of E is denoted by the symbol $\mathfrak{P}(E)$. For the projections π and ρ we define $\pi \leq \rho$ if and only if $\pi x \leq \rho x$ for all $0 \leq x \in E$.

THEOREM. *Let E be an arbitrary K-space. Projecting onto bands defines an isomorphism $K \mapsto [K]$ of Boolean algebras $\mathfrak{B}(E)$ and $\mathfrak{P}(E)$. If there exists a unit in E then the mappings $\pi \mapsto \pi 1$ from $\mathfrak{P}(E)$ into $\mathfrak{E}(E)$ and $e \mapsto \{e\}^{\perp\perp}$ from $\mathfrak{E}(E)$ into $\mathfrak{B}(E)$ are isomorphisms of Boolean algebras.*

1.5. The projection π_u onto the principal band $\{u\}^{\perp\perp}$, where $0 \leq u \in E$, can be computed by means of a simpler rule than is indicated in 1.2:

$$\pi_u x = \sup\{x \wedge (nu) : n \in \mathbb{N}\}.$$

In particular, a K_σ-space contains the projection of any element onto every principal band.

Let E be a K-space with unit 1. The projection of the unit onto the band $\{x\}^{\perp\perp}$ is called the *trace* of the element x and is denoted by the symbol e_x. Thus, $e_x = \sup\{1 \wedge (n|x|) : n \in \mathbb{N}\}$. The trace e_x can be used both as a unit in $\{x\}^{\perp\perp}$ and as a unit element in E. For any real number λ, e_λ^x denotes the trace of the positive part of the element $\lambda 1 - x$, i.e., $e_\lambda^x = e_{(\lambda 1 - x)^+}$. The associated function $\lambda \mapsto e_\lambda^x$ is called the *spectral function* or the *characteristic of* x.

1.6. An ordered space E over F is called an *ordered algebra* over F, if it is an algebra over F and, moreover, the following condition is satisfied: if $x, y \in E$ with $x \geq 0$ and $y \geq 0$, then $xy \geq 0$. In order to characterize the positive cone E_+ of an ordered algebra E, another property should be added to those mentioned in 1.1: $E_+ \cdot E_+ \subset E_+$. We say that E is a lattice ordered algebra if E is a vector lattice and an ordered algebra, simultaneously. A lattice ordered algebra is called an *f-algebra* if for any $a, x, y \in E_+$ from the condition $x \wedge y = 0$ it follows that $(ax) \wedge y = 0$ and $(xa) \wedge y = 0$. An f-algebra is called *faithful* if for any two elements x and y the equality $xy = 0$ implies $x \perp y$. It is not difficult to show that an f-algebra is faithful if and only if it has no nonzero nilpotent elements. The faithfulness of an f-algebra is also equivalent to the absence of strictly positive elements with zero square.

---

[5] *Translator's note.* This is the Russian terminology for a Dedekind complete vector lattice.

[6] *Translator's note.* Literally translated from the Russian; in the French translation of [13] this is called "achevé", and in [30] the phrase "universally complete" is used for this notion.

1.7. The complexification $E \oplus iE$ (i is the imaginary unit) of a real vector lattice E is called a *complex vector lattice*. In addition, it is often required that

$$|z| := \sup \left\{ \operatorname{Re}(e^{i\theta} z) \colon 0 \leq \theta \leq \pi \right\}$$

for any element $z \in E \oplus iE$. In the case of a K-space or an arbitrary Banach lattice this requirement is automatically fulfilled. Thus, a complex K-space is the complexification of a real K-space. Speaking of order properties of a complex vector lattice $E \oplus iE$, we have in mind its real part E. The notions of a sublattice, ideal, bands of projection, etc. are naturally extended to the case of a complex vector lattice with the help of suitable complexifications.

1.8. Various types of convergence are related with the order relation in a vector lattice. Let $(A, \leq)$ be an upwards-filtered set (i.e., filtered with respect to increase). A net $(x_\alpha) := (x_\alpha)_{\alpha \in A}$ in E is said to be increasing (decreasing) if $x_\alpha \leq x_\beta$ $(x_\beta \leq x_\alpha)$ for $\alpha \leq \beta$ $(\alpha,\ \beta \in A)$.

It is said that a net (x_α) o-converges to an element $x \in E$ if there exists a decreasing net $(e_\alpha)_{\alpha \in A}$ in E with the properties $\inf\{e_\alpha : \alpha \in A\} = 0$ and $|x_\alpha - x| \leq e_\alpha$ $(\alpha \in A)$. If this is the case, x is called an o-limit of the net (x_α) and this is denoted by $x = o\text{-}\lim x_\alpha$ or $x_\alpha \xrightarrow{(o)} x$. In a K-space E, the upper and lower o-limits for an order bounded net are introduced by the following formulas:

$$\limsup_{\alpha \in A} x_\alpha := \overline{\lim_{\alpha \in A}}\, x_\alpha := \inf_{\alpha \in A} \sup_{\beta \geq \alpha} x_\beta;$$

$$\liminf_{\alpha \in A} x_\alpha := \underline{\lim_{\alpha \in A}}\, x_\alpha := \sup_{\alpha \in A} \inf_{\beta \geq \alpha} x_\beta.$$

There is an evident relation between these objects:

$$x = o\text{-}\lim_{\alpha \in A} x_\alpha \leftrightarrow \limsup_{\alpha \in A} x_\alpha = x = \liminf_{\alpha \in A} x_\alpha.$$

It is said that the net $(x_\alpha)_{\alpha \in A}$ regulator converges to $x \in E$ if there exist an element $0 \leq u \in E$, which is called a regulator of convergence, and a net of numbers $(\lambda_\alpha)_{\alpha \in A}$ with the properties $\lim \lambda_\alpha = 0$ and $|x_\alpha - x| \leq \lambda_\alpha u$ $(\alpha \in A)$. In addition, x is called a r-limit of the net (x_α) and this is denoted as $x = r\text{-}\lim_{\alpha \in A} x_\alpha$ or $x_\alpha \xrightarrow{(r)} x$. Clearly, regulator convergence is convergence in the normed space $(I(u),\ \|\cdot\|_u)$.

The presence of o-convergence in a K-space allows us to define the sum of an infinite family $(x_\xi)_{\xi \in \Xi}$. Indeed, let $A := \mathfrak{P}_{\text{fin}}(\Xi)$ be the set of all finite subsets of Ξ. We write $y_\alpha := x_{\xi_1} + \ldots + x_{\xi_n}$ for $\alpha := \{\xi_1, \ldots, \xi_n\} \in A$, obtaining a net $(y_\alpha)_{\alpha \in A}$, where A is naturally ordered by inclusion. If $x := o\text{-}\lim_{\alpha \in A} y_\alpha$ exists, then the element x is called the *o-sum* of the family (x_ξ) and it is denoted as $x = o\text{-}\sum_{\xi \in \Xi} x_\xi$ or simply $x = \sum_{\xi \in \Xi} x_\xi$. It is clear that for $x_\xi \geq 0$ $(\xi \in \Xi)$ the o-sum of the family (x_ξ) exists if and only if the family $(y_\alpha)_{\alpha \in A}$ is order bounded; moreover,

$$o\text{-}\sum_{\xi \in \Xi} x_\xi = \sup_{\alpha \in A} y_\alpha.$$

If the elements of the family (x_ξ) are pairwise disjoint, then

$$o\text{-}\sum_{\xi \in \Xi} x_\xi = \sup_{\xi \in \Xi} x_\xi^+ - \sup_{\xi \in \Xi} x_\xi^-.$$

Any K-space is o-complete in the following sense. If a net $(x_\alpha)_{\alpha\in A}$ in E satisfies the condition $\limsup |x_\alpha - x_\beta| := \inf_{\gamma\in A} \sup_{\alpha,\beta\geq\gamma} |x_\alpha - x_\beta| = 0$, then there exists an element $x \in E$ such that $x = o\text{-}\lim x_\alpha$.

1.15. Comments. (a) The history of functional analysis in ordered vector spaces is usually related to the contributions of G. Birkhoff, L. V. Kantorovich, M. G. Krein, H. Nakano, F. Riesz, H. Freudenthal, etc. At the present time, the theory and applications of ordered vector spaces forms an extensive domain in mathematics which, essentially, is one of the main branches of contemporary functional analysis. The field is well presented in monographs; see **[1, 7, 13, 14, 16, 18, 20, 25, 26, 29, 31, 32, 36–39, 41, 50, 51]**. We should also mention the surveys with extensive bibliographies **[2–4]**. The necessary information on the theory of Boolean algebras can be found in **[6, 24, 30]**.

(b) L. V. Kantorovich singled out the most important class of ordered vector spaces—the order complete vector lattices, i.e., K-spaces. They were introduced in Kantorovich's first fundamental work on the subject **[11]**, where he wrote: "In this note I define a new type of spaces which I call linear semi-ordered spaces. The introduction of these spaces allows us to study linear operations of one general class (operations with values belonging to such a space) as linear functionals."

In the same paper Kantorovich formulated an important methodological principle—a heuristic transfer principle for the K-spaces. As an example of application of this principle one can take Theorem 3 from **[11]** which is also called the Hahn-Banach theorem. It states that the Kantorovich principle can be realized in the case of the classical theorem on the majorized extension of a linear functional, i.e., the real numbers in the Hahn-Banach-Kantorovich theorem can be replaced by the elements of an arbitrary K-space, and the linear functionals by linear operators with values in this K-space.

LECTURE 2. Boolean-valued models

This lecture presents a short survey of necessary information from the theory of Boolean-valued models. The details can be found in **[10, 18, 21, 40, 45–47]**.

The main feature of the method of Boolean-valued models lies in the comparative analysis of two models—standard and nonstandard (Boolean-valued)—using a certain technique of descents and ascents. In addition, a syntactic comparison of formal strings has often to be applied. Therefore, before starting our study of the technique of descents and ascents we need to have a more precise idea about the status of mathematical objects within the framework of formalized set theory.

2.1. At present, the Zermelo-Frenkel set theory is the most widely used axiomatic basis of mathematics. We recall briefly some of the concepts of this theory, concentrating on the details that will be necessary below. It should be noted that regarding formal set theory we shall use (since it is unavoidable) the level of rigor that is accepted in mathematics; we shall introduce abbreviations with the help of the definition operator := and we will not go into the concomitant details.

(1) The alphabet of the Zermelo-Frenkel theory (shortened ZF or ZFC) consists of symbols for variables; parentheses (,); propositional connectives (= operations of the propositional algebra); $\vee$, $\wedge$, $\to$, $\leftrightarrow$, $\rceil$; the quantifiers $\forall$, $\exists$; the sign of equality $=$; and a symbol for a special two-place predicate $\in$. Conceptually, the range of the variables of ZF is conceived as the world (universe) of sets. In other words, the

universe of ZF has no other objects but sets. Instead of $\in(x, y)$, one writes $x \in y$ and says that x is an element of y.

(2) The formulas of ZF are defined by the usual procedure. In other words, the ZF formulas are finite strings obtained from the atomic formulas of the form $x = y$ and $x \in y$, where x, y are variables of ZF, with the help of reasonable placements of parentheses, quantifiers, and propositional connectives. Natural meaning is given to the terms of free and bound variables (or, equivalently, to the concept of the action area of a quantifier).

(3) When studying ZF theory, it is convenient to use the expressions that are absent in its formal language. In particular, it is appropriate to use the notions of a class and of a definable class, and also corresponding symbols for classes of the form $A\varphi := A_{\varphi(\cdot)} := \{x : \varphi(x)\}$ and $A\psi := A_{\psi(\cdot, y)} := \{x : \psi(x, y)\}$, where φ, ψ are formulas from ZF, and y is a selected set of variables. If one wishes to make the resulting notations more precise (or eliminate them) one can assume that the use of classes and classifiers is only related to the usual conventions about the introduction of abbreviations. This convention, sometimes called the Church scheme, is postulated as follows:

$$z \in \{x : \varphi(x)\} \leftrightarrow \varphi(z),$$
$$z \in \{x : \psi(x, y)\} \leftrightarrow \psi(z, y).$$

When working with ZF, abbreviations widely used in mathematics are involved. Some of them are:

$\cup x := \{z : (\exists y \in x)\, z \in y\}$;
$\cap x := \{z : (\forall y \in x)\, z \in y\}$;
$x \subset y := (\forall z)\,(z \in x \to z \in y)$;
$\mathscr{P}(x) :=$ the class of all subsets of $x := \{z : z \subset x\}$;
$V :=$ the class of all sets $:= \{x : x = x\}$.

We note that in further discussions more complicated descriptions, in which a lot is implicitly understood, are allowed:

$\mathrm{Funct}(f) := f$ is a function;
$\mathrm{dom}(f) :=$ the domain of definition of f;
$\mathrm{im}(f) :=$ the range of values of f;
$\varphi \vDash \psi := \langle\langle \psi$ is derivable from $\varphi \rangle\rangle$;
the class A is a set $:= A \in V := (\exists x)(\forall y)\, y \in x \leftrightarrow y \in A$.

Similar simplifications without special stipulations are used in writing down complicated concepts and formulas. For instance, the following formulas of ZF, which are rather large in the language of ZF itself, are simply written down as:

$f : x \to y := f$ is a function from x to y;
E is a K-space.

2.2. In ZF set theory we accept the usual axioms and inference rules for first-order theories with equality, which fix the standard methods of classical reasoning (syllogisms, law of the excluded middle, modus ponens, generalization, etc.). Besides, the following special and characteristic axioms are assumed.

(1) **Axiom of extensionality**:

$$(\forall x)(\forall y) \quad (x \subset y \wedge y \subset x \to x = y).$$

(2) **Axiom of union**:

$$(\forall x) \quad \cup x \in V.$$

(3) **Axiom of power set**:

$$(\forall x) \quad \mathscr{P}(x) \in V.$$

(4) **Axiom scheme of replacement**:

$$(\forall x)(((\forall y)(\forall z)(\forall u)\varphi(y,z) \wedge \varphi(y,u) \to z = u)$$
$$\to \{z : (\exists y \in x)\varphi(y,z)\} \in V).$$

(5) **Axiom of foundation**:

$$(\forall x)(x \neq \varnothing \to (\exists y \in x)(y \cap x = \varnothing)).$$

(6) **Axiom of infinity**:

$$(\exists \omega)((\varnothing \in \omega) \wedge (\forall x \in \omega)(x \cup \{x\} \in \omega)).$$

(7) **Axiom of choice**:

$$(\forall F)(\forall x)(\forall y)((x \neq \varnothing \wedge F : x \to \mathscr{P}(y))$$
$$\to ((\exists f) f : x \to y \wedge (\forall z \in x) \quad f(z) \in F(z))).$$

The precise concept of the class of all sets as the von Neumann universe V is based on the presented axiomatics. The initial object in this construction is the empty set $\varnothing$. A simple step for introducing new sets consists in forming the union of sets or in taking subsets of the sets already constructed. The transfinite repetition of such steps exhausts the class of all sets. More precisely, it is assumed that $V := \bigcup_{\alpha \in \mathrm{Or}} V_\alpha$, where Or is the class of all ordinals and

$$V_0 := \varnothing; \qquad V_{\alpha+1} := \mathscr{P}(V_\alpha); \qquad V_\beta := \bigcup_{\alpha<\beta} V_\alpha \quad (\beta \text{ is a limit ordinal}).$$

The class V is the standard model of ZF theory.

2.3. Now we shall describe the construction of a Boolean-valued universe. Let B be a complete Boolean algebra. For each ordinal α we set

$$V_\alpha^{(B)} := \{x : \mathrm{Funct}(x) \wedge (\exists \beta)(\beta < \alpha \wedge \mathrm{dom}(x) \subset V_\beta^{(B)} \wedge \mathrm{im}(x) \subset B)\}.$$

Thus, in more detailed notation:
$V_0^{(B)} := \varnothing$;
$V_{\alpha+1}^{(B)} := \{x : x$ is a function with domain in $V_\alpha^{(B)}$ and with range in $B\}$:
$V_\alpha^{(B)} := \bigcup_{\beta<\alpha} V_\beta^{(B)}$.

The following class is considered as the Boolean-valued universe $V^{(B)}$:
$V^{(B)} := \bigcup_{\alpha \in Or} V_\alpha^{(B)}$.

Elements of the class $V^{(B)}$ are called *B-valued sets*. It is worth noting that $V^{(B)}$ consists only of functions. In particular, $\varnothing$ is the function with the domain $\varnothing$ and the range $\varnothing$.

2.4. Suppose we have an arbitrary formula $\varphi = \varphi(u_1, \dots, u_n)$ from ZF theory. By replacing the variables $u_1, \dots, u_n$ with elements $x_1, \dots, x_n \in V^{(B)}$ we obtain a certain statement about the objects $x_1, \dots, x_n$, the validity of which we try to estimate. The desired truth-value $[\![\varphi]\!]$ should be an element from the algebra B. In the process we wish the ZF theorems to be judged valid, i.e., the largest truth value in B, namely one, to be ascribed to them.

The assignment of truth-values is carried out by a double induction which takes into account the character of constructing formulas from the atomic ones, and assigns the truth-values for $[\![x \in y]\!]$ and $[\![x = y]\!]$, where $x, y \in V^{(B)}$, based on the method of construction of $V^{(B)}$.

It is clear that if φ and ψ are already estimated ZF formulas, and $[\![\varphi]\!] \in B$ and $[\![\psi]\!] \in B$ are their truth-values, then one should put

$$[\![\varphi \wedge \psi]\!] := [\![\varphi]\!] \wedge [\![\psi]\!];$$
$$[\![\varphi \vee \psi]\!] := [\![\varphi]\!] \vee [\![\psi]\!];$$
$$[\![\varphi \to \psi]\!] := [\![\varphi]\!] \Rightarrow [\![\psi]\!];$$
$$[\![\urcorner \varphi]\!] := [\![\varphi]\!]^*;$$
$$[\![(\forall x)\varphi(x)]\!] := \bigwedge_{x \in V^{(B)}} [\![\varphi(x)]\!];$$
$$[\![(\exists x)\varphi(x)]\!] := \bigvee_{x \in V^{(B)}} [\![\varphi(x)]\!],$$

where on the right-hand sides we have the Boolean operations corresponding to the logical connectives and quantifiers from the left sides: $\wedge$ is the infimum, $\vee$ the supremum, $*$ the complement, $\bigwedge$ and $\bigvee$ denote the infimum and supremum of arbitrary sets, and the operation $\Rightarrow$ is introduced in the following way: $a \Rightarrow b := a^* \vee b$ $(a, b \in B)$. Only these definitions allow us to obtain the unit truth-value for the classical tautologies.

Now we pass to the estimation of atomic formulas $x \in y$ and $x = y$ for $x, y \in V^{(B)}$. The intuitive idea is that a B-valued set is a "fuzzy (lattice) set", i.e., a "set that contains an element z from $\operatorname{dom}(y)$ with the probability $y(z)$". Taking into account this idea as well as the goal to save both the logical truth $x \in y \leftrightarrow (\exists z \in y)x = z$ and the extensionality axiom, we are compelled to use the following recursive definition:

$$[\![x \in y]\!] := \bigvee_{z \in \operatorname{dom}(y)} y(z) \wedge [\![z = x]\!],$$

$$[\![x = y]\!] := \bigwedge_{z \in \operatorname{dom}(x)} x(z) \Rightarrow [\![z \in y]\!] \wedge \bigwedge_{z \in \operatorname{dom}(y)} y(z) \Rightarrow [\![z \in x]\!].$$

2.5. Now, we are already in a position to give meaning to formal expressions of the form $\varphi(x_1, \dots, x_n)$, where $x_1, \dots, x_n \in V^{(B)}$ and φ is a ZF formula, i.e., to give precise meaning to the expression: "a set-theoretic expression $\varphi(u_1, \dots, u_n)$ holds for the elements $x_1, \dots, x_n \in V^{(B)}$". Namely, we say that the formula $\varphi(x_1, \dots, x_n)$ is true inside $V^{(B)}$ or that the elements $x_1, \dots, x_n$ possess the property φ in $V^{(B)}$ if $[\![\varphi(x_1, \dots, x_n)]\!] = 1$. This is denoted by $V^{(B)} \models \varphi(x_1, \dots, x_n)$.

It is not difficult to be convinced that the axioms and theorems of the first-order predicate calculus with equality are valid in $V^{(B)}$. In particular,

(1) $[\![x = x]\!] = 1$,

(2) $[\![x = y]\!] = [\![y = x]\!]$,
(3) $[\![x = y]\!] \wedge [\![y = z]\!] \leq [\![x = z]\!]$,
(4) $[\![x = y]\!] \wedge [\![z \in x]\!] \leq [\![z \in y]\!]$,
(5) $[\![x = y]\!] \wedge [\![x \in z]\!] \leq [\![y \in z]\!]$.

It is useful to note that, in general, for every formula φ the following holds:

$$V^{(B)} \models x = y \wedge \varphi(x) \to \varphi(y),$$

i.e., written out explicitly

(6) $[\![x = y]\!] \wedge [\![\varphi(x)]\!] \leq [\![\varphi(y)]\!]$.

2.6. In a Boolean-valued universe the relation $[\![x = y]\!] = 1$ does not mean at all that the functions x and y (considered as elements of V) coincide. For instance, the zero function on any layer $V_\alpha^{(B)}$, where $\alpha \geq 1$, plays the role of the empty set in $V^{(B)}$. This makes some constructions more difficult. In view of this, we can introduce a separated Boolean-valued universe $\bar{V}^{(B)}$, which is often denoted by the same symbol $V^{(B)}$; i.e., one sets $V^{(B)} := \bar{V}^{(B)}$. In order to define $\bar{V}^{(B)}$ in the class $V^{(B)}$, the relation $\{(x, y) : [\![x = y]\!] = 1\}$ is considered, which evidently is an equivalence. Selecting an element (a representative of the smallest rank) from every equivalence class one arrives at the separated universe $\bar{V}^{(B)}$. It should be noted that for every formula φ of ZF theory and any elements $x, y \in V^{(B)}$ the following holds:

$$[\![x = y]\!] = 1 \to [\![\varphi(x)]\!] = [\![\varphi(y)]\!].$$

Therefore, in a separated universe the truth-value of formulas can be computed independently of the method of selecting representatives. In general, when dealing with a separated universe, the equivalence class is often replaced (with the necessary precaution) by a certain representative, as it is done, for instance, in the case of function spaces.[7]

2.7. The most important properties of a Boolean-valued universe are contained in the following three principles.

Transfer principle. All the theorems of the ZF theory are true in $V^{(B)}$; i.e., the transfer principle, symbolically written down as

$$V^{(B)} \models \text{ ZF theorem},$$

is valid.

The transfer principle is established by a rather laborious test showing that all ZF axioms have the truth-value 1, and that the derivation rules preserve the validity of formulas. The transfer principle is sometimes expressed by saying that $V^{(B)}$ is a Boolean-valued model of ZF set theory.

Maximum principle. For every formula φ of ZF theory there is $x_0 \in V^{(B)}$ such that

$$[\![(\exists x)\varphi(x)]\!] = [\![\varphi(x_0)]\!].$$

In particular, if it is true that in $V^{(B)}$ there exists x for which $\varphi(x)$ holds, then, in

[7] *Translator's note.* Such as $L^2(\mathbb{R})$, where one usually thinks in terms of true functions instead of equivalence classes of functions that differ on sets of measure zero.

fact, in $V^{(B)}$ (in the sense of V!) there can be found an element x_0 such that $\varphi(x_0)$. Symbolically,

$$V^{(B)} \models (\exists x)\varphi(x) \to (\exists x_0)\, V^{(B)} \models \varphi(x_0).$$

In other words, for any formula φ of the ZF theory the maximum principle holds:

$$(\exists x_0 \in V^{(B)})[\![\varphi(x_0)]\!] = \bigvee_{x\in V^{(B)}} [\![\varphi(x)]\!].$$

The latter equality also explains the origin of the term "maximum principle". The proof of this principle is a simple application of the following mixing principle.

Mixing principle. Let $(b_\xi)_{\xi\in\Xi}$ be a partition of unity in B, i.e., a family of elements from the Boolean algebra B such that $\xi \neq \eta \to b_\xi \wedge b_\eta = 0$ and $\bigvee_{\xi\in\Xi} b_\xi = 1$.

For any family of elements $(x_\xi)_{\xi\in\Xi}$ of the universe $V^{(B)}$ and any partition of unity $(b_\xi)_{\xi\in\Xi}$ there exists a (unique) mixing (x_ξ) with the probabilities (b_ξ); i.e., there exists an element x of the separated universe $V^{(B)}$ such that for all $\xi \in \Xi$ the following holds: $[\![x = x_\xi]\!] \geq b_\xi$.

The mixing x of the set (x_ξ) with respect to (b_ξ) is denoted as

$$x := \operatorname*{mix}_{\xi\in\Xi}(b_\xi x_\xi) = \operatorname{mix}\{b_\xi x_\xi \ : \ \xi \in \Xi\}.$$

2.8. The comparative analysis which was discussed at the beginning of this lecture is possible due to a close interrelation of the worlds V and $V^{(B)}$. In other words, it is necessary to have a rigorous mathematical technique that would permit us to determine interrelations between interpretations of the same fact in the models V and $V^{(B)}$. The basis of this technique consists in introducing the operations of canonical imbedding, descent, and ascent, which will be defined below. Let us begin with the canonical imbedding of the von Neuman universe. For $x \in V$ the standard name of x in $V^{(B)}$ is $x^\wedge$, so that we have the following recursion scheme:

$$\varnothing^\wedge := \varnothing, \quad \operatorname{dom}(x^\wedge) := \{y^\wedge \ : y \in x\}, \quad \operatorname{im}(x^\wedge) := \{1\}.$$

We note the necessary properties of the mapping $x \mapsto x^\wedge$.

(1) For every $x \in V$ and formula φ the following holds

$$[\![(\exists y \in x^\wedge)\varphi(y)]\!] = \bigvee\{[\![\varphi(z^\wedge)]\!] \colon z \in x\};$$

$$[\![(\forall y \in x^\wedge)\varphi(y)]\!] = \bigwedge\{[\![\varphi(z^\wedge)]\!] \colon z \in x\}.$$

(2) If x, y are elements from V, then using the transfinite induction, one establishes the following:

$$x \in y \leftrightarrow V^{(B)} \models x^\wedge \in y^\wedge,$$

$$x = y \leftrightarrow V^{(B)} \models x^\wedge \in y^\wedge.$$

In other words, the standard name can be considered as an imbedding of V into $V^{(B)}$. Moreover, the standard name maps V onto $V^{(2)}$, as can be noticed from the following fact.

(3) The following statement holds:

$$(\forall u \in V^{(2)})(\exists!\, x \in V)\, V^{(B)} \models u = x^\wedge.$$

(4) A formula is called *bounded* if all occurring variables are included into it under the signs of bounded quantifiers, i.e., quantifiers which range over some set.

The latter phrase means that any occurring variable x is found in the domain of action of a quantifier of the form $(\forall x \in y)$ or $(\exists x \in y)$, for some y.

The principle of bounded transfer. For any bounded formula φ of ZF theory and for any family $x_1, \dots, x_n \in V$ the following equivalence holds:

$$\varphi(x_1, \dots, x_n) \leftrightarrow V^{(B)} \models \varphi(x_1^\wedge, \dots, x_n^\wedge).$$

Let us agree, in the study of a separated universe $\bar{V}^{(B)}$ below, to keep the symbol $x^\wedge$ for a selected element of the class corresponding to x.

(5) Let us note, as an example, that the principle of bounded transfer implies

$$(\Phi \text{ is a correspondence from } x \text{ to } y)$$
$$\leftrightarrow (V^{(B)} \models \Phi \text{ is a correspondence from } x^\wedge \text{ to } y^\wedge);$$
$$(f \text{ is a function from } x \text{ to } y)$$
$$\leftrightarrow (V^{(B)} \models f^\wedge \text{ is a function from } x^\wedge \text{ to } y^\wedge).$$
$$\text{Moreover, } f(a)^\wedge = f^\wedge(a^\wedge) \text{ for every } a \in x.$$

Thus, the standard name can be regarded as a covariant functor from the category of sets (or correspondences) of V into a suitable subcategory $V^{(2)}$ of the separated universe $V^{(B)}$.

2.9. For an arbitrary element x from the (separated) Boolean-valued universe $V^{(B)}$ the descent $x\downarrow$ of this element is defined by the formula

$$x\downarrow := \{y \in V^{(B)} : [\![y \in x]\!] = 1\}.$$

Let us mention the main properties of the descent procedure.

(1) The class $x\downarrow$ is a set; i.e., $x\downarrow \in V$ for every $x \in V^{(B)}$. If $[\![x \neq \varnothing]\!] = 1$, then $x\downarrow$ is a nonempty set.

(2) Let $z \in V^{(B)}$ and $[\![z \neq \varnothing]\!] = 1$. Then for any formula φ of ZF theory the following holds:

$$[\![(\forall x \in z)\, \varphi(x)]\!] = \bigwedge\{[\![\varphi(x)]\!] : x \in z\downarrow\};$$
$$[\![(\exists x \in z)\, \varphi(x)]\!] = \bigvee\{[\![\varphi(x)]\!] : x \in z\downarrow\}.$$

In addition, there exists $x_0 \in z\downarrow$ such that $[\![\varphi(x_0)]\!] = [\![(\exists x \in z)\varphi(x)]\!]$.

(3) Let Φ be a correspondence from X into Y in $V^{(B)}$. Thus Φ, X, Y are elements of $V^{(B)}$ and, moreover, $[\![\Phi \subset X \times Y]\!] = 1$. Then there exists a unique correspondence $\Phi\downarrow$ from $X\downarrow$ into $Y\downarrow$ such that for any nonempty subset A of the set X in $V^{(B)}$ the following holds:

$$\Phi\downarrow(A\downarrow) = \Phi(A)\downarrow.$$

The correspondence $\Phi\downarrow$ from $X\downarrow$ into $Y\downarrow$ that appears in this statement is called the *descent of the correspondence* Φ from X into Y in $V^{(B)}$.

(4) The descent of the composition of correspondences in $V^{(B)}$ is the composition of the descents of these correspondences:

$$(\Psi \circ \Phi)\downarrow = \Psi\downarrow \circ \Phi\downarrow.$$

(5) If Φ is a correspondence in $V^{(B)}$, then $(\Phi^{(-1)})\downarrow = \Phi\downarrow^{-1}$.

(6) Let I_X be the identity mapping of a set $X \in V^{(B)}$. Then $(I_X)\downarrow = I_{X\downarrow}$.

(7) Let f, X, $Y \in V^{(B)}$ be such that $[\![f : X \to Y]\!] = 1$; i.e., f is a mapping from X into Y in $V^{(B)}$. Then $f\downarrow$ is the unique mapping from $X\downarrow$ into $Y\downarrow$ for which

$$[\![f\downarrow(x) = f(x)\downarrow]\!] = 1 \quad (x \in X\downarrow).$$

As illustrated by statements (1)–(7), the operation of descent can be regarded as a functor from B-valued sets and mappings (correspondences) into the category of usual (i.e., in the sense of V) sets and mappings (correspondences).

(8) For $x_1, \dots, x_n \in V^{(B)}$ we denote the corresponding ordered n-tuple in $V^{(B)}$ by $(x_1, \dots, x_n)^B$. Let us assume that P is an n-ary relation on X in $V^{(B)}$; i.e., X, $P \in V^{(B)}$ and $[\![P \subset X^n]\!] = 1$ $(n \in \omega)$. Then there exists an n-ary relation P' on $X\downarrow$ such that

$$(x_1, \dots, x_n) \in P' \leftrightarrow [\![(x_1, \dots, x_n)^B \in P]\!] = 1.$$

The relation P' is also denoted by the symbol $P\downarrow$ and is called the descent of P.

2.10. Let $x \in V$ and $x \subset V^{(B)}$; i.e., x is a set consisting of B-valued sets or, in other words, $x \in \mathscr{P}(V^{(B)})$. We put $\varnothing\uparrow := \varnothing$ and

$$\operatorname{dom}(x\uparrow) := x, \quad \operatorname{im}(x\uparrow) := \{1\}$$

if $x \neq \varnothing$. The element $x\uparrow$ (of the separated universe $V^{(B)}$, i.e., the selected representative of the class $\{y \in V^{(B)} \cdot [\![y = x\uparrow]\!] = 1\}$) is called the *ascent* of x.

(1) For any $x \in \mathscr{P}(V^{(B)})$ and any formula φ the following equalities hold:

$$[\![(\forall z \in x\uparrow)\,\varphi(z)]\!] = \bigwedge_{y \in x} [\![\varphi(y)]\!];$$

$$[\![(\exists z \in x\uparrow)\,\varphi(z)]\!] = \bigvee_{y \in x} [\![\varphi(y)]\!].$$

To introduce the ascent of a correspondence $\Phi \subset X \times Y$ it is necessary to exercise some caution related to the distinction between the source domain X and the domain of definition $\operatorname{dom}(\Phi) := \{x \in X : \Phi(x) \neq \varnothing\}$. For our further investigations this distinction is not essential; therefore, we can assume that in talking about the ascent we always consider everywhere defined correspondences, i.e., such that $\operatorname{dom}(\Phi) = X$.

(2) Let X, Y, $\Phi \in \mathscr{P}(V^{(B)})$. Further, let Φ be a correspondence from X into Y. There exists a (unique) correspondence $\Phi\uparrow$ from $X\uparrow$ into $Y\uparrow$ such that for every subset A of the set X the following is fulfilled:

$$\Phi\uparrow(A\uparrow) = \Phi(A)\uparrow$$

if and only if Φ is extensional, i.e., satisfies the condition

$$y_1 \in \Phi(x_1) \to [\![x_1 = x_2]\!] \leq \bigvee_{y_2 \in \Phi(x_2)} [\![y_1 = y_2]\!]$$

for x_1, $x_2 \in \operatorname{dom}(\Phi) = X$. Further, $\Phi\uparrow = \Phi'\uparrow$, where $\Phi' := \{(x, y)^B : (x, y) \in \Phi\}$. The element $\Phi\uparrow$ is called the *ascent of the initial correspondence* Φ.

(3) The composition of extensional correspondences is extensional. Moreover, the ascent of a composition is equal to the composition of ascents (in $V^{(B)} : V^{(B)} \models (\Psi \circ \Phi)\uparrow = \Psi\uparrow \circ \Phi\uparrow$).

It should be noted that if Φ and Φ^{-1} are extensional, then $(\Phi\uparrow)^{-1} = (\Phi^{-1})\uparrow$. However, the extensionality of Φ does not guarantee the extensionality of Φ^{-1}.

(4) It is worth noting that if an extensional correspondence f is a function from X into Y, then the ascent $f\uparrow$ is a function from $X\uparrow$ into $Y\uparrow$. Here, the extensionality f can be formulated in the following way:

$$[\![x_1 = x_2]\!] \leq [\![f(x_1) = f(x_2)]\!] \quad (x_1, x_2 \in X).$$

For a set $X \subset V^{(B)}$ the symbol mix(X) denotes the set of all the mixings of the form mix$(b_\xi x_\xi)$, where $(x_\xi) \subset X$ and (b_ξ) is an arbitrary partition of unity. The following statements are called the *rules of reducing arrows* or the *rules of "descent-ascent"* and *"ascent-descent"*.

(5) Let X and X' be subsets from $V^{(B)}$ and let $f : X \to X'$ be an extensional mapping. Let Y, Y', $g \in V^{(B)}$ be such that $[\![Y \neq \varnothing]\!] = [\![g : Y \to Y']\!] = 1$. Then the following relations hold:

$$X\uparrow\downarrow = \mathrm{mix}\,(X), \quad f\uparrow\downarrow = f, \quad Y\downarrow\uparrow = Y, \quad g\downarrow\uparrow = g.$$

2.11. Let $X \in V$, $X \neq \varnothing$; i.e., X is a nonempty set. We denote by the letter $\imath$ the imbedding $x \mapsto x^\wedge$ $(x \in X)$. Then $\imath(X)\uparrow = X^\wedge$ and $X = \imath^{-1}(X^\wedge\downarrow)$. Using these relations we can extend descent and ascent to the case when Φ is a correspondence from X into $Y\downarrow$ and $[\![\Psi$ is a correspondence from $X^\wedge$ into $Y]\!] = 1$, where $Y \in V^{(B)}$. Namely, we set $\Phi\upharpoonleft := (\Phi \circ \imath)\uparrow$ and $\Psi\downharpoonleft := \Psi\downarrow \circ \imath$. We call $\Phi\upharpoonleft$ the *modified ascent* of the correspondence Φ, and $\Psi\downharpoonleft$ the *modified descent* of the correspondence Ψ. (If the context precludes misunderstandings, then one can consider just ascent and descent, and use ordinary arrows.) It is not difficult to see that $\Phi\upharpoonleft$ is the only correspondence in $V^{(B)}$ satisfying the following relation:

$$[\![\Phi\upharpoonleft(x^\wedge) = \Phi(x)\uparrow]\!] = 1 \qquad (x \in X).$$

Analogously, $\Psi\downharpoonleft$ is the only correspondence from X into $Y\downarrow$ satisfying the following equality:

$$\Psi\downharpoonleft(x) = \Psi(x^\wedge)\downarrow \qquad (x \in X).$$

If $\Phi := f$ and $\Psi := g$ are functions, then these defining relations take the form:

$$[\![f\upharpoonleft(x^\wedge) = f(x)]\!] = 1, \quad g\downharpoonleft(x) = g(x^\wedge) \quad (x \in X).$$

2.12. (1) A pair (X, d), where $X \in V$, $X \neq \varnothing$, and d is a mapping from $X \times X$ into a Boolean algebra B is called a *Boolean set* or a *B-set*, or just a *set with B-structure*, if for any x, y, $z \in X$ it satisfies the following conditions:

(a) $d(x, y) = 0 \leftrightarrow x = y$;

(b) $d(x, y) = d(y, x)$;

(c) $d(x, y) \leq d(x, z) \vee d(z, y)$.

As an example of a B-set we can take any $\varnothing \neq X \subset V^{(B)}$ by setting $d(x, y) := [\![x \neq y]\!] = [\![x = y]\!]^*$ $(x, y \in X)$. As another example, take a nonempty sets X with the "discrete B-metric" d; i.e., $d(x, y) = 1$ if $x \neq y$ and $d(x, y) = 0$ if $x = y$.

(2) Let (X, d) be some B-set. There exist an element $\mathscr{X} \in V^{(B)}$ and an injection $\imath : X \to X' := \mathscr{X}\downarrow$ such that $d(x, y) = [\![\imath x \neq \imath y]\!]$ $(x, y \in X)$ and any element $x' \in X'$ admits a representation $x' = \mathrm{mix}_{\xi\in\Xi}(b_\xi \imath x_\xi)$, where $(x_\xi)_{\xi\in\Xi} \subset X$ and $(b_\xi)_{\xi\in\Xi}$ is a partition of unity in B.

The element $\mathscr{X} \in V^{(B)}$ is called the *Boolean realization* of X. If X is a discrete B-set, then $\mathscr{X} = X^\wedge$ and $\imath x = x^\wedge$ $(x \in X)$. If $X \subset V^{(B)}$, then $\imath\uparrow$ is an injection from $X\uparrow$ into $\mathscr{X}$ (in $V^{(B)}$).

A mapping f from a B-set (X, d) into a B-set (X', d') is called *nonexpanding* if $d(x, y) \geq d'(f(x), f(y))$, for all $x, y \in X$.

(3) Let X and Y be some B-sets, $\mathscr{X}$ and $\mathscr{Y}$ their Boolean realizations, and $\imath$ and κ the corresponding injections $X \to \mathscr{X}\downarrow$ and $Y \to \mathscr{Y}\downarrow$. If $f : X \to Y$ is a nonexpanding correspondence, then there exists a unique element $g \in V^{(B)}$ such that $[\![g : \mathscr{X} \to \mathscr{Y}]\!] = 1$ and $f = \kappa^{-1} \circ g\downarrow \circ \imath$.

A similar statement holds for correspondences.

(4) We shall present an example of a B-set which is important below. Let E be a vector lattice and $B := \mathfrak{B}(E)$. We put

$$d(x, y) := \{|x - y|\}^{\perp\perp} \qquad (x, y \in E).$$

It is not difficult to verify that d satisfies the conditions (b), (c) in (1), while condition (a) in (1) is fulfilled only for an Archimedean E (see 1.3).

Thus, (E, d) is a B-set if and only if the vector lattice E is Archimedean.

2.13. Starting from the results of 2.9 we can define the descent of an algebraic system. For simplicity we shall restrict ourselves to the case of a finite signature. Let $\mathfrak{A}$ be an algebraic system of finite signature in $V^{(B)}$. This means that there exist elements A, $f_1, \dots, f_n$, $P_1, \dots, P_m \in V^{(B)}$ and natural numbers $a(f_1), \dots, a(f_n)$, $a(P_1), \dots, a(P_m)$ such that the following conditions (all in $V^{(B)}$) are fulfilled:

$$A \neq \varnothing, \quad P_k \subset A^{a(P_k)^\wedge} \quad (k := 1, \dots, m);$$

$$f_l : A^{a(f_l)^\wedge} \to A \quad (l := 1, \dots, n);$$

$$\mathfrak{A} := (A, f_1, \dots, f_n, P_1, \dots, P_m).$$

Having obtained the descent of the set A, of the functions $f_1, \dots, f_n$, and of the relations $P_1, \dots, P_m$ according to the rules 1.9, we obtain an algebraic system $\mathfrak{A}\downarrow = (A\downarrow, f_1\downarrow, \dots\dots, f_n\downarrow, P_1\downarrow, \dots, P_m\downarrow)$ which is called the *descent of* $\mathfrak{A}$. Thus the descent of the algebraic system $\mathfrak{A}$ is the descent of the base set A together with the descended operations and relations.

2.14. Comments. (a) As was noted above in 1.15(b), the heuristic transfer principle introduced by Kantorovich in connection with the concept of a K-space subsequently found many confirmations in the investigations of Kantorovich himself and his followers. Essentially, this principle is one of those ideas, which as the organizing and direction-giving idea in a new field, finally brought about a profound and complete theory of K-spaces, rich with various applications. Already at the beginning of the development of this theory attempts were made to formalize these heuristic arguments. There were also so-called theorems on preservation of relations, which state that if a certain proposition containing a finite number of functional relations is proved for the real numbers, then a similar fact is automatically valid for elements of a K-space (see **[7, 14]**).

However, the intrinsic mechanism controlling the phenomenon of preservation of relations, the bounds of applicability of such statements, and also the general reasons for a number of analogies and parallels with classical function theory were still obscure. The depth and the universal character of the Kantorovich principle became apparent in the framework of Boolean-valued analysis.

(b) The part of functional analysis which uses a special model-theoretic technique, the Boolean-valued models of the set theory, is called *Boolean-valued analysis*. It is of interest to note that the construction of Boolean-valued models was not related to the theory of ordered vector spaces. The necessary language and technical tools were already forged within mathematical logic by the 1960s. However, there was at that time no general idea to give life to this mathematical apparatus and to lead to progress in model theory. This idea only came with the discovery of P. Cohen, who established the absolute unsolvability (in a precise mathematical sense) of the classical continuum problem. Indeed, it was in connection with Cohen's method of forcing that there emerged Boolean-valued models of set theory, whose creation is associated with the names of P. Vopenka, D. Scott, and R. Solovay (see **[43, 45, 48, 49]**).

(c) The method of forcing is naturally divided into two parts: general and special. The general part is the technique of Boolean-valued models of set theory, i.e., the construction of a Boolean-valued universe $V^{(B)}$ and the interpretation of set-theoretic statements in it. Here the complete Boolean algebra B is totally arbitrary. The special part consists in the construction of a specific Boolean algebra B that provides necessary (rather frequently pathological, exotic) properties of objects (for instance, of a K-space) obtained from B. Both parts are of independent interest, but the most effective results are obtained by combining them. In this section, as in most investigations in Boolean-valued analysis, only the general part of the method of forcing is applied. The special part is most actively used in proofs of independence or consistency (see **[10, 27, 47]**). Further progress in Boolean-valued analysis probably will be connected with full application of the forcing method.

(d) The material in 2.1–2.8 is standard and a detailed description of it can be found in **[18, 21, 27, 47]**; see also **[10, 23]**. Various versions of the methods presented in 2.9–2.11 are widely applied in investigations of Boolean-valued models. In **[17, 22]** the descent and ascent technique is given in a form that is better adapted to problems of analysis. Indeed, in this form they are studied in **[21]**. The imbedding (2.10) of sets with Boolean structure into a Boolean-valued universe was introduced in [17]. The basis of such an imbedding is the Solovay-Tennenbaum method, proposed earlier for imbeddings of complete Boolean algebras **[44]**.

LECTURE 3. Vector lattices and numerical systems

Boolean-valued analysis can be traced back to the statement by Scott and Solovay that the image of the field of real numbers in a Boolean model is an extended K-space. Depending on what Boolean algebra B (algebra of measurable sets, or of regular open sets, or of projections in a Hilbert space) is used as a base in constructing a Boolean-valued model $V^{(B)}$, different K-spaces are obtained (spaces of measurable functions, or of semicontinuous functions, or of selfadjoint operators). Thereby there arises a remarkable possibility of transferring what is known about numbers to many classical objects of analysis. This will be discussed in the present lecture.

3.1. By the field of real numbers we understand an algebraic system in which the axioms of an Archimedean ordered field (with different zero and unit element) and the axiom of completeness are fulfilled. We recall two well-known statements.

(1) The field $\mathbb{R}$ of real numbers exists and is unique up to isomorphism.

(2) If P is an Archimedean ordered field, then there exists an isomorphic imbedding h of the field P into $\mathbb{R}$ such that the image $h(P)$ is a subfield of $\mathbb{R}$ containing the subfield of rational numbers. In particular, $h(P)$ is dense in $\mathbb{R}$.

Applying to (1) consecutively the transfer and maximum principles, we can find an element $\mathscr{R} \in V^{(B)}$, for which $[\![\mathscr{R}$ is a field of real numbers $]\!] = 1$. Moreover, if any $\mathscr{R}' \in V^{(B)}$ satisfies the condition $[\![\mathscr{R}'$ is a field of real numbers $]\!] = 1$, then the condition $[\![$ the ordered fields $\mathscr{R}$ and $\mathscr{R}'$ are isomorphic $]\!] = 1$ is also satisfied. In other words, in the model $V^{(B)}$ there exists a field of real numbers $\mathscr{R}$ which is unique up to isomorphism.

Let us also note that the formula $\varphi(\mathbb{R})$ representing the formal description of the axioms of an Archimedean ordered field is bounded and, therefore, $[\![\varphi(\mathbb{R}^\wedge)]\!] = 1$; i.e., $[\![\mathbb{R}^\wedge$ is an Archimedean ordered field $]\!] = 1$. "Having passed" the statement (2) through the transfer principle we can the conclude that $[\![\mathbb{R}^\wedge$ is isomorphic to a dense subfield of the field $\mathscr{R}]\!] = 1$. Based on this fact, we will assume below that $\mathscr{R}$ is a field of real numbers in the model $V^{(B)}$ and that $\mathbb{R}^\wedge$ is a dense subfield in it. As is easy to see, the elements $0 := 0^\wedge$ and $1 := 1^\wedge$ are the zero and the unit element of the field $\mathscr{R}$.

It should be emphasized that in the general case the equality $\mathscr{R} = \mathbb{R}^\wedge$ does not hold. Indeed, the completeness axiom for $\mathbb{R}$ is not a bounded formula, and it might fail for $\mathbb{R}^\wedge$ in $V^{(B)}$.

Now we shall consider the descent $\mathscr{R}\downarrow$ of the algebraic system $\mathscr{R}$. In other words, the descent of the carrier set of the system $\mathscr{R}$ is regarded together with the descended operations and order. For simplicity we shall denote the operations and order relation in $\mathscr{R}$ and $\mathscr{R}\downarrow$ by the same symbols $+$, $\cdot$, $\leq$. Thus, to be more precise, the addition and multiplication, and the relation of order in $\mathscr{R}\downarrow$ are introduced by the following formulas:

$$z = x + y \leftrightarrow [\![z = x + y]\!] = 1;$$

$$z = x \cdot y \leftrightarrow [\![z = x \cdot y]\!] = 1;$$

$$x \leq y \leftrightarrow [\![x \leq y]\!] = 1;$$

$$(x, y, z \in \mathscr{R}\downarrow).$$

Multiplication by real numbers can also be introduced in $\mathscr{R}\downarrow$ by the rule:

$$y = \lambda x \leftrightarrow [\![\lambda^\wedge x = y]\!] = 1 \qquad (\lambda \in \mathbb{R};\ x, y \in \mathscr{R}\downarrow).$$

3.2. Theorem (Gordon). *Let $\mathbb{R}$ be an ordered field of real numbers in the model $V^{(B)}$. Then $\mathscr{R}\downarrow$ (with operations and order descended) is an extended K-space with unit* 1. *Moreover, there exists an isomorphism χ of the Boolean algebra B onto the base $\mathscr{P}(\mathscr{R}\downarrow)$ such that the following equivalences are valid*:

$$\chi(b)x = \chi(b)y \leftrightarrow b \leq [\![x = y]\!],$$

$$\chi(b)x \leq \chi(b)y \leftrightarrow b \leq [\![x \leq y]\!]$$

for all $x, y \in \mathscr{R}\downarrow$ and $b \in B$.

3.3. The extended K-space $\mathscr{R}\downarrow$ is at the same time a faithful f-algebra with ring unit 1, where for any $b \in B$ the projection $\chi(b)$ is the operator of multiplication by the unit element $\chi(b)1$.

From what was said above it is clear that the mapping $b \mapsto \chi(b)1$ $(b \in B)$ is a Boolean isomorphism between B and the algebra of unit elements in $\mathfrak{E}(\mathscr{R}\downarrow)$. This isomorphism is denoted by the same letter χ. Thus, depending on the context, $x \mapsto \chi(b)x$ is either a band projection or the operator of multiplication by the unit element $\chi(b)$.

3.4. Everywhere below $\mathscr{R}$ is the field of real numbers in the model $V^{(B)}$. Let us explain the meaning of exact bounds and order limits in the K-space $\mathscr{R}\downarrow$.

(1) Let $(b_\xi)_{\xi\in\Xi}$ be a partition of unity in B, and $(x_\xi)_{\xi\in\Xi}$ a set in $\mathscr{R}\downarrow$. Then

$$\underset{\xi\in\Xi}{\operatorname{mix}}(b_\xi x_\xi) = o\text{-}\sum_{\xi\in\Xi}\chi(b_\xi)x_\xi.$$

(2) For a nonempty set $A \subset \mathscr{R}\downarrow$ and arbitrary $a \in \mathscr{R}$, $b \in B$ the following equivalences are valid:

$$b \le [\![a = \sup(A\uparrow)]\!] \leftrightarrow \chi(b)a = \sup(b)A;$$

$$b \le [\![a = \inf(A\uparrow)]\!] \leftrightarrow \chi(b)a = \inf\chi(b)A.$$

(3) Let A be an upwards filtered set and let $s : A \to \mathscr{R}\downarrow$ be a net in $\mathscr{R}\downarrow$. Then $A^\wedge$ is upwards filtered and $\sigma := s\uparrow : A^\wedge \to \mathscr{R}$ is a net in $\mathscr{R}$ (in $V^{(B)}$); moreover, for any $x \in \mathscr{R}\downarrow$ and $b \in B$ we have

$$b \le [\![x = \lim\sigma]\!] \leftrightarrow \chi(b)x = o\text{-}\lim\chi(b)\circ s.$$

(4) Let the elements A and $\sigma \in V^{(B)}$ be such that $[\![A$ is upwards filtered and $\sigma : A \to \mathscr{R}]\!] = 1$. Then $A\downarrow$ is an upwards filtered set and thus the mapping $s := \sigma\downarrow : A\downarrow \to \mathscr{R}\downarrow$ is a net in $\mathscr{R}\downarrow$. Besides, for any $x \in \mathscr{R}\downarrow$ and $b \in B$, the following is satisfied:

$$b \le [\![x = \lim\sigma]\!] \leftrightarrow \chi(b)x = o\text{-}\lim\chi(b)\circ s.$$

(5) Let f be a mapping from a nonempty set Ξ into $\mathscr{R}\downarrow$ and $g := f\uparrow$. Then for any $x \in \mathscr{R}\downarrow$ and $b \in B$ the following holds:

$$b \le [\![z = \sum_{\xi\in\Xi^\wedge} g(\xi)]\!] \leftrightarrow \chi(b)x = \sum_{\xi\in\Xi}\chi(b)f(\xi).$$

3.5. For every element $x \in \mathscr{R}\downarrow$ the following relations hold:

$$e_x = \chi([\![x = 0]\!]), \quad e_\lambda^x = \chi([\![x < \lambda^\wedge]\!]) \quad (\lambda \in \mathbb{R}).$$

A real number t is not equal to zero if and only if the supremum of the set $\{1 \wedge (n|t|) : n \in \omega\}$ is equal to 1. Consequently, according to the transfer principle, for $x \in \mathscr{R}\downarrow$ we have $b := [\![x \ne 0]\!] = [\![\sup A = 1]\!]$, where $A \in V^{(B)}$ is defined by the formula $A := \{1 \wedge (n|x|) : n \in \omega^\wedge\}$. If $C := \{1 \wedge (n|x|) : n \in \omega\}$, then, by using the second formula from 2.10(1) and the representation $\omega^\wedge = (\iota\omega)\uparrow$ from 2.11, we shall prove that $[\![C\uparrow = A]\!] = 1$. So $[\![\sup(A) = \sup(C\uparrow)]\!] = 1$. Invoking 3.4(2), we derive

$$b = [\![\sup(C\uparrow) = 1]\!] = [\![\sup(C) = 1]\!] = [\![e_x = 1]\!].$$

On the other hand, $[\![e_x = 0]\!] = [\![e_x = 1]\!]^* = b^*$. According to 3.2, we can write

$$\chi(b)e_x = \chi(b)1 = \chi(b); \qquad \chi(b^*)e_x = 0 \to \chi(b)e_x = e_x.$$

Finally, $\chi(b) = e_x$.

Let us take $\lambda \in \mathbb{R}$ and define $y := (\lambda 1 - x)^+$. Since $[\![\lambda^\wedge = \lambda 1]\!] = 1$, we have $[\![y = (\lambda^\wedge - x)^+]\!] = 1$. Consequently, $e_\lambda^x = e_y = \chi([\![y = 0]\!])$. It remains to note that in $V^{(B)}$ the number $y = (\lambda^\wedge - x) \vee 0$ is not equal to zero only if $\lambda^\wedge - x > 0$, i.e., $[\![y \neq 0]\!] = [\![x < \lambda^\wedge]\!]$.

3.6. THEOREM. *Let E be an Archimedean vector lattice, $\mathscr{R}$ a field of real numbers in the model $V^{(B)}$, and j an isomorphism of B onto the base $\mathfrak{B}(E)$. There exists an element $\mathscr{E} \in V^{(B)}$ satisfying the following conditions:*

(1) *$V^{(B)} \models \langle\langle \mathscr{E}$ is a vector sublattice of the field $\mathscr{R}$ regarded as a vector lattice over $\mathbb{R}^\wedge\rangle\rangle$;*

(2) *$E' := \mathscr{E}\downarrow$ is a vector sublattice of $\mathscr{R}\downarrow$ invariant under every projection $\chi(b)$ $(b \in B)$ and such that any set of the positive pairwise disjoint sets has a supremum;*

(3) *there exists a 0-continuous lattice isomorphism $\imath\colon E \to E'$ such that $\imath(E)$ is a minorant sublattice in $\mathscr{R}\downarrow$;*

(4) *for every $b \in B$ the operator of projection onto a band generated in $\mathscr{R}\downarrow$ by the set $\imath(j(b))$ coincides with $\chi(b)$.*

Let us set $d(x, y) := j^{-1}(\{|x - y|\}^{\perp\perp})$. Let $\mathscr{E}$ be a Boolean realization of a B-set (E, d) and $E' := \mathscr{E}\downarrow$ (see 2.12(4)). By 2.12(2), we can say without loss of generality that $E \subset E'$, $d(x, y) = [\![x \neq y]\!]$ $(x, y \in E)$, and $E' = \operatorname{mix}(E)$. Further, in the set E' we can introduce the structure of vector lattice. For that we take a number $\lambda \in \mathbb{R}$ and elements $x, y \in E'$ of the form $x := \operatorname{mix}(b_\xi x_\xi)$, $y := \operatorname{mix}(b_\xi y_\xi)$, where $(x_\xi) \subset E$, $(y_\xi) \subset E$, and (b_ξ) is a partition of unity in B, and define

$$x + y := \operatorname{mix}(b_\xi(x_\xi + y_\xi));$$
$$\lambda x := \operatorname{mix}(b_\xi(\lambda x_\xi));$$
$$x \le y \leftrightarrow x = \operatorname{mix}(b_\xi(x_\xi \wedge y_\xi)).$$

Inside $V^{(B)}$ we define addition $\oplus$, multiplication $\odot$, and order relation $\circledcirc$ on the set $\mathscr{E}$ as ascents of the corresponding operations from E'. More precisely, the operations $\oplus\colon \mathscr{E} \times \mathscr{E} \to \mathscr{E}$ and $\odot\colon \mathbb{R}^\wedge \times \mathscr{E} \to \mathscr{E}$ and the relation $\circledcirc \subset \mathscr{E} \times \mathscr{E}$ are defined by

$$[\![x \oplus y = x + y]\!] = 1;$$
$$[\![\lambda^\wedge \odot y = \lambda x]\!] = 1 \quad (x, y \in E',\ \lambda \in \mathbb{R}),$$
$$[\![x \circledcirc y]\!] = \bigvee\{[\![x = x']\!] \wedge [\![y = y']\!]\colon x', y' \in E',\ x' \le y'\}.$$

Then we can claim that $\mathscr{E}$ is a vector lattice over the field $\mathbb{R}^\wedge$ and, in particular, it is a lattice ordered group in $V^{(B)}$. It is also clear that the Archimedean axiom is valid for $\mathscr{E}$ because the lattice E' is Archimedean.

Note that if $x \in E_+$, then $\{x\}^{\perp\perp} = d(x, 0) = [\![x \neq 0]\!]$; i.e., $\{x\}^\perp = [\![x = 0]\!]$. Consequently, for disjoint $x, y \in E$ we get $[\![x = 0]\!] \vee [\![y = 0]\!] = \{x\}^\perp \vee \{y\}^\perp = 1_B$. From this it is easy to derive that $[\![\mathscr{E}$ is linearly ordered $]\!] = 1$, since

$$[\![(\forall x \in \mathscr{E})(\forall y \in \mathscr{E})(|x| \wedge |y| = 0 \to x = 0 \vee y = 0)]\!] = 1.$$

It is well known that the Archimedean linearly ordered groups are isomorphic to additive subgroups of the field of real numbers. Applying this statement to $\mathscr{E}$ in $V^{(B)}$, it can be assumed without any loss of generality that $\mathscr{E}$ is an additive subgroup of the field $\mathscr{R}$. In addition, we shall assume that $1 \in \mathscr{E}$; otherwise $\mathscr{E}$ can be replaced by the group $e^{-1}\mathscr{E}$, $0 < e \in \mathscr{E}$, which is isomorphic to $\mathscr{E}$. The multiplication $\odot$ represents a $\mathbb{R}^\wedge$-bilinear continuous mapping from $\mathbb{R}^\wedge \times \mathscr{E}$ into $\mathscr{E}$. Let $\beta\colon \mathscr{R} \times \mathscr{R} \to \mathscr{R}$ be its extension by continuity. Then β is $\mathscr{R}$-bilinear and $\beta(1, 1) = 1^\wedge \odot 1 = 1$. Consequently,

β coincides with the usual multiplication in $\mathscr{R}$; i.e., $\mathscr{E}$ is a vector sublattice of the field $\mathscr{R}$, regarded as a vector lattice over $\mathbb{R}^\wedge$. Thereby $E' \subset \mathscr{R}\downarrow$.

The minorant property of E' in $\mathscr{R}\downarrow$ evidently follows from the fact that $[\![\mathscr{E}$ is dense in $\mathscr{R}]\!] = 1$. We shall prove that E is minorant in E'.

From the properties of the isomorphism χ (see 3.2) it is clear that

$$\chi(b)\imath x = 0 \leftrightarrow j(b) \leq \{x\}^\perp \leftrightarrow x \in j(b)^\perp$$

for any $b \in B$ and $x \in E_+$. Thus $\chi(b)$ is the projection onto the band generated in $\mathscr{R}\downarrow$ by the set $\imath(j(b))$. Besides, if $\chi(b)x = 0$ for all $x \in E_+$, then $b = \{0\}$. So, for any $b \in B$ there can be found a strictly positive element $y \in E$, for which $y = \chi(b)y$. Now we shall take $0 < z \in E'$. The representation $z = \operatorname{mix}(b_\xi x_\xi)$, where (b_ξ) is a partition of unity in B and $(x_\xi) \subset E_+$, holds. Apparently, $\chi(b_\xi)x_\xi \neq 0$ for at least one index ξ. Let $\pi := \chi(b_\xi) \circ \chi([\![x_\xi \neq 0]\!])$ and y be a strictly positive element in E, for which $y = \pi y$. Then for $x_0 := y \wedge x_\xi$ we have $0 < x_0 \leq \pi x_\xi \leq \chi(b_x i)x_\xi \leq z$ and $x_0 \in E$. Therefore, E is minorant in E'.

3.7. The element $\mathscr{E} \in V^{(B)}$ from Theorem 3.6 is called the *Boolean realization* of E. Thus, vector sublattices of the field of real numbers $\mathscr{R}$ regarded as vector lattices over the field $\mathbb{R}^\wedge$ serve as Boolean realizations of Archimedean vector lattices.

Now we shall note several corollaries from 3.2 and 3.6 keeping the same notations $B, E, E', \mathscr{E}, \imath, \mathscr{R}$.

(1) For every $x' \in E'$ there exists a set $(x_\xi) \subset E$ and a partition of unity (π_ξ) in $\mathscr{P}(\mathscr{R}\downarrow)$ such that

$$x' = o\text{-}\sum_{\xi \in \Xi} \pi_\xi \imath x_\xi.$$

(2) For any $x \in \mathscr{R}\downarrow$ and $\varepsilon > 0$ there exists $x_\xi \in E'$ such that $|x - x_\xi| \leq \varepsilon 1$.

(3) If $h\colon E \to \mathscr{R}\downarrow$ is a lattice isomorphism and for every $b \in B$ the projection onto the band generated by the set $h(j(b))$ in $\mathscr{R}\downarrow$ coincides with $\chi(b)$, then there exists an $a \in \mathscr{R}\downarrow$, for which $hx = a \cdot \imath(x)$ $(x \in E)$.

(4) If E contains the order unit $\mathbf{1}$, then the isomorphism $\imath$ is uniquely defined by the additional requirement $\imath\mathbf{1} = 1$.

(5) If E is a K-space, then $\mathscr{E} = \mathscr{R}$, $E' = \mathscr{R}\downarrow$, and $\imath(E)$ is a foundation of the K-space $\mathscr{R}\downarrow$. Moreover, $\imath^{-1} \circ \chi(b) \circ \imath$ is the projection onto the band of $j(b)$ for every $b \in B$.

(6) The image $\imath(E)$ coincides with all of $\mathscr{R}\downarrow$ if and only if E is an extended K-space.

(7) Extended K-spaces are isomorphic if and only if their bases are isomorphic.

(8) Let E be an extended K-space with unit $\mathbf{1}$. Then the there exists a unique multiplication in E such that E is a faithful f-algebra with the multiplication unit $\mathbf{1}$.

3.8. We shall dwell on questions of extension and completion of Archimedean vector lattices.

By a *maximal extension* of an Archimedean vector lattice E we understand a K-space $mE := \mathscr{R}\downarrow$, where $\mathscr{R}$ is a field of real numbers in the model $V^{(B)}$, $B := \mathfrak{B}(E)$. It is clear from the Theorem 3.6 that there exists an isomorphism $\imath\colon E \to mE$; moreover, the sublattice $\imath(E)$ is minorant in mE and $\imath(E)^{\perp\perp} = mE$. The maximal extension is defined up to isomorphism by these properties. To be more precise, the following statements are valid.

(1) Let E be an Archimedean vector lattice and F an extended K-space. We assume that an isomorphism h from E onto a minorant sublattice of F is given and $h(E)^{\perp\perp} = F$. Then there exists an isomorphism κ from F onto mE such that $\imath = \kappa \circ h$.

From the above conditions it is easy to derive that $j\colon b \mapsto j(b) := h(b)^{\perp\perp}$ is an isomorphism from $B := \mathfrak{B}(E)$ into $\mathfrak{B}(F)$. According to 3.6(5),(6) there exists an isomorphism k from F onto mE, for which $k^{-1} \circ \chi(b) \circ k$ is the projection onto the band $j(b)$ (for each $b \in B$). We shall apply 3.6(3) to $F_0 := h(E)$ and $g := \imath \circ h^{-1}\colon F_0 \to \mathscr{R}\downarrow$. There can be found an element $a \in \mathscr{R}\downarrow$ such that $g(x) = a \cdot k(x)$ $(x \in F_0)$. Now set $\kappa(x) := a \cdot k(x)$ $(x \in F)$. Then $\imath = \kappa \circ h$.

(2) For any Archimedean vector lattice E there exists a K-space ${}_{\circ}E$, unique up to isomorphism and an o-continuous lattice isomorphism $\imath\colon E \to {}_{\circ}E$ such that

$$\sup\{\imath x\colon x \in E,\ \imath x \leq y\} = y = \inf\{\imath x\colon x \in E,\ \imath x \geq y\}$$

for every element $y \in {}_{\circ}E$.

Let F be a K-space and $A \subset F$. We denote by dA the set of all $x \in F$ that can be represented in the form $o\text{-}\sum_{\xi\in\Xi} \pi_\xi a_\xi$, where $(a_\xi) \subset A$ and (π_ξ) is a partition of unity in $\mathfrak{P}(F)$. Let rA be the set of all elements $x \in F$ of the form $x = r\text{–}\lim_n a_n$, where (a_n) is an arbitrary regular convergent sequence in A.

(3) For an Archimedean vector lattice E the formula ${}_{\circ}E = rdE$ holds.

3.9. Interpreting the notion of convergent numerical net in $V^{(B)}$ and invoking 3.4(3), 3.7(5) one can obtain useful tests for o-convergence in a K-space E with unit $\mathbf{1}$.

THEOREM. *Let $(x_\alpha)_{\alpha\in A}$ be an order bounded net in E and $x \in E$. The following statements are equivalent:*

(1) *the net (x_α) o-converges to the element x;*

(2) *for any number $\varepsilon > 0$ the net of unit elements $(e_\varepsilon^{y(\alpha)})_{\alpha\in A}$, where $y(\alpha) := |x - x_\alpha|$, o-converges to zero;*

(3) *for any number $\varepsilon > 0$ there exists a partition of unity $(\pi_\alpha)_{\alpha\in A}$ in the Boolean algebra $\mathfrak{P}(E)$ such that*

$$\pi_\alpha|x - x_\beta| < \varepsilon\mathbf{1} \qquad (\alpha, \beta \in A);$$

(4) *for any number $\varepsilon > 0$ there exists an increasing net of projections $(\rho_\alpha)_{\alpha\in A} \subset \mathfrak{P}(E)$ such that*

$$\rho_\alpha|x - x_\beta| < \varepsilon\mathbf{1} \qquad (\alpha, \beta \in A;\ \beta \geq \alpha).$$

3.10. Comments. (a) The Boolean status of K-spaces is established by the Gordon Theorem 3.2 (see **[8]**). This fact can be formulated in the following way: an extended K-space is an interpretation of the field of real numbers in a suitable Boolean-valued model. In addition, it turns out that any theorem (within the framework of ZF theory) on real numbers has its analogue in the corresponding K-space. Conversion of one kind of theorems into others is realized by certain precisely defined procedures: ascent, descent, canonical imbedding; i.e., as a matter of fact it is realized algorithmically. Therefore, the Kantorovich statement "the elements from a K-space are generalized numbers" finds a precise mathematical formulation in Boolean-valued analysis. On the other hand, Boolean-valued analysis turns the heuristic transfer principle, which played an auxiliary guiding role in most investigations of the pre-Boolean theory of the K-spaces, into a precise research method.

(b) If in 3.2 B is the σ-algebra of measurable sets modulo sets of measure zero for a measure μ, then $\mathscr{R}\downarrow$ is isomorphic to the extended K-space of measurable functions $L^0(\mu)$. This fact (for the Lebesgue measure on the interval) was already known to Scott and Solovay (see **[43]**). If B is the complete Boolean algebra of projections in a Hilbert space, then $\mathscr{R}\downarrow$ is isomorphic to the space of those self-adjoint operators which have a spectral function acting in B. The two special cases of the Gordon theorem noted above were effectively used by G. Takeuti; see **[45]**, and also the bibliography in **[21]**. The object $\mathscr{R}\downarrow$ for general Boolean algebras was also considered by T. Jech **[33, 34]** who essentially rediscovered the Gordon theorem. The difference is that in **[33]** a (complex) extended K-space with unit is defined by another system of axioms and is called a complete Stone algebra. The interconnections from 3.4, 3.5 between properties of numerical objects and corresponding objects in a K-space $\mathscr{R}\downarrow$ were obtained essentially by Gordon **[8, 9]**.

(c) The Realization Theorem 3.6 was obtained by Kusraev **[19]**. There is a closely related result (formulated in other terms) in **[35]**, where a Boolean interpretation of the theory of linearly ordered sets is developed. Corollaries 3.7(7),(8) are well known (see **[7, 14]**). The concept of maximal extension for a K-space was introduced by Pinsker in a different way. He also proved the existence of a maximal extension unique up to isomorphism, for an arbitrary K-space. Theorem 2.8(2) on order completion of an Archimedean vector lattice was stated by Yudin. Corresponding references are in **[7, 14]**. Statement 2.6(3) was obtained by Veksler **[5]**.

The tests for o-convergence 3.9(2) and 3.9(4) (for sequences) were established by Kantorovich and Vulikh, respectively (see **[14]**). In 3.8 it is shown that these tests are, essentially, just interpretations of convergence properties of numerical nets (sequences).

(d) As was noted in 2.14(a), the first attempts to formalize the heuristic Kantorovich principle led to theorems on preservation of relations (see **[7, 14]**). Modern forms of theorems on preservation of relations, which use the method of Boolean-valued models, can be found in **[9, 35]** (see also **[21]**).

(e) Boolean realizations (not only of Archemedean vector spaces) provide subsystems of the field $\mathscr{R}$ (see 3.6(1)). For example, the following statements are formulated in **[19]**: (1) a Boolean realization of an Archimedean lattice ordered groups is a subgroup of the additive group of $\mathscr{R}$; (2) an Archimedean f-ring contains two mutually complementary bands, one of which has zero multiplication and is realized as (1), and the other is realized as a subring of $\mathscr{R}$; (3) an Archimedean f-algebra contains two mutually complementary bands, one of which is realized as in 3.6, and the other is a sublattice and a subalgebra of the field $\mathscr{R}$ considered as a lattice ordered algebra over the field $\mathbb{R}^\wedge$ (see also **[35]**).

References

1. G. P. Akilov and S. S. Kutateladze, *Ordered vector spaces*, "Nauka", Novosibirsk, 1978. (Russian)
2. A. V. Bukhvalov, *Order-bounded operators in vector lattices and spaces of measurable functions*, Itogi Nauki i Tekhniki: Mat. Anal, vol. 26, VINITI, Moscow, 1988, pp. 3–63; English transl. in J. Soviet Math. **54** (1991).
3. A. V. Bukhvalov, A. I. Veksler, and V. A. Geiler, *Normed lattices*, Itogi Nauki i Tekhniki: Mat. Anal, vol. 18, VINITI, Moscow, 1980, pp. 125–184; English transl. in J. Soviet Math. **18** (1982).
4. A. V. Bukhvalov, A. I. Veksler, and G. Ya. Lozanovski, *Banach lattices–some Banach aspects of the theory*, Uspekhi Mat. Nauk **34** (1979), no. 2, 137–183; English transl. in Russian Math. Surveys **34** (1979).

5. A. I. Veksler, *A new construction of Dedekind completion of vector lattices and l-groups with division*, Sibirsk. Mat. Zh. **10** (1969), 1206–1213; English transl. in Siberian Math. J. **10** (1969).
6. D. A. Vladimirov, *Boolean algebras*, "Nauka", Moscow, 1969. (Russian)
7. B. Z. Vulikh, *Introduction to the theory of partially ordered spaces*, Fizmatgiz, Moscow, 1961; English transl., Noordhoff, Groningen, 1967.
8. E. I. Gordon, *Real numbers in Boolean-valued models of set theory and K-spaces*, Dokl. Akad. Nauk SSSR **237** (1977), no. 4, 773–775; English transl. in Soviet Math. Dokl. **18** (1977).
9. ———, *On theorems on the preservation of relations in K-spaces*, Sibirsk. Mat. Zh. **23** (1982), no. 3, 55–65; English transl. in Siberian Math. J. **23** (1982).
10. T. J. Jech, *Lectures in set theory, with particular emphasis on the method of forcing*, Lecture Notes in Math., vol. 217, Springer-Verlag, Berlin and New York, 1971.
11. L. V. Kantorovich, *On semiordered linearly spaces and their applications in the theory of linear operations*, Dokl. Akad. Nauk SSSR **4** (1935), 11–14. (Russian)
12. ———, *Materials and bibliographies of Soviet scientists*, Ser. Mat., vol. 18, "Nauka", Moscow, 1989. (Russian)
13. L. V. Kantorovich and B. Z. Akilov, *Functional analysis*, 3rd ed., "Nauka", Moscow, 1984; English transl., Pergamon Press, Oxford, 1982.
14. L. V. Kantorovich, B. Z. Vulikh, and A. G. Pinsker, *Functional analysis in partially ordered spaces*, GITTL, Moscow, 1950. (Russian)
15. P. Cohen, *Set theory and the continuum hypothesis*, Benjamin, New York, 1966.
16. M. A. Krasnosel′skiĭ, *Positive solutions of operator equations*, "Fizmatgiz", Moscow, 1962; English transl., Noordhoff, Groningen, 1964.
17. A. G. Kusraev, *Some categories and functors of Boolean-valued analysis*, Dokl. Akad. Nauk SSSR **271** (1983), no. 2, 281–286; English transl. in Soviet Math. Dokl. **28** (1983).
18. ———, *Vector duality and its applications*, "Nauka", Novosibirsk, 1985. (Russian)
19. ———, *Numerical systems in Boolean models of the theory of sets*, Soviet Conference of Math. Logic, Moscow, 1986. (Russian)
20. A. G. Kusraev and S. S. Kutateladze, *Subdifferential calculus*, "Nauka", Moscow, 1987. (Russian)
21. ———, *Nonstandard methods of analysis*, "Nauka", Novosibirsk, 1990; English transl., KAP (to appear).
22. S. S. Kutateladze, *Descents and ascents*, Dokl. Akad. Nauk SSSR **272** (1983), no. 3, 521–524; English transl. in Soviet Math. Dokl. **28** (1983).
23. Yu. I. Manin, *The provable and nonprovable*, "Soviet Radio", Moscow, 1979. (Russian)
24. R. Sikorski, *Boolean algebras*, 3rd ed., Springer-Verlag, Berlin and New York, 1969.
25. C. D. Aliprantis and O. Burkinshaw, *Locally solid Riesz spaces*, Academic Press, New York, 1978.
26. ———, *Positive operators*, Academic Press, New York, 1985.
27. T. L. Bell, *Boolean-valued models and independence proofs in set theory*, Clarendon Press, Oxford, 1979.
28. P. G. Dodds and D. H. Fremlin, *Compact operators in Banach lattices*, Israel J. Math **34** (1979), 287–320.
29. D. H. Fremlin, *Toplogical Riesz spaces and measure theory*, Cambridge Univ. Press, New York, 1974.
30. P. R. Halmos, *Lectures on Boolean algebras*, Van Nostrand, New York, 1963.
31. E. de Jonge and A. C. M. van Rooij, *Introduction to Riesz spaces*, Mathematisch Centrum, Amsterdam, 1977.
32. G. J. O. Jameson, *Ordered linear spaces*, Lecture Notes in Math., vol. 141, Springer-Verlag, Berlin and New York, 1970.
33. T. Jech, *Abstract theory of abelian operator algebras: an application of forcing*, Trans. Amer. Math. Soc. **289** (1985), 133–162.
34. ———, *First order theory of complete Stonean algebras*, Canad. Math. Bull. **30** (1987), 385–392.
35. ———, *Boolean linear spaces*, Adv. Math. **81** (1990), 117–197.
36. H. E. Lacey, *The isometric theory of classical Banach spaces*, Springer-Verlag, Berlin and New York, 1974.
37. J. Lindenstrauss and L. Tzafriri, *Classical Banach spaces. II. Function spaces*, Springer-Verlag, Berlin and New York, 1979.
38. W. A. J. Luxemburg and A. C. Zaanen, *Riesz spaces*, Vol. I, North-Holland, Amsterdam, 1971.
39. A. L. Peressini, *Ordered topological vector spaces*, Harper and Row, New York, 1967.
40. J. B. Rosser, *Simplified independence proofs. Boolean valued models of set theory*, Academic Press, New York, 1969.
41. H. H. Schaefer, *Banach lattices and positive operators*, Springer-Verlag, Berlin and New York, 1974.

42. H.-U. Schwarz, *Banach lattices and operators*, Teubner, Leipzig, 1984.
43. R. M. Solovay, *A model of set-theory in which every set of reals is Lebesgue measurable*, Ann. of Math. (2) **92** (1970), 1–56.
44. R. M. Solovay and S. Tennenbaum, *Iterated Cohen extensions and Souslin's problem*, Ann. of Math. (2) **94** (1971), 201–245.
45. G. Takeuti, *Two applications of logic to mathematics*, Princeton Univ. Press, Princeton, NJ, 1978.
46. G. Takeuti and W. M. Zaring, *Introduction to axiomatic set theory*, Springer-Verlag, Berlin and New York, 1971.
47. ______, *Axiomatic set theory*, Springer-Verlag, Berlin and New York, 1973.
48. P. Vopenka, *General theory of ∇-models*, Comment. Math. Univ. Carolin. **8** (1967), 145–170.
49. P. Vopenka and P. Hajek, *The theory of semisets*, Academia, Prague, 1972.
50. Y.-Ch. Wong and K.-F. Ng, *Partially ordered topological vector spaces*, Claredon Press, Oxford, 1973.
51. A. C. Zaanen, *Riesz spaces*, Vol. II, North-Holland, Amsterdam, 1983.

Translated by D. ŽVIRÉNAITÉ

Amer. Math. Soc. Transl.
(2) Vol. **163**, 1995

The Whitehead Groups of Algebraic Groups and Applications to Some Problems of Algebraic Group Theory

A. P. Monastyrnyĭ and V. I. Yanchevskiĭ

Let G be an algebraic group defined over an infinite field k ($\operatorname{char} k \neq 2$). Let G be absolutely simple and isotropic over k, let $G(k)$ be the group of all elements of G that are rational over k, and $G(k)^+$ the subgroup of $G(k)$ generated by all unipotent elements of $G(k)$ that are contained in the unipotent radical of all minimal parabolic subgroups of G, defined over k. The group $W(G,K) = G(k)/G(k)^+$ is called the *Whitehead group* of G. It was proved by J. Tits **[1]** that every subgroup normalized by $G(k)^+$ either is central in G, or contains $G(k)^+$. In particular the quotent of $G(k)^+$ by its center is a simple group. The well-known Kneser-Tits conjecture was formulated in connection with this result as follows: *if G is simply connected, then $W(G,K)$ is trivial.*

The main topics of our report are the following:

1. the Kneser-Tits conjecture;
2. the structure and calculation of Whitehead groups;
3. the applications of Whitehead groups to algebraic group theory (the problem of the k-rationality, the weak approximation problem, R-equivalence on algebraic groups).

First of all, it is useful to give some historical remarks on the Kneser-Tits conjecture. Of course, by the time it was posed the conjecture had been proved in all known particular cases, for instance, for classical groups over a field k **[2]**. Later, for local fields k it was proved by V. P. Platonov **[3]**. In the case when $G(k)$ is the classical group over a skew field over k, the first results on this problem were obtained by T. Nakayama and Y. Matsushima in 1943 **[4]** and by S. Wang in 1950 **[5]** in the context of the Tannaka-Artin problem. They proved that $W(G,K) = 1$, if $G(k) = \mathrm{SL}_n(D)$, $n \geq 2$, where $\mathrm{SL}_n(D) = \{a \in \mathrm{GL}_n(D) \mid \mathrm{Nrd}(a) = 1\}$, Nrd is the reduced norm homomorphism, and D is a finite-dimensional division algebra over a p-adic or algebraic number field. For a long time nothing has been known about the validity of this conjecture for groups of classical type in the following three cases:

1. $G(k) = \mathrm{SL}_n(D)$, D is a skew field central and finite-dimensional over k,

2. $G(k) = \mathrm{SU}(\Phi_n, D)$, D is a skew field central and finite over a field K with involution τ of the second kind, such that $k = \{a \in K \mid a^\tau = a\}$, Φ_n is a skew-Hermitian sesquilinear isotropic form on the n-dimensional vector space over D,

1991 *Mathematics Subject Classification*. Primary 19B99.

3. $G(k) = \mathrm{Spin}(\Phi_n, D)$ is the universal covering of the group $U(\Phi_n, D)$, D is a skew field central and finite-dimensional over k with involution τ of the first kind of the second type, Φ_n is a skew-Hermitian sesquilinear isotropic form on an n-dimensional vector space over D.

But in 1975 V. P. Platonov resolved the Tannaka-Artin problem **[6]** in the negative and, consequently, settled the general Kneser-Tits conjecture in the negative as well. In his reduced K-theory **[6–8,10]** he developed methods for the calculation of $W(G, K)$ in the case (1) and solved, as applications of his investigations, the problem of the k-rationality of the varieties of simply connected algebraic groups, the problem of a weak approximation, and others. Some modifications of the calculations of $\mathrm{SK}_1(D)$ for Henselian skew fields D were suggested by P. Draxl **[11]** and Yu. L. Ershov **[12]**. After the construction of reduced K-theory the question arose: are these results of a general nature or not? More precisely, are similar results valid for other types of groups, in particular, for groups of classical type? In the case (2) V. I. Yanchevskiĭ developed a reduced unitary K-theory and obtained similar results **[13, 14]**. The case (3) has been a difficult problem for a long time. But at present we have developed methods for the calculation of the Whitehead groups in this case. The latest results in the spinor situation **[15]** enable us to observe the situation for algebraic groups of the classical type in general. Now we present these results and show how we can obtain complete statements in the general situation.

Let D be a central skew field over k, $(D : k) = m^2$, and let m be its index, $m = \mathrm{ind}(D)$. Let τ be an involution on D and $S_\tau = \{d \in D \mid d^\tau = d\}$. Suppose that $S_\tau \cap k = k$ and $\dim_k S_\tau = m(m-1)/2$; i.e. τ is an involution of the first kind of the second type. It is well known (the Artin theorem) that $m = 2^s$, $s \geq 0$; in this case. Let Φ_n be a sesquilinear skew-Hermitian nondegenerate isotropic form on an n-dimensional vector space over D, $U(\Phi_n, D)$ the unitary group of Φ_n. It is easy to see that $U(\Phi_n, D) = \mathrm{SU}(\Phi_n, D)$.

PROPOSITION 1. *Let G be an algebraic group defined over k and such that $G(k) = U(\Phi_n, D)$. Then there exist an algebraic group H defined over k and such that there exists an exact sequence*

$$1 \to k^* \to H(k) \to G(k) \to 1.$$

There exist an algebraic group $\widetilde{G}$ and homomorphism

$$\Theta_{\mathrm{red}} \colon G(k) \to k^*/k^{*2}$$

defined over k such that the following sequence is exact:

$$1 \to \{-1, 1\} \to \widetilde{G}(k) \to G(k) \xrightarrow{\Theta_{\mathrm{red}}} k^*/k^{*2} \to N \to 1,$$

where $N < k^*/k^{*2}$.

DEFINITION 1. One calls the group $H(k)$ the Clifford group of the form Φ_n and denotes it by $\mathrm{Clif}(\Phi_n, D)$. One calls the group $\widetilde{G}(k)$ the spinor group of the form Φ_n and denotes it by $\mathrm{Spin}(\Phi_n, D)$. The homomorphism

$$\Theta_{\mathrm{red}} \colon U(\Phi_n, D) \to k^*/k^{*2}$$

is called the reduced spinor norm.

It should be pointed out that Θ_{red} is a generalization of the classical Eichler's spinor norm (see **[2]**).

The first problem in the spinor case is to describe the group $W(G,k)$ for $G(k) = \operatorname{Spin}(\Phi_n, D)$ in terms of the ground skew field D only. Let $\Sigma_\tau(D)$ be the multiplicative subgroup of D^*, which is generated by the nonzero elements of S_τ, and let $[D^*, D^*]$ be the commutator subgroup of D^*.

THEOREM 1. *Let $G(k) = \operatorname{Spin}(\Phi_n, D)$. There exist exact sequences*

$$\{-1,1\} \to W(G,k) \to R(D)/\Sigma_\tau(D)[D^*,D^*] \to 1 \qquad \text{for} \quad n > 2,$$
$$\{-1,1\} \to W(G,k) \to R(D)/\Sigma_\tau(D) \to 1 \qquad \text{for} \quad n = 2,$$

where $R(D) = \{\, d \in D^* \mid \operatorname{Nrd}(d) \in k^{*2} \,\}$.

DEFINITION 2. One calls the group $R(D)/\Sigma_\tau(D)[D^*,D^*]$ the spinor Whitehead group and denotes it by $K_1 \operatorname{Spin}(D)$.

First of all we would like to say few words about the idea of the proof of Theorem 1. There exists the well-known Wall spinor norm homomorphism (see **[16]**)

$$\Theta_W\colon\ \operatorname{SU}(\Phi_n, D) \to D^*/\Sigma_\tau(D)[D^*,D^*] \qquad (n > 2),$$

and there exists an exact sequence

$$\{-1,1\} \to W(G,k) \to \ker\Theta_{\text{red}}/\ker\Theta_W \to 1,$$

where $G(k) = \operatorname{Spin}(\Phi_n, D)$. We prove the following important result.

PROPOSITION 2. *The diagram*

$$\begin{array}{ccc} & U(\Phi_n, D) & \\ {\scriptstyle\Theta_W}\swarrow & & \searrow{\scriptstyle\Theta_{\text{red}}} \\ D^*/\Sigma_\tau(D)[D^*,D^*] & \xrightarrow{\operatorname{Nrd}} & k^*/k^{*2} \end{array}$$

is commutative.

And after these considerations we prove Theorem 1.

The result of Theorem 1 is similar to the corresponding result of J. Dieudonné: if $G(k) = \operatorname{SL}_n(D)$, then $W(G,k) \cong \operatorname{SL}_1(D)/[D^*,D^*] \overset{\text{def}}{=} \operatorname{SK}_1(D)$, and the result of G. E. Wall **[16]**: if $G(k) = \operatorname{SU}(\Phi_n, D)$ then $W(G,k) \cong \Sigma'_\tau(D)/\Sigma_\tau(D) \overset{\text{def}}{=} \operatorname{SK}_1 U(D)$ ($\Sigma'_\tau = \{\, d \in D^* \mid \operatorname{Nrd}(d) \in k = K \cap S_\tau \,\}$), which were established in the fifties.

From the point of view of Theorem 1 the group $K_1 \operatorname{Spin}(\Phi_n, D)$, which we shall call *the spinor Whitehead group* of D, is of great importance for the calculation of the group $W(G,k)$, $G(k) = \operatorname{Spin}(\Phi_n, D)$. It is well known that if $\operatorname{ind}(D) = 2$, then the group $W(G,k)$ is trivial (see **[9]**). Therefore we are interested only in the case where $\operatorname{ind}(D) = 2^s$ and $s \geq 2$. In the case where $s = 2$ we have

THEOREM 2. *Let $G(k) = \operatorname{Spin}(\Phi_n, D)$, $n \geq 2$, $\operatorname{ind}(D) = 4$. Then $W(G,k) = 1$.*

It is evident that if the group $K_1 U(D) = D^*/\Sigma_\tau(D)[D^*,D^*]$ is trivial then $K_1 \operatorname{Spin}(D)$ as its subgroup is trivial too. The group $K_1 U(D)$, where D is a Henselian skew field of exponent 2, was calculated by Platonov and Yanchevskiĭ **[17]**. It follows from this calculation that $K_1 \operatorname{Spin}(D)$ is trivial for a wide class of Henselian skew fields. Nevertheless the answer, in general, on the Kneser-Tits conjecture for spinor groups is negative. We would like to emphasize the central role of Henselian skew fields for obtaining this result. The following theorem gives us a good formula for the computation of $K_1\,\mathrm{Spin}(D)$.

THEOREM 3. *Let* $k = K\langle x_1\rangle\dots\langle x_t\rangle$, $a_1,\dots,a_t \in K^*$, $L = K(\sqrt{a_1},\dots,\sqrt{a_t})$, $D = (a_1,x_1)\otimes_k\dots\otimes_k(a_t,x_t)$ *a skew field (note that* D *is a skew field if and only if* $[L:K] = 2^t$*),* $R(L) = \{l \in L^* \mid N_{L|K}(l) \in K^{*2}\}$. *Then* $K_1\,\mathrm{Spin}(D) = R(L)/\prod_{\sigma\in\mathrm{Gal}(L/K)\setminus\{1\}}(L^\sigma)^*$, *where* L^σ *is the field of invariants of the cyclic group* $\langle\sigma\rangle$.

By means of Theorem 3 we have constructed the first counterexample for the Kneser-Tits conjecture for spinor groups:

THEOREM 4. *Let* $K = \mathbb{Q}(x)$, *let* p *be a prime such that* $p \neq 3$ *and* $p \neq a^2 + b^2$, *where* $a,b \in \mathbb{Q}$, $L = K(\sqrt{x},\sqrt{p},\sqrt{z})$, *where* $z = (1+px)/(1+p)$, *and let* D *be a skew field as in Theorem* 3. *Then* $K_1\,\mathrm{Spin}(D) \neq 1$.

Theorem 4 provides us with an infinite series of counterexamples.

It is easy to see that there exists a connection between the Kneser-Tits conjecture and the normal structure of algebraic groups. Furthermore the Whitehead groups of algebraic groups are closely connected with some other problems of algebraic group theory. We shall begin with the problem of the stability of $W(G,k)$ under pure transcendental extensions of the ground field.

THEOREM 5. *Let* F/k *be a pure transcendental extension. Then the natural homomorphism* $K_1\,\mathrm{Spin}(D) \to K_1\,\mathrm{Spin}(D\otimes_k F)$ *is an isomorphism.*

In order to prove Theorem 5 we need the following important statement.

PROPOSITION 3. *Let* $D[t]$ *be the ring of skew (noncommutative) polynomials,* $D(t)$ *the skew field of noncommutative rational functions. Let* $P \in D(t)$ *and* $\mathrm{Nrd}(P) \in k(t)^{*2}$. *Then* $P \in \Sigma_\tau(D(t))[D(t)^*, D(t)^*]$.

The stability under pure transcendental extensions of the group $W(\mathrm{SL}_n(D),k)$ was established first by Platonov [**10**] and later by means of different methods by Platonov and Yanchevskiĭ [**18**]. For $W(\mathrm{SU}(\Phi_n,D),k)$ it was established by Yanchevskiĭ [**14**]. Now we can formulate this fact for all simply connected groups of classical type.

THEOREM 6. *Let* G *be an algebraic group of classical type defined over* k *and isotropic over* k. *Suppose* G *is simply connected and* $G(k) \neq \mathrm{Spin}(\Phi_2,D)$. *Let* F/k *be a pure transcendental extension. Then there exists a natural isomorphism* $W(G,k) \simeq W(G,F)$.

It should be pointed out that in the case $G(k) = \mathrm{Spin}(\Phi_2,D)$ the group

$$\mathrm{Spin}(\Phi_2,D)/[\mathrm{Clif}(\Phi_2,D),\mathrm{Clif}(\Phi_2,D)]$$

is stable under pure transcendental extensions of k.

Note that this result from algebraic group theory was proved by means of methods involving skew fields. In each case the stability of $W(G,k)$ under pure transcendental extensions of k was proved separately for $G(k) = \mathrm{SL}_n(D)$, $G(k) = \mathrm{SU}_n(\Phi_n,D)$, and $G(k) = \mathrm{Spin}(\Phi_n,D)$. It is unknown how to prove this fact by means of methods of algebraic group theory.

The important application of the above theorem is the following result on the R-equivalence on algebraic groups. We shall begin with some definitions (see **[19]**):

Let X be an irreducible affine algebraic variety defined over k. Two points $x, y \in X(k)$ are said to be *strictly R-equivalent* if and only if there exists a k-morphism $f : \mathbb{P}^1 \to X$ such that $f(a) = x$, $f(b) = y$, $a, b \in \mathbb{P}^1(k)$. Two points $x, y \in X(k)$ are said to be *R-equivalent* if and only if there exist points $x = x_1, x_2, \dots, x_n = y \in X(k)$ such that the points x_i, x_{i+1} are strictly R-equivalent for $i = 1, \dots, n-1$. It is easy to see that the group structure of the algebraic group G defined over k induces a group structure on the set of equivalence classes of $G(k)$ under R-equivalence. This group is denoted by $G(k)/R$.

For cubic surfaces, this notion was introduced by Yu. I. Manin. The theory of R-equivalence on algebraic groups was developed by J.-L. Colliot-Thelene and J.-J. Sansuc **[20]**. The following interesting result is due to V. E. Voskresenskiĭ **[19]**, who proved that

$$W(G, k) \cong G(k)/R \cong \mathrm{SL}_1(D)/R \cong \mathrm{SK}_1(D)$$

for $G(k) = \mathrm{SL}_n(D)$. Subsequently, a similar result on $G(k) = \mathrm{SU}(\Phi_n, D)$ was established by Yanchevskiĭ. The stability of the group $W(G, k)$ under pure transcendental extensions of k is of great importance for proving these results. Similarly we use Theorem 5 in order to prove

THEOREM 7. *Let $G(k) = \mathrm{Spin}(\Phi_n, D)$, where the Witt index v of the form Φ_n is greater than* 1. *Then $W(G, k) \cong G(k)/R$.*

In the case $v = 1$ the following exact sequence exists:

$$\{-1, 1\} \to W(G, k) \to G(k)/R \to 1 \qquad \text{for} \quad n > 2.$$

Now we can formulate this fact for all simply connected algebraic groups of classical type.

THEOREM 8. *Let G be an algebraic group of classical type defined over k and isotropic over k. Suppose G is simply connected and $G(k) \neq \mathrm{Spin}(\Phi_n, D)$, where the Witt index of Φ_n is* 1. *Then $W(G, k) \cong G(k)/R$.*

COROLLARY 1. *$W(G, k)$ is a birational invariant of G.*

The fact that the variety of $\mathrm{SL}_n(D)$ is not rational in general was established by Platonov in 1977 **[21]**. It is easy to see that the group $G(k)/R$ of a rational group G is trivial. It follows from Theorem 7 and Theorem 3 that

COROLLARY 2. *The variety of the group $\mathrm{Spin}(\Phi_n, D)$ is not rational in general.*

THEOREM 9. *Let Φ_n be a form of the Witt index* 1. *Then there exists a decomposition $\Phi_n = \Phi_2 \oplus \Phi_{n-2}$, where Φ_2 is a form of Witt index* 1 *and*

$$\mathrm{Spin}(\Phi_n, D)/\big[\mathrm{Clif}(\Phi_2, D), \mathrm{Clif}(\Phi_2, D)\big]\,\mathrm{Spin}(\Phi_n, D)^+ \cong \mathrm{Spin}(\Phi_n, D)/R.$$

The method developed here gives us an opportunity to establish some conditions for the element -1 to belong to the subgroup $\mathrm{Spin}(\Phi_n, D)^+$ (see Theorem 1).

THEOREM 10. *Let Φ_n be a form of Witt index ≥ 2 over a skew field D, and let $M = \{T \mid k \subset T \subset D, (T : k) = \mathrm{ind}(D)/2\}$ be the set of all subfields of codimension*

2 in the skew field D. If $-1 \in \mathrm{Spin}(\Phi_n, D)^+$, then $-1 \in \prod_{T\in M} N_{T/k}(T)$. Suppose that $\mathrm{SK}_1(D) = 1$; then $-1 \in \mathrm{Spin}(\Phi_n, D)^+$ if and only if $-1 \in \prod_{T\in M} N_{T/k}(T)$.

It seems that the calculation of the group $\overline{R(L)} \overset{\mathrm{def}}{=} R(L)/\prod_{\Sigma\in \mathrm{Gal}(L|K)\setminus\{1\}}(L^{\sigma})^*$ (see Theorem 3) is the most realistic way for the investigation of $W(G,k)$. For instance we have the following theorem.

THEOREM 11. *Let K be a global field* ($\mathrm{char}\, K \neq 2$), $L = K(\sqrt{a_1}, \dots, \sqrt{a_t})$, $\Gamma = \mathrm{Gal}(L/K)$, *$P(K)$ the set of all places of K. Then the group $\overline{R(L)}$ is finite. If there exists $v = P(K)$ such that $\Gamma_v = \Gamma$ then $\overline{R(L)} = 1$. If $|\Gamma_v| \leq 2$ for every $v \in P(K)$, then $|R(L)| \geq 2$, where $m \geq 2 - t(t+1)/2 - 2$.*

In order to prove Theorem 11 we define and consider the algebraic torus

$$R^{(2)}_{L/K} = \{ (\alpha, \beta) \in R_{L/K}(G_m) \times G_{m,K} \mid N_{L/K}(\alpha) = \beta^2 \}$$

and establish that $\overline{R(L)}$ coincides, in essence, with the group of R-equivalences $R^{(2)}_{L/K}/R$ of $R^{(2)}_{L/K}$. Calculations of $R^{(2)}_{L/K}$ lead us to the result of Theorem 11.

Let k be an algebraic number field. Then for all groups G of classical types $W(G,k) = 1$. In fact in the case $G(k) = \mathrm{SL}_n(D)$ this was proved by S. Wang **[5]**, in the case $G(k) = \mathrm{SU}(\Phi_n, D)$ by Platonov and Yanchevskiĭ **[22]** and independently by C. T. C. Wall, and in the case $G(k) = \mathrm{Spin}(\Phi_n, D)$ by E. A. M. Seip-Hornix (see also **[3]**). On the other hand, Platonov showed that in the case when k is of transcendence degree greater than 1 this is not true in general. The question arises: is the Kneser-Tits conjecture true in the case when k is a function field of an arithmetic curve? At present, this an open question. We are concerned here with the following particular situation.

Let k be an algebraic number field, $P(k)$ the set of all its places, k_v the completion of k at a place $v \in P(k)$, X a smooth geometrically irreducible algebraic curve defined over k, and K its function field. A finite-dimensional simple central algebra over K is said to be *completely splitting* if its class in the Brauer group $\mathrm{Br}(K)$ belongs to the kernel of the homomorphism $\mathrm{Br}(K) \to \oplus_{v\in P(k)} \mathrm{Br}(K_v)$, where K_v is the function field of the curve $X \otimes_k k_v$. In the above notation we have

THEOREM 12 ((Yanchevskiĭ)). *If $G(k)$ is a classical group over a completely splitting algebra B and the index of B is free from cubes, then $G(k) = G(k)^+$ and consequently $G(k)/R = 1$.*

A field k is called the *C_2^0-field* if it satisfies the following conditions: for any finite field extension K/k and for any skew field D central and finite-dimensional over K the homomorphism $\mathrm{Nrd}_{D/K}\colon D^* \to K^*$ is surjective. It was proved by Yanchevskiĭ in **[23, 24]** that if k is C_2^0-field then $W(G,k)$ is trivial for $G(k) = \mathrm{SL}_n(D)$ and $G(k) = \mathrm{SU}(\Phi_n, D)$. A similar result for $G(k) = \mathrm{Spin}(\Phi_n, D)$ is established by means of Theorem 1 and Theorem 10 as follows.

THEOREM 13 ((Yanchevskiĭ)). *Let k be a C_2^0-field. If D is a skew field over k with involution of the first kind, then $K_1\,\mathrm{Spin}(D) = 1$. Suppose G is a simply connected algebraic group of classical type defined over k and $G(k) \neq \mathrm{Spin}(\Phi_n, D)$, where the Witt index of Φ_n is 1. Then $W(G,k) = 1$.*

Finally we formulate the following result on weak approximation in spinor groups. Let v be a valuation of k. We consider the group $G(k_v)$ as a topological group; let

$\overline{G(k)}$ be the closure of $G(k)$ in $G(k_v)$. One says that G possesses the *property of weak approximation with respect to* v iff $G(k_v)/\overline{G(k)} = 1$.

THEOREM 14. *For any* $r \in \mathbb{N}$ *there exists a group* $G(k) = \mathrm{Spin}(\Phi_n, D)$ *such that* $|G(k_v)/\overline{G(k)}| > r$.

Similar results for the cases $G(k) = \mathrm{SL}_n(D)$ and $G(k) = \mathrm{SU}_n(D)$ were obtained earlier by Platonov and Yanchevskiĭ, respectively, **[10, 14]**.

Summarizing we may say that at present time we have a good reduced K-theory for all algebraic groups of classical type.

References

1. J. Tits, *Algebraic and abstract simple groups*, Ann. of Math. (2) **80** (1964), 313–329.
2. J. A. Dieudonné, *La géométrie des groupes classiques*, Ergeb. Math. Grenzgeb. (3), Band 5, 3rd ed., Springer-Verlag, Berlin and New York, 1971.
3. V. P. Platonov, *The strong approximation problem and the Kneser-Tits conjecture for algebraic groups*, Izv. Akad. Nauk SSSR Ser. Mat. **33** (1969), no. 6, 1211–1219; English transl. in Math. USSR-Izv. **3** (1969).
4. T. Nakayama and Y. Matsushima, *Über die multiplicative Gruppe einer p-adischen Divisionsalgebra*, Proc. Imper. Acad. Japan **19** (1943), 622–628.
5. S. Wang, *On the commutator group of a simple algebra*, Amer. J. Math. **72** (1950), 323–324.
6. V. P. Platonov, *On the Tannaka-Artin problem*, Dokl. Akad. Nauk SSSR **221** (1975), no. 5, 1038–1041; English transl. in Soviet Math. Dokl. **16** (1975).
7. ———, *The Tannaka-Artin problem and reduced K-theory*, Izv. Akad. Nauk SSSR Ser. Mat. **40** (1976), no. 2, 227–261; English transl. in Math. USSR-Izv. **10** (1976).
8. ———, *Algebraic groups and reduced K-theory*, Proc. Internat. Congr. Math. (Helsinki, 1978), Acad. Sci. Fennica, Helsinki, 1980, pp. 311–317.
9. J. Tits, *Groupes de Whitehead de groupes algébriques simples sur un corps (d'apres V.P.Platonov et al.)*, Séminaire Bourbaki 293 annee (1976/1977), Exp. No. 505, Lecture Notes in Math., vol. 677, Springer-Verlag, Berlin and New York, 1978, pp. 218–236.
10. V. P. Platonov, *Reduced K-theory and approximation in algebraic groups*, Trudy Mat. Inst. Akad. Steklov. **142** (1976), 198–207; English transl. in Proc. Steklov Inst. Math. **1979**, no. 3.
11. P. Draxl and M. Kneser (eds.), *SK_1 von Schiefkörpern. SK_1 of division algebras*, Lecture Notes in Math., vol. 778, Springer-Verlag, Berlin and New York, 1980.
12. Yu. L. Ershov, *Henselian valuations of division rings and the group SK_1*, Mat. Sb. **117** (1982), no. 1, 60–68; English transl. in Math. USSR-Sb. **45** (1983).
13. V. I. Yanchevskiĭ, *The reduced unitary K-theory and skew fields over Henselian discrete-valued fields*, Izv. Akad. Nauk SSSR Ser. Mat. **42** (1978), no. 4, 879–918; English transl. in Math. USSR-Izv. **13** (1979).
14. ———, *The reduced unitary K-theory. Applications to algebraic groups*, Mat. Sb. **110** (1979), no. 4, 579–596; English transl. in Math. USSR-Sb. **13** (1981).
15. A. P. Monastyrnyi and V. I. Yanchevskiĭ, *On the Whitehead groups and the Kneser-Tits conjecture for spinor groups*, Dokl. Akad. Nauk SSSR **307** (1989), no. 1, 31–35; English transl. in Soviet Math. Dokl. **40** (1991).
16. G. E. Wall, *The structure of a unitary factor group*, Inst. Hautes Études Sci. Publ. Math. **1** (1959).
17. V. P. Platonov and V. I. Yanchevskiĭ, *The Dieudonné conjecture on the structure of unitary groups over skew field and Hermitian K-theory*, Izv. Akad. Nauk SSSR Ser. Mat. **48** (1984), no. 6, 1266–1294; English transl. in Math. USSR-Izv. **25** (1985).
18. ———, *On stability in reduced K-theory*, Dokl. Akad. Nauk SSSR **242** (1978), no. 4, 769–772; English transl. in Soviet Math. Dokl. **19** (1978).
19. V. E. Voskresenskiĭ, *Algebraic tori*, "Nauka", Moscow, 1977. (Russian)
20. J. L. Colliot-Thelene and J.-J Sansuc, *La R-equivalence sur les tores*, Ann. Sci. École Norm. Sup.(4) **10** (1977), 175–229.
21. V. P. Platonov, *Birational properties of reduced Whitehead group*, Dokl. Akad. Nauk BSSR **21** (1977), no. 3, 197–198. (Russian)

22. V. P. Platonov and V. I. Yanchevskiĭ, *Structure of unitary groups and commutant of a simple algebra over global fields*, Dokl. Akad. Nauk SSSR **208** (1973), no. 3, 541–554; English transl. in Soviet Math. Dokl. **14** (1973).
23. V. I. Yanchevskiĭ, *Commutants of simple algebras with surjective reduced norm*, Dokl. Akad. Nauk SSSR **221** (1975), no. 5, 1056–1058; English transl. in Soviet Math. Dokl. **16** (1975).
24. ———, *Simple algebras with involutions and unitary groups*, Mat. Sb. **93** (1974), no. 3, 368–380; English transl. in Math. USSR-Sb. **22** (1974).

INSTITUTE OF MATHEMATICS, BYELORUSSIAN ACADEMY OF SCIENCES, MINSK, BELARUS

Amer. Math. Soc. Transl.
(2) Vol. **163**, 1995

Diffeomorphicity Criteria for Simply Connected Manifolds

N. Yu. Netsvetaev

§0. Introduction

All manifolds below are smooth, connected, oriented and (if not specified otherwise) compact and, in this section, closed, i.e., without boundary. The symbol $\cong$ denotes the diffeomorphism relation.

How can we know whether two manifolds given in some or other way are diffeomorphic? From the viewpoint of the general theory, which works by reducing differential-topological problems to algebraic-topological ones, in the even-dimensional simply connected case a quite adequate answer is given by the Novikov Theorem 0.2 below. This theorem will be our starting point and the object of extension.

0.1. Notation. Let M_0, M_1 be simply connected manifolds of dimension m, X a simply connected CW-complex, ξ a vector bundle over X of sufficiently large dimension ($\geq m+1$). Let $f_i\colon M_i \to X$, $i = 0, 1$, be two normal maps (see 1.1) and let f_0 and f_1 be normally (co)bordant.

0.2. Theorem (S. P. Novikov). *With the notation* 0.1 *assumed, let m be even, $m \geq 6$. If f_0, f_1 are homotopy equivalences, then $M_0 \cong M_1$.*

This theorem follows from the three Theorems 0.3–0.5 below. The first is the classification theorem of Novikov.

0.3. Theorem (S. P. Novikov, see **[1, 2]**). *With the notation* 0.1 *assumed, let $m \geq 4$. If f_0, f_1 are homotopy equivalences, then M_0 is h-cobordant to the connected sum $M_1 \# \Sigma$ of M_1 with some homotopy sphere Σ bounding a parallelizable manifold.*

0.4. Theorem (Kervaire-Milnor, see **[3, 4]**). *If a homotopy sphere Σ of even dimension $m = 2n$ bounds a parallelizable manifold, then Σ is h-cobordant to the standard sphere S^{2n}. (Symbolically,* $\mathrm{b}P_{2n+1} = 0$.)

0.5. Theorem (Smale **[5]**). *If two simply connected manifolds of dimension ≥ 5 are h-cobordant, then they are diffeomorphic.*

Note that the h-cobordism Theorem 0.5 can be reformulated in the following way.

1991 *Mathematics Subject Classification*. Primary 57R80.

0.6. THEOREM $(=0.5')$. *Let W be a simply connected $(m+1)$-dimensional manifold (with boundary) whose boundary has two components N_0 and N_1. Let $m \geq 5$. If the inclusions $N_i \subset W$, $i = 0, 1$, are homotopy equivalences, then $N_0 \cong N_1$.*

It is easily seen that Theorem 0.2 generalizes, for m even, Theorem 0.5 in the form 0.6. On the other hand its application in certain concrete cases is hampered by the fact that it requires an *a priori* homotopy equivalence from M_0 and M_1, in contrast to 0.6, where, because of duality, the $[(m+1)/2]$-connectedness of the inclusions $N_i \subset W$ suffices. Equally, in a series of interesting situations taking place in differential topology and algebraic geometry, the hypotheses 0.1 with even m are fulfilled but f_0, f_1 are only $m/2$-connected. M. Freedman has obtained corresponding generalizations of the Novikov Theorem 0.2 which are most effective for $m = 4k+2$.

0.7. THEOREM (Freedman **[6, 7]**). *With the notation* 0.1 *assumed, let $m = 2n = 4k+2$ and let f_0, f_1 be n-connected. If the middle Betti numbers of M_0 and M_1 are the same* $(\operatorname{rk} H_n M_0 = \operatorname{rk} H_n M_1)$, *then M_0 and M_1 are diffeomorphic*: $M_0 \cong M_1$.

For the dimensions divisible by 4 he obtained only a weaker result.

0.8. THEOREM (Freedman **[6, 7]**). *With the notation* 0.1 *assumed, let $m = 2n = 4k > 4$ and let f_0, f_1 be n-connected. If the quadratic $\mathbb{Z}$-modules (with respect to intersection number)* $\ker f_{i,*}^{(H_n)}$, $i = 0, 1$, *are unimodular and isometric to each other, then $M_0 \cong M_1$.*

In particular, it follows from Theorems 0.7 and 0.8 that in Theorem 0.2 it suffices to demand the $(n+1)$-connectedness of f_0, f_1 instead of homotopy equivalence (here $n = m/2$).

The object of the present paper is, first, to obtain an extension of Theorem 0.8 sufficient for rather interesting applications. To this purpose Theorems 1.4, 1.5, 1.7, 1.8, 1.9 are devoted. Furthermore, we extend Theorem 0.7 to the nonempty boundary case, and our proof does not depend on the original one by Freedman.

For applications see **[8, 9]**.

§1. Main results

1.1. Terminology: normal maps and normal bordism. Let X be a CW-complex, ξ an oriented vector bundle over X, M a manifold with stable normal bundle ν. A normal map $f\colon (M,\nu) \to (X,\xi)$ is a map $f\colon M \to X$ covered by a fiber isomorphism $\nu \to \xi$. Normal bordism between normal mappings and normal bordism relative to the boundary are defined in the natural way. Normal bordism and diffeomorphism relative to the boundary are denoted by the symbol $\operatorname{rel} \partial$.

1.2. Notation and assumptions: X, M_0, M_1, f_0, f_1. In what follows X is a connected CW-complex, ξ is an oriented vector bundle over X. M_0, M_1 are compact manifolds of dimension $2n$ and $f_i\colon (M_i, \nu_i) \to (X, \xi)$, $i = 0, 1$, are normally bordant $\operatorname{rel} \partial M_i$ normal mappings (in particular, a diffeomorphism $\partial M_0 \cong \partial M_1$ is fixed).

We assume that f_0, f_1 are n-connected.

Let $\operatorname{rk} H_n M_0 \geq \operatorname{rk} H_n M_1$, for definiteness. Since M_0 and M_1 are cobordant relative to the boundary and f_0, f_1 are n-connected, $\delta := \frac{1}{2}(\operatorname{rk} H_n M_0 - \operatorname{rk} H_n M_1)$ is a nonnegative integer.

1.3. THEOREM. *With the assumptions* 1.2 *fulfilled, let n be odd. Then $M_0 \cong_{\operatorname{rel}\partial} M_1 \mathbin{\#} \delta(S^n \times S^n)$. In particular if* $\operatorname{rk} H_n M_0 = \operatorname{rk} H_n M_1$, *then $M_0 \cong_{\operatorname{rel}\partial} M_1$.*

In the closed case this implies Freedman's Theorem 0.7.

1.4. THEOREM. *With the assumptions* 1.2 *fulfilled, let n be even,* $n \geq 4$. *If* $\delta > 0$ *then* $M_0 \cong_{\mathrm{rel}\,\partial} M_1 \# \delta(S^n \times S^n)$. *If* $\delta = 0$, *i.e.,* $\operatorname{rk} H_n M_0 = \operatorname{rk} H_n M_1$, *then* $M_0 \# S^n \times S^n \cong_{\mathrm{rel}\,\partial} M_1 \# S^n \times S^n$.

In view of 1.4, the case where n is even and $\operatorname{rk} H_n M_0 = \operatorname{rk} H_n M_1$ is of special interest. We consider it in the rest of this section.

1.5. THEOREM. *With the assumptions* 1.2 *fulfilled, let n be even and* $\operatorname{rk} H_n M_0 = \operatorname{rk} H_n M_1$. *If there are classes* $u, v \in \ker f_{0,*}^{(H_n)}$ *with* $u^2 = 0$, $uv = 1$, *then* $M_0 \cong_{\mathrm{rel}\,\partial} M_1$.

1.6. Lefschetz normal maps. Let M be a closed manifold of dimension $2n$ and $f: (M, \nu) \to (X, \xi)$ a normal map. We call f a Lefschetz map, if f is n-connected and the intersection number form restricted to $\ker f_*^{(H_n)}/\text{torsion}$ is nondegenerate (i.e., of $\det \neq 0$). This notion generalizes the notion of Lefschetz submanifold introduced in **[10]**, where the connection with Lefschetz is explained.

The following Theorem 1.7 widely extends Freedman's Theorem 0.8.

1.7. THEOREM. *With the assumptions* 1.2 *fulfilled, let n be even,* $n \geq 4$, *and* M_0, M_1 *closed. Let* f_0, f_1 *be Lefschetz and* $\operatorname{rk} H_n M_0 = \operatorname{rk} H_n M_1$. *In addition let* $\operatorname{tors}(H_n X) = f_*(\operatorname{tors}(H_n M_i))$. *Then every isometry* φ *making the following diagram commutative*

$$\begin{array}{ccc} H_n M_0/\text{torsion} & \xrightarrow{\varphi} & H_n M_0/\text{torsion} \\ & {\scriptstyle f_{0,*}}\searrow \quad \swarrow{\scriptstyle f_{1,*}} & \\ & H_n X/\text{torsion} & \end{array}$$

is realized by a diffeomorphism $g: M_0 \to M_1$.

1.8. THEOREM. *If under the conditions of* 1.7 *the inequality*

$$\operatorname{rk} H_n M_i \geq 2 \operatorname{rk} H_n X + 2$$

holds, and furthermore the quadratic $\mathbb{Z}$*-module* $\ker f_{0,*}^{(H_n)}$ *is indefinite, then the desired* φ *exists. In particular* $M_0 \cong M_1$.

Theorem 1.7 can be transferred to the nonclosed case (and, moreover, the Lefschetz condition can be eliminated) at the expense of more intricate formulation (cf. 1.10.b). We restrict ourselves to the following partial result.

1.9. THEOREM. *With the assumption* 1.2 *fulfilled, let n be even,* $n \geq 4$, *and let the quadratic* $\mathbb{Z}$*-module* $\ker f_{i,*}^{(H_n)}/$*torsion be unimodular,* $i = 1, 2$. *If these modules are isometric to each other, then* $M_0 \cong_{\mathrm{rel}\,\partial} M_1$.

(The mentioned modules are *a priori* isometric if they are indefinite and of the same rank. But then Theorem 1.5 is also applicable.) In the closed case we once more obtain Freedman's Theorem 0.8.

1.10. REMARKS. a) If $M_0 = M_1 =: M$ is a regular complete intersection in $\mathbb{C}P^N =: X$, then Theorem 1.7 implies that every automorphism of the (graded) integer cohomology ring H^*M is realized by an autodiffeomorphism of M. This is important when studying complete intersections with isolated singularities. For other applications to algebraic geometry see **[8]**.

b) Let M be a parallelizable $4k$-dimensional manifold ($k \neq 1$) homotopy equivalent to a bouquet of $2k$-spheres and with a simply connected boundary; i.e., let M

be a handlebody from $H(4k, 2k)$. Let $\mathrm{Diff}(M \text{ rel}\,\partial M)$ be the group of all autodiffeomorphisms of M that are the identity on ∂M. Then we have a sequence of groups and homomorphisms,

$$\mathrm{Diff}(M \text{ rel}\,\partial M) \to \mathrm{Aut}(H_{2n}M) \to \mathrm{Aut}(H_{2n-1}\partial M),$$

where $\mathrm{Aut}(H_{2n}M)$ is the group of autoisometries. Similarly to 1.7, we can prove that this sequence is exact. For the case $\partial M = S^{4k-1}$ this was proved by Kreck in **[11]**. (As Wall **[12]** proved, $\mathrm{Diff}\, M \to \mathrm{Aut}(H_{2n}M)$ is onto.)

§2. Normal n-connected maps

The following assertion is proved in a standard way (and, in fact, $(n-1)$-connectedness is sufficient).

2.1. PROPOSITION. *Let $f : M \to X$ be an n-connected map where X is simply connected. Then the Hurewicz homomorphism gives an epimorphism*

$$\ker f_*^{(\pi_n)} \to \ker f_*^{(H_n)}.$$

2.2. COROLLARY. *Let $f : (M, \nu) \to (X, \xi)$ be a normal k-connected map from a $2n$-dimensional manifold M to a simply connected CW-space X, where $n \geq 3$. Then $\ker f_*^{(H_n)}$ is generated by spherical classes which can be realized by embedded n-spheres with stably trivial normal bundles. In particular if n is even, then $L := \ker f_*^{(H_n)}/\text{torsion}$ is an even lattice (this holds also for $n = 2$).*

2.3. THEOREM ((Freedman **[7]**, Kreck **[11]**)). *With the notation of 1.2 assumed, M_0 and M_1 are stably diffeomorphic $\text{rel}\,\partial M_i$; i.e., $M_0 \# a_0(S^n \times S^n) \cong_{\text{rel}\,\partial} M_1 \# a_1(S^n \times S^n)$ for certain a_0, $a_1 \geq 0$. (This holds also without the simple connectedness assumption.) More explicitly, there is a $2n$-manifold M and a normal n-connected map $f : (M, \nu) \to (X, \xi)$ normally bordant to f_i, $i = 0, 1$, and obtainable from f_i by a_i trivial normal surgeries.*

§3. Proofs of Theorems 1.3, 1.4, 1.5

3.1. Theorems 1.3 and 1.4 immediately follow from the Freedman-Kreck Theorem 2.3 and the parts (a) and (b) of the following proposition.

3.2. PROPOSITION. *Let K_0 and K_1 be compact simply connected $2n$-manifolds, $2n \geq 6$, stably diffeomorphic $\text{rel}\,\partial K_i$. Let, e.g., $\text{rk}\, H_nK_0 \geq \text{rk}\, H_nK_1$. Set $\delta = \frac{1}{2}(\text{rk}\, H_nK_0 - \text{rk}\, H_nK_1)$. Evidently, δ is integer. Further,*
(a) *Let n be odd. Then $K_0 \cong_{\text{rel}\,\partial} K_1 \# \delta(S^n \times S^n)$. In particular, if $\delta = 0$, then $K_0 \cong_{\text{rel}\,\partial} K_1$.*
(b) *Let n be even, $n \neq 2$. If $\delta > 0$, then $K_0 \cong_{\text{rel}\,\partial} K_1 \# \delta(S^n \times S^n)$. If $\delta = 0$ then $K_0 \# S^n \times S^n \cong_{\text{rel}\,\partial} K_1 \# S^n \times S^n$.*
(c) *Let n be even and $\delta = 0$. If $S^n \times S^n \setminus \text{pt}$ embeds in K_0, then $K_0 \cong_{\text{rel}\,\partial} K_1$.*

PROOF. This easily follows from the main results of **[9]**, notably 5.1 and 6.1. □

Note that on the other hand we can deduce 3.2(a) from 1.3.

3.3. Theorem 1.5 follows from 3.2(c) and the following lemma.

3.4. LEMMA. *With the hypotheses of 1.5 fulfilled, $S^n \times S^n \setminus \text{pt}$ embeds in M_0.*

PROOF. By 2.2, v^2 is even. Thus, replacing v by $v-(v^2/2)u$ if necessary, we can assume $v^2=0$. Then, again by 2.2, u and v are realized by embedded spheres with trivial normal bundles. Using the Whitney trick we can make these spheres intersect transversally in a single point. Then the regular neighborhood of their union is the desired $S^n\times S^n$ without a point embedded in M_0. □

§4. Auxiliary algebraic results

4.0. The aim of this section is to prove Corollaries 4.3 and 4.7, needed for the proofs of Theorems 1.7 and 1.8. We use the terminology of **[14]**. Let H be a unimodular integer lattice, A a free abelian group, $\alpha\colon H\to A$ a group epimorphism, $L:=\ker\alpha$. We assume that the lattice L is nondegenerate (i.e., of $\det\neq 0$) and even. If $a\in H$, $a^2=\pm 2$, then s_a denotes the reflection $H\to H$ defined as $s_a(x)=x\mp(x,a)a$. We will need the following easy lemma.

4.1. LEMMA. *Let $\varphi\colon H\to H$ be an automorphism of the lattice H such that $\alpha\circ\varphi=\alpha$. Then it leaves the sublattice L invariant and it is uniquely determined by the restriction to L. On the other hand, an automorphism $\vartheta\colon L\to L$ extends to an automorphism $\psi\colon H\to H$ with $\alpha\circ\psi=\alpha$ if and only if ϑ induces the identity on the discriminant group $G_L:=L^{\#}/L$ of L.*

4.2. ASSERTION. *Let the Witt $\mathbb{Z}$-index of L be ≥ 3. Then an automorphism of L decomposes into a product of reflections s_a with $a^2=\pm 2$ if and only if it induces the identity of G_L.*

PROOF. This follows from Theorem 2.7 in **[15]**. □

4.3. COROLLARY. *If the Witt $\mathbb{Z}$-index of L is ≥ 3 then every automorphism $\varphi\colon H\to H$ with $\alpha\circ\varphi=\alpha$ decomposes into a product of reflections s_a with $a\in L$, $a^2=\pm 2$.*

PROOF. This immediately follows from 4.1 and 4.2. □

4.4. LEMMA. *$l(G_L)\leq \operatorname{rk} A$, where l is the minimal number of generators.*

4.5. DEFINITION. U_{2p} denotes the lattice $\mathbb{Z}^{2p}$ with the form $\left(\begin{smallmatrix}0&E\\E&0\end{smallmatrix}\right)$ on it. An even lattice S is *stable*, if every isometric embedding: $U_{2p}\to S\oplus U_{2p}$ extends to an automorphism $\vartheta\colon S\oplus U_{2p}\to S\oplus U_{2p}$.

4.6. ASSERTION. *If $\operatorname{rk} S\geq 2+l(G_S)$ and S is indefinite, then S is stable.*

PROOF. This easily follows from the results of **[14]**; see Theorems 1.13.2 and 1.14.2 there. □

4.7. COROLLARY. *Let $\mathrm{in}_i\colon U_{2a}\to L$, $i=0,1$, be isometric embeddings. If $\operatorname{rk} H\geq \operatorname{rk} A+2a+2$ and the orthogonal complement of the sublattice $\mathrm{in}_0\,U_{2a}$ in L is indefinite, then there is an automorphism $\psi\colon H\to H$ with $\alpha\circ\psi=\alpha$ such that $\psi\circ\mathrm{in}_0=\mathrm{in}_1$.*

PROOF. This follows easily from 4.4, 4.6, and 4.1. □

§5. Proofs of Theorems 1.7, 1.8, and 1.9

5.1. First, we prove Theorem 1.7 in the special case $M_0=M_1=:M$.

So let $f\colon(M,\nu)\to(X,\xi)$ be a Lefschetz normal map. We introduce the following notation. Let $H:=H_nM/\text{torsion}$, $L:=\ker f_*^{(H_n)}/\text{torsion}$, $A:=H_nX/\text{torsion}$. Then L is a nondegenerate primitive even sublattice of the unimodular lattice H; L coincides with the kernel of $\alpha = f_*\colon H \to A$. In addition assume that the Witt $\mathbb{Z}$-index of L is ≥ 3.

5.2. ASSERTION. *The automorphism $\varphi\colon H \to H$ in 1.7 is realized by an autodiffeomorphism of M.*

PROOF. It follows from 4.3 that φ decomposes into a product of reflections $s_{a(i)}$ with $a(i) \in L$, $a(i)^2 = \pm 2$. Since every $a(i)$ is realized by an embedded n-sphere with stably trivial normal bundle (see 2.2) and thus isomorphic to the tangent bundle, it follows that $s_{a(i)}$ is realized by an autodiffeomorphism of M, namely by the Dehn-Lickorish twist along this n-sphere (cf. [9]). □

5.3. Now we prove 1.7 in the general case. Consider the manifold M from the Freedman-Kreck Theorem 2.3 diffeomorphic to $M_i \# a(S^n \times S^n)$, $i = 0, 1$, with $a = a_0 = a_1$, and the corresponding Lefschetz normal map $f\colon (M, \nu) \to (X, \xi)$. Set $H_i := H_n M_i/\text{torsion}$, $i = 0$, 1. Clearly, $H = H_i \oplus U_{2a,i}$, $i = 0$, 1. Assume for the moment that $a \geq 3$. We obtain an automorphism $\psi = \varphi \oplus \mathrm{id}(U_{2a})\colon H \to H$. By 5.2 it is realized by an autodiffeomorphism $h\colon M \to M$. For the sake of simplicity, in the considerations below let $a = 1$. (Although we assumed $a \geq 3$ somewhat earlier. This is no contradiction.) Cf. [**9,** §§5, 6].

Let W_i be a cobordism between M and M_i obtained by gluing a trivial n-handle $\mathbb{H}_i$ to $M \times I$. Let m_i, $c_i \in H_n M$ be the classes realized by the fibres of $S^n \times S^n$. Let m_i be the comeridian sphere of $\mathbb{H}_i$, $i = 0, 1$. The automorphism ψ carries the pair m_0, c_0 to the pair m_1, c_1. We can assume, e.g., that $\psi(m_0) = c_1$ and $\psi(c_0) = m_1$. Now, if we glue the cobordisms W_0 and $(-W_1)$ with the help of the diffeomorphism h, then the handles $\mathbb{H}_0$ and $\mathbb{H}_1$ cancel and we obtain a trivial cobordism between M_0 and M_1. It is easily seen that the corresponding diffeomorphism induces the isomorphism φ. This proves Theorem 1.7.

5.4. PROOF OF THEOREM 1.8. To construct the desired isomorphism $\varphi\colon H_0 \to H_1$ it is necessary and sufficient to find an automorphism $\psi\colon H \to H$ with $\alpha \circ \psi = \alpha$ carrying $U_{2a,0}$ to $U_{2a,1}$. But the existence of such ψ follows from 4.7. □

5.5. PROOF OF THEOREM 1.9. This is is similar to that of 1.7 and is even easier, since we need not control the discriminant groups; they are trivial because of the unimodularity of the lattices involved. We prove that every isometry in 1.9 is realized by a diffeomorphism $M_0 \to M_1$ as in the proof of Theorem 1.7, by reduction to the case $M_0 = M_1 =: M$. □

References

1. S. P. Novikov, *Diffeomorphisms of simply connected manifolds*, Dokl. Akad. Nauk SSSR **143** (1962), 1046–1049; English transl. in Soviet Math. Dokl. **3** (1962).
2. ______, *Homotopically equivalent smooth manifolds*. I, Izv. Akad. Nauk SSSR Ser. Mat. **28** (1964), no. 2, 365–474; English transl. in Amer. Math. Soc. Transl. Ser. 2 **48** (1965).
3. M. A. Kervaire and J. W. Milnor, *Groups of homotopy spheres*. I, Ann. of Math (2) **77** (1963), 504–537.
4. C. T. C. Wall, *Killing the middle homotopy groups of odd dimensional manifolds*, Trans. Amer. Math. Soc. **103** (1962), 421–433.
5. S. Smale, *On the structure of manifolds*, Amer. J. Math. **84** (1962), 387–399.
6. M. H. Freedman, *On the classification of taut submanifolds*, Bull. Amer. Math. Soc. **81** (1975), 1067–1068.
7. ______, *Uniqueness theorems for taut submanifolds*, Pacific J. Math. **62** (1976), 379–387.

8. N. Yu. Netsvetaev, *Diffeomorphism criteria for smooth manifolds and algebraic varieties*, Proc. Internat. Conf. on Algebra, Part 3 (Novosibirsk, 1989), Contemp. Math., vol. 131, Amer. Math. Soc., Providence, RI, 1992, pp. 453–459.
9. ———, *Diffeomorphism and stable diffeomorphism of simply connected manifolds*, Algebra i Analiz **2** (1990), no. 2, 112–120; English transl. in Leningrad Math. J. **2** (1991).
10. ———, *Decomposition of complex projective manifolds into a connected sum*, Dokl. Akad. Nauk SSSR **277** (1984), no. 2, 299–303; English transl. in Soviet Math. Dokl. **30** (1984).
11. M. Kreck, *Isotopy classes of diffeomorphisms of $(k-1)$-connected almost parallelizable $2k$-manifolds*, Algebraic Topology (Aarhus, 1978), Lecture Notes in Math., vol. 763, Springer-Verlag, Berlin and New York, 1979, pp. 643–646.
12. C. T. C. Wall, *Classification of $(n-1)$-connected $2n$-manifolds*, Ann. of Math. (2) **75** (1962), 163–189.
13. M. Kreck, *Duality and surgery: An extension of results of Browder, Novikov and Wall about surgery on compact manifolds*, Vieweg Verlag (to appear).
14. V. V. Nikulin, *Integer symmetric bilinear forms and some of their geometric applications*, Izv. Akad. Nauk SSSR Ser. Mat. **43** (1979), no. 1, 111–177; English transl. in Math. USSR-Izv. **14** (1979).
15. M. Kneser, *Erzeugung ganzzahliger orthogonaler Gruppen durch Spiegelungen*, Math. Ann. **255** (1981), 453–462.

Amer. Math. Soc. Transl.
(2) Vol. **163**, 1995

On the K-theory of Generalized Fibre Bundles and Some of Their Twisted Forms

I. A. Panin

Introduction

D. Quillen constructed the K-theory of projective fibre bundles and their twisted forms (Severi-Brauer schemes) in his fundamental paper **[10]** on higher algebraic K-theory. R. Swan **[12]** constructed the K-theory of nonsingular projective quadrics.

A new approach to the construction of the higher K-theory of algebraic varieties is developed in the present paper. In particular, we prove a formula relating K-groups of twisted forms of generalized flag fibre bundles to those of the associated Azumaya algebra and its powers.

In §0 some preliminary information is discussed, §§1 and 3 are devoted to a description of the method in the paper (see 1.8 and 3.8). Theorem 1.8 deals with the "untwisted" case, Proposition 3.8 enables to reduce the "twisted" case to an "untwisted" one (see the proof of 4.9). In §§2 and 4 Quillen's results are reproved as an application of this method (see 2.1, 4.9). Finally in §5 some results which may be obtained in this way are formulated without proof. Proofs may be found in **[14]**.

The author thanks A. S. Merkur′ev, A. A. Suslin, A. L. Smirnov, A. A. Beilinson, and M. M. Kapranov for useful discussions and attention to the present paper. Theorem 5.2 in the characteristic zero case was obtained together with A. S. Merkur′ev.

Results similar to ours have been obtained independently by M. Levine together with V. Srinivas. Recently A. A. Suslin and independently M. Levine computed the K-theory of $SL_{1,D}$, where D is an arbitrary simple algebra.

§0. Preliminaries

In this paper all schemes are supposed to be quasicompact (but not necessarily Noetherian or separated). The following proposition will be used in §1 to define the direct image homomorphism (1.1.1).

0.1. Proposition. *Let $f\colon T \to S$ be a flat projective morphism of schemes and let $\mathscr{M}$ be a locally free coherent sheaf on T. Then there exists an exact sequence $0 \to \mathscr{M} \to \mathscr{N} \to \mathscr{P} \to 0$ of locally free coherent sheaves on T such that $R^i f_*(\mathscr{N}) = 0$ for $i > 0$.*

Proof. Since f is projective, there exists a projective fibre bundle $p\colon \mathbb{P}_S(\mathscr{E}) \to S$ and a closed embedding $i\colon T \hookrightarrow \mathscr{P}_S(\mathscr{E})$ such that $f = p \circ i$. Denote by $\mathscr{O}(1)$ the

1991 *Mathematics Subject Classification*. Primary 19D10.

canonical invertible sheaf on $\mathbb{P}_S(\mathscr{E})$ and put $\mathscr{L} = i^*\mathscr{O}(1)$. Since S is the union of finitely many open affines, there exists an n_0 such that $R^i f_*(\mathscr{L}^{\otimes n}) = 0$ and $R^i f_*(\mathscr{L}^{\otimes n} \otimes \mathscr{M}) = 0$ for $i > 0$, $n \geq n_0$ (see **[5,** (III.5.2)]). It follows from (**[9,** (II.5.2)]) that the sheaf $f_*(\mathscr{L}^{\otimes n})$ is locally free for such n. Consider the natural homomorphisms

$$(*,\ m) \qquad \alpha_m \colon f^* f_*(\mathscr{L}^{\otimes m}) \longrightarrow \mathscr{L}^{\otimes m}.$$

As S is the union of finitely many open affines, there exists an m_0 such that α_m is an epimorphism for $m \geq m_0$ (**[5,** (III.8.8)]). Dualizing $(*, m)$ and tensoring with $\mathscr{L}^{\otimes m}$, we obtain an exact sequence of locally free coherent sheaves

$$0 \longrightarrow \mathscr{O}_T \longrightarrow f^*\big(f_*(\mathscr{L}^{\otimes m})^V\big) \otimes \mathscr{L}^{\otimes m} \longrightarrow \mathscr{K}_m \longrightarrow 0.$$

Tensoring it with the sheaf $\mathscr{M}$ we obtain a sequence

$$(**,\ m) \qquad 0 \longrightarrow \mathscr{M} \longrightarrow \mathscr{N}_m \longrightarrow \mathscr{P}_m \longrightarrow 0.$$

It will be shown now that $R^i f_*(\mathscr{N}_m) = 0$ for $i > 0$, $m \gg 0$. Indeed, $R^i f_*(\mathscr{N}_m) = R^i f_*\Big(f^*\big(f_*(\mathscr{L}^{\otimes m})^V\big) \otimes \mathscr{L}^{\otimes m} \otimes \mathscr{M}\Big) = f_*(\mathscr{L}^{\otimes m})^V \otimes R^i f_*(\mathscr{L}^{\otimes m} \otimes \mathscr{M})$ since $R^i f_*(\mathscr{L}^{\otimes m})$ is locally free as noted above. Since $R^i f_*(\mathscr{L}^{\otimes m} \otimes \mathscr{M}) = 0$ for $i > 0$ and $m \geq n_0$ (see the beginning of this proof), the sequence $(**, m)$ satisfies the conclusion of the proposition. □

§1. $K_0(T \times_S T)$ and endomorphisms of $K_*(T)$

The main result of this section is Theorem 1.8. We fix a quasicompact scheme S and a flat projective morphism $h \colon T \to S$ up to the end of this section.

1.1. Consider the following categories:

$\mathscr{P}(S)$, the category of locally free coherent sheaves on S;

$\mathscr{P}(T)$, the category of locally free coherent sheaves on T;

$\mathscr{P}(h, T)$, the full subcategory in $\mathscr{P}(T)$ consisting of such $\mathscr{F}$ that $R^i f_*(\mathscr{F}) = 0$ for $i > 0$.

If $\mathscr{F} \in \mathscr{P}(h, T)$, then $h_*(\mathscr{F}) \in \mathscr{P}(S)$ (see **[9,** (II.5.2)]). Therefore, one has a well-defined exact functor of direct images

$$h_* : \mathscr{P}(h, T) \longrightarrow \mathscr{P}(S).$$

This functor defines homomorphisms $K_i(\mathscr{P}(h, T)) \to K_i(S)$ $(i \geq 0)$. In view of **[10,** (7.2.7)]) and Proposition 0.1 the natural inclusion $\mathscr{P}(h, T) \hookrightarrow \mathscr{P}(T)$ induces isomorphisms $K_i(\mathscr{P}(h, T)) \xrightarrow{\sim} K_i(T)$ $(i \geq 0)$. Composing this isomorphism with the homomorphism $K_i(\mathscr{P}(h, T)) \to K_i(S)$ defined above, we obtain a homomorphism which will be denoted

$$(1.1.1) \qquad h_* : K_i(T) \longrightarrow K_i(S).$$

1.2. The functor $h^* : \mathscr{P}(S) \to \mathscr{P}(T)$ is exact; hence it induces homomorphisms $h^* : K_i(S) \longrightarrow K_i(T)$. The homomorphism h_* is a homomorphism of $K_0(S)$-modules; i.e.,

$$(1.2.1) \qquad h_*\big(h^*(a) \cdot x\big) = a \cdot h_*(x)$$

for any $a \in K_0(S)$, $x \in K_i(T)$. If the square

$$\begin{array}{ccc} T' & \xrightarrow{h'} & S' \\ {\scriptstyle g'}\downarrow & & \downarrow{\scriptstyle g} \\ T & \xrightarrow{h} & S \end{array}$$

is Cartesian and g is a flat morphism, then the base change formula holds:

(1.2.2) $$g^* \circ h^* = h'_* \circ (g')^* : K_i(T) \longrightarrow K_i(S').$$

1.3. DEFINITION. Define a $K_0(S)$-linear pairing $\langle \, , \, \rangle$ on $K_0(T)$ as follows:

$$\langle a, b \rangle = h_*(a \cdot b).$$

1.4. DEFINITION. Each element $z \in K_0(T \times_S T)$ defines homomorphisms

$$Z_* \colon K_i(T) \longrightarrow K_i(T)$$

by the formula $z_*(x) = p_{1,*}(p_2^*(x) \cdot z)$, where $p_1, p_2 \colon T \times_S T \to T$ are natural projections, $x \in K_i(T)$ (see also **[8]**).

1.5. It is clear that $(z_1 + z_2)_* = z_{1,*} + z_{2,*}$. It follows from (1.2.1) that the correspondence $z \to z_*$ defines a group homomorphism

$$K_0(T \times_S T) \longrightarrow \mathrm{End}_{K_0(S)}\big(K_*(T)\big).$$

1.6. LEMMA. *If $a, b \in K_0(T)$ and $a \otimes b \overset{\text{def}}{=} p_1^*(a) \cdot p_2^*(b) \in K_0(T \times_S T)$, then the homomorphism*

$$(a \boxtimes b)_* \colon K_i(T) \longrightarrow K_i(T)$$

is calculated by the formula

(1.6.2) $$(a \boxtimes b)_*(x) = h^* \big[h_*(b \cdot x)\big] \cdot a.$$

If $x \in K_0(T)$, then this formula may be rewritten in the following way:

(1.6.2) $$(a \boxtimes b)_*(x) = \langle b, x \rangle \cdot a.$$

PROOF. The statements of the lemma follow from (1.2.1), (1.2.2), and 1.3. □

1.7. REMARK. Suppose that the structure sheaf $\mathcal{O}_{\Delta(T/S)}$ of the diagonal $\Delta(T/S) \subset T \times_S T$ has a finite locally free resolution on $T \times_S T$. Then the sheaf $\mathcal{O}_{\Delta(T/S)}$ defines an element of $K_0(T \times_S T)$, which will be denoted $\big[\mathcal{O}_{\Delta(T/S)}\big]$. It is easy to see that for $i \geq 0$

$$\big[\mathcal{O}_{\Delta(T/S)}\big]_* = \mathrm{id}_{K_i(T)} \, .$$

1.8. THEOREM. *Let $f : X \to Y$ be a flat projective morphism of schemes. Suppose the structural sheaf $\mathcal{O}_{\Delta(X/Y)}$ of the diagonal $\Delta(X/Y) \subset X \times_Y X$ has a finite locally free resolution on $X \times_Y X$. Suppose the following relation holds in $K_0(X \times_Y X)$:*

$$\big[\mathcal{O}_{\Delta(X/Y)}\big] = \sum_{i=1}^{n} a_i \boxtimes b_i$$

for some $a_i, b_i \in K_0(X)$. Then the following hold:

a) *$K_0(X)$ is a finitely generated projective $K_0(Y)$-module;*

b) *each one of the families* $\{a_i\}_{i=1}^n$ *and* $\{b_j\}_{j=1}^n$ *generates the* $K_0(Y)$*-module* $K_0(X)$;

c) *the homomorphism* α *induced by the pairing* $\langle\ ,\ \rangle$ (*see* 1.3) *is an isomorphism*; $\alpha\colon K_0(X) \to \operatorname{Hom}_{K_0(Y)}(K_0(X), K_0(Y))$;

d) *the natural homomorphisms* $K_0(X) \otimes_{K_0(Y)} K_p(Y) \to K_p(X)$ *are isomorphisms.*

PROOF. Consider the following diagram

$$\begin{array}{ccc} K_*(Y)^n & \underset{\psi}{\overset{\varphi}{\rightleftarrows}} & K_*(X) \\ \wr\uparrow{\scriptstyle\rho'} & & \uparrow{\scriptstyle\rho} \\ K_0(Y)^n \otimes_{K_0(Y)} K_*(Y) & \underset{\psi_0\otimes \mathrm{id}}{\overset{\varphi_0\otimes \mathrm{id}}{\rightleftarrows}} & K_0(X)^n \otimes_{K_0(Y)} K_*(Y) \end{array}$$

where ρ' and ρ are natural, and φ and ψ are homomorphisms of $K_*(Y)$-modules defined in the following way: $\varphi(e_i) = a_i$, where $e_i = (0, \dots, 1, \dots, 0) \in K_*(Y)^n$, $\psi(x) = \sum_i f_*(b_i \cdot x) \cdot e_i$, finally $\varphi_0 = \varphi|_{K_0(Y)^n}$, $\psi_0 = \psi|_{K_0(X)}$.

It follows from the definitions of ρ, ρ', φ, ψ, φ_0, ψ_0 that $\psi \circ \rho = \rho' \circ (\psi_0 \otimes \mathrm{id})$, $\varphi \circ \rho' = \rho \circ (\varphi_0 \otimes \mathrm{id})$. We have $\varphi \circ \psi = \mathrm{id}$ and, hence, $\varphi_0 \circ \psi_0 = \mathrm{id}$. Indeed (see 1.5, 1.6.2, and 1.7),

$$(\varphi \circ \psi)(x) = \varphi\Big(\sum_i f_*(b_i \cdot x) \cdot e_i\Big) = \sum_i f^* f_*(b_i \cdot x) \cdot a_i$$
$$= \sum (a_i \boxtimes b_i)_*(x) = \left[\mathscr{O}_{\Delta(X/Y)}\right]_*(x) = x.$$

Let us prove 1.8a), 1.8b), and 1.8d). Statements 1.8a), 1.8b) follow from the equality $\varphi_0 \circ \psi_0 = \mathrm{id}$. It follows from the equalities $\varphi \circ \psi = \mathrm{id}$, $\varphi_0 \circ \psi_0 = \mathrm{id}$, $\psi \circ \rho = \rho' \circ (\psi_0 \otimes \mathrm{id})$, $\varphi \circ \rho' = \rho \circ (\varphi_0 \otimes \mathrm{id})$, and the bijectivity of ρ' that ρ is an isomorphism.

It remains to prove 1.8c). For each $x \in K_0(X)$ we have

$$(*) \qquad x = \left[\mathscr{O}_{\Delta(X/Y)}\right]_*(x) = \Big(\sum_i a_i \boxtimes b_i\Big)_*(x) = \sum_i \langle b_i, x\rangle \cdot a_i,$$

$$(**) \qquad x = \left[\mathscr{O}_{\Delta(X/Y)}\right]_*(x) = \Big(\sum_i b_i \boxtimes a_i\Big)_*(x) = \sum_i \langle a_i, x\rangle \cdot b_i.$$

Define $\beta\colon \operatorname{Hom}_{K_0(Y)}(K_0(X), K_0(Y)) \to K_0(Y)$ by $\beta(\xi) = \sum_i \xi(b_i) \cdot a_i$ and show that $\alpha \circ \beta = \mathrm{id}$ and $\beta \circ \alpha = \mathrm{id}$.

Now

$$(\beta \circ \alpha)(x) = \sum_i \langle b_i, x\rangle \cdot a_i = x$$

(see $(*)$) and

$$[(\alpha \circ \beta)(\xi)](x) = \Big[\alpha\Big(\sum_i \xi(b_i) \cdot a_i\Big)\Big](x) = \Big\langle \sum_i \xi(b_i) \cdot a_i, x \Big\rangle$$
$$= \sum_i \xi(b_i) \cdot \langle a_i, x\rangle = \sum_i \xi(b_i \cdot \langle a_i, x\rangle) = \xi\Big(\sum_i \langle a_i, x\rangle \cdot b_i\Big) = \xi(x).$$

Here the $K_0(Y)$-linearity of ξ and $\langle\ ,\ \rangle$ and the equality $(**)$ are used. This proves that α is an isomorphism. $\square$

1.9. REMARK. The statement 1.8d) was conjectured by A. A. Suslin. The proof of the statement 1.8c) presented here was communicated to the author by A. A. Beilinson. Projectivity of $K_0(X)$ over $K_0(Y)$ was remarked by O. T. Izboldin.

1.10. COROLLARY. *If under the conditions of Theorem* 1.8 *the family* $\{a_i\}_{i=1}^n$ *is a basis of* $K_0(X)$ *over* $K_0(Y)$, *then*

$$\langle a_i, b_i\rangle = \delta_{ij} = \begin{cases} 0, & i \neq j, \\ 1, & i = j, \end{cases}$$

and the family $\{b_j\}_{j=1}^n$ *is another basis of* $K_0(X)$ *over* $K_0(Y)$.

PROOF. In view of the equality $(*)$ from the proof of the theorem, we have $a_i = \sum_j \langle b_j, a_i\rangle \cdot a_j$. Since $\{a_i\}_{i=1}^n$ is a basis of $K_0(X)$ over $K_0(Y)$, it follows that $\langle b_j, a_i\rangle = \delta_{ij}$. As the family $\{b_j\}_{j=1}^n$ generates $K_0(X)$ over $K_0(Y)$ (see 1.8b)), $\{b_j\}_{j=1}^n$ is a basis. □

§2. K-theory of projective fibre bundles

In this section we reprove the following result of Quillen.

2.1. THEOREM (**[10]**). *Let* Y *be a quasicompact scheme,* $\mathscr{E}$ *a locally free sheaf of rank* n *on* Y, *and* $\mathbb{P}_Y(\mathscr{E})$ *the projective fibre bundle associated with* $\mathscr{E}$. *Then the natural homomorphism*

$$K_0(\mathbb{P}_Y(\mathscr{E})) \otimes_{K_0(Y)} K_*(Y) \longrightarrow K_*(\mathbb{P}_Y(\mathscr{E}))$$

is an isomorphism and $K_0(\mathbb{P}_Y(\mathscr{E})) = \overset{n-1}{\underset{i=0}{\oplus}} K_0(Y) \cdot [\mathscr{O}(-i)]$, *where* $\mathscr{O}(-1)$ *is the tautological locally free sheaf of rank* 1 *on* $\mathbb{P}_Y(\mathscr{E})$.

PROOF. Let $p\colon \mathbb{P}_Y(\mathscr{E}) \to Y$ be the natural projection and let $\alpha\colon p^*(\mathscr{E}^*) \to \mathscr{O}(1)$ be the canonical epimorphism. Put $J = \ker(\alpha)$. It is well known that $p_*\mathscr{O}(1) = \mathscr{E}^*$ and $p_*(J^*) = \mathscr{E}$. Denote $\mathbb{P}_Y(\mathscr{E})$ by $\mathbb{P}$. Then we have

$$H^0(\mathbb{P} \times_Y \mathbb{P}, \mathscr{O}(1) \boxtimes J^*) = H^0(Y, \mathscr{E}^* \otimes \mathscr{E}) = \mathrm{End}_Y(\mathscr{E}).$$

The section $s \in H^0(\mathbb{P} \times_Y \mathbb{P}, \mathscr{O}(1) \boxtimes J^*)$, corresponding to the unit operator, vanishes exactly along the diagonal $\Delta(\mathbb{P}/Y)$ and defines the Koszul resolution

$$(2.1.1) \qquad \{0 \longrightarrow \mathscr{O}(-n+1) \boxtimes \Lambda^{n-1} J \longrightarrow \dots \longrightarrow \mathscr{O}(-1) \boxtimes \Lambda^1 J \longrightarrow \mathscr{O} \boxtimes \mathscr{O}\}$$

of the sheaf $\mathscr{O}_{\Delta(\mathbb{P}/Y)}$. The resolution (2.1.1) together with Theorem 1.8d) proves the first statement.

To calculate $K_0(\mathbb{P}_Y(\mathscr{E}))$, consider the homomorphisms

$$\varphi\colon K_0(Y)^n \longrightarrow K_0(\mathbb{P}_Y(\mathscr{E})), \qquad \psi\colon K_0(\mathbb{P}_Y(\mathscr{E})) \longrightarrow K_0(Y)^n$$

defined in the following way: $\varphi(e_i) = [\mathscr{O}(-i+1)], \quad \psi(x) = \sum_{i=1}^n p_*(x \cdot [\mathscr{O}(i-1)])$.

We show that the homomorphism φ is an isomorphism.

The surjectivity follows from (2.1.1) and 1.8b). To prove the injectivity of φ we recall the following result:

2.2. LEMMA (**[5, (III, 5.1)]**). *Under the notation of Theorem* 2.1 *we have*

1) $\forall m > 0 \ \forall i \geq 0 \quad R^m p_*\mathscr{O}(i) = 0$;

2) $\forall i \geq 0 \quad p_*\mathscr{O}(i) = S^i\mathscr{E}^*$ (*ith symmetric power*);
3) $\forall m \geq 0 \ \forall i \quad \bigl(-(n-1) \leq i \leq -1 \Rightarrow R^m p_*\mathscr{O}(i) = 0\bigr)$.

The definition of the homomorphism $p_*\colon K_0(\mathbb{P}_Y(\mathscr{E})) \to K_0(Y)$ (see (1.1.1)), the projection formula (1.2.1), and Lemma 2.2 show now that the homomorphism $(\psi \circ \varphi)_{ij}\colon K_0(Y) \to K_0(Y)$ coincides with the multiplication by $[S^{i-j}\mathscr{E}^*]$ for $i \geq j$, and $(\psi \circ \varphi)_{ij} = 0$ for $i < j$. Hence the homomorphism $\psi \circ \varphi$ is determined by an $n \times n$ upper triangular matrix with units on the diagonal. In particular, $\psi \circ \varphi$ is an isomorphism and φ is a monomorphism. □

2.3. COROLLARY. *Under the conditions of Theorem* 2.1, *the family* $\{(-1)^k[\Lambda^k J]\}_{k=0}^{n-1}$ *is a basis of* $K_0(\mathbb{P}_Y(\mathscr{E}))$ *over* $K_0(Y)$, *dual to the basis* $\{[\mathscr{O}(-k)]\}_{k=0}^{n-1}$ *relatively the form* $\langle\ ,\ \rangle$ *from* 1.3.

PROOF. This follows from 2.1, (2.1.1), and 1.10. □

2.4. REMARK. On the category of regular sheaves on $\mathbb{P}_Y(\mathscr{E})$ the functor T_i from Quillen's calculation (see **[10,** §8]) coincides with the functor $\mathscr{F} \mapsto p_*(\mathscr{F} \otimes \Lambda^i J)$ and the canonical resolution of the regular sheaf $\mathscr{F}$ from **[10,** (8.1.11)] coincides with the complex $p_{1,*}\bigl(p_2^*(\mathscr{F}) \otimes (2.1.1)\bigr)$.

§3. A technical result

The main result of this section is Proposition 3.8. We fix some quasicompact scheme Y and a flat projective morphism $f\colon X \to Y$ up to the end of this section.

3.1. Let A be a sheaf of algebras on Y that is locally isomorphic to the sheaf of $M_n(\mathscr{O}_Y)$ on Y for the étale topology on Y; i.e., A is an Azumaya algebra on Y. Consider the following categories:

$\mathscr{P}(Y, A)$; its objects are sheaves of left A-modules, which are locally free coherent $\mathscr{O}_Y$-modules; its morphisms are morphisms of left A-modules;

$\mathscr{P}(X, A)$; its objects are sheaves of left f^*A-modules, which are locally free coherent $\mathscr{O}_X$-modules; its morphisms are morphisms of left f^*A-modules;

$\mathscr{P}(f, A)$; the full subcategory of $\mathscr{P}(X, A)$ consisting of $\mathscr{F}$ such that $R^i f_*(\operatorname{Res}\mathscr{F}) = 0$ for $i > 0$, where $\operatorname{Res}\colon \mathscr{P}(X, A) \to \mathscr{P}(X)$ is the forgetful functor.

3.2. DEFINITION. Put

$$K_*(X, A) = K_*\bigl(\mathscr{P}(X, A)\bigr), \qquad K_*(Y, A) = K_*\bigl(\mathscr{P}(Y, A)\bigr).$$

3.3. If $\mathscr{F}$ is a left f^*A-module, then the sheaf $f_*\mathscr{F}$ is equipped with a natural left A-module structure. In view of **[9,** (II.5.2)] $f_*(\operatorname{Res}\mathscr{F})$ is a locally free sheaf on Y. Hence, we have the functor

$$f_*^A\colon \mathscr{P}(f, A) \longrightarrow \mathscr{P}(Y, A).$$

It is evidently exact, so it defines homomorphisms

$$K_i\bigl(\mathscr{P}(f, A)\bigr) \longrightarrow K_i\bigl(\mathscr{P}(Y, A)\bigr) \qquad (i \geq 0).$$

As in 1.1, it may be shown that the natural inclusion $\mathscr{P}(f, A) \hookrightarrow \mathscr{P}(X, A)$ induces isomorphisms $K_i(\mathscr{P}(f, A)) \xrightarrow{\sim} K_i(X, A)$. Composing this isomorphism with the homomorphism $K_i(\mathscr{P}(f, A)) \to K_i(Y, A)$ defined above, we obtain a homomorphism which will be denoted

$$f_*^A \colon K_i(X, A) \longrightarrow K_i(Y, A).$$

The functor $f^* \colon \mathscr{P}(X, A) \to \mathscr{P}(X, A)$ is exact, so it defines a homomorphism, which will be denoted

$$f_A^* \colon K_i(Y, A) \longrightarrow K_i(X, A).$$

3.4. The following functors

$$\begin{aligned} \mathscr{P}(X, A^{\mathrm{op}}) \times \mathscr{P}(X, A) &\longrightarrow \mathscr{P}(X), \quad (M, N) \mapsto M \otimes_A N, \\ \mathscr{P}(X) \times \mathscr{P}(X, A) &\longrightarrow \mathscr{P}(X, A), \quad (\mathscr{F}, \mathscr{Y}) \mapsto \mathscr{F} \otimes \mathscr{Y} \end{aligned}$$

are biexact. Therefore, they induce products

$$\begin{aligned} K_0(X, A^{\mathrm{op}}) \otimes K_i(X, A) &\longrightarrow K_i(X) \qquad (i \geq 0), \\ K_0(X) \otimes K_i(X, A) &\longrightarrow K_i(X, A) \qquad (i \geq 0), \end{aligned}$$

which satisfy, in particular, the following projection formulas:

$$f_*\left(f_{A^{\mathrm{op}}}^*(\alpha) \otimes_{[A]} x\right) = \alpha \otimes_{[A]} f_*^A(x), \tag{3.4.1}$$

$$f_*^A(\beta \cdot f_A^*(y)) = f_*(\beta) \cdot y, \tag{3.4.2}$$

where $\alpha \in K_0(Y, A^{\mathrm{op}})$, $x \in K_i(X, A)$, $\beta \in K_0(X)$, $y \in K_i(Y, A)$.

3.5. Lemma. *Let the following diagram*

$$\begin{array}{ccc} X' & \xrightarrow{g'} & X \\ \downarrow{\scriptstyle f'} & & \downarrow{\scriptstyle f} \\ Y' & \xrightarrow{g} & Y \end{array}$$

be Cartesian with flat projective morphisms f, g. Then the morphisms f', g' are also flat projective and the following base change formula holds:

$$f_A^* \circ g_*^A = (g')_*^A \circ (f')_A^* \colon K_i(Y', A) \longrightarrow K_i(X, A) \qquad (i \geq 0).$$

3.6. Definition. Let $\alpha \in K_0(X, A^{\mathrm{op}})$ (see 3.1). Under the conditions of 3.5 define homomorphisms

$$\varphi_\alpha \colon K_*(Y, A) \longrightarrow K_*(X), \qquad \varphi'_\alpha \colon K_*(Y', A) \longrightarrow K_*(X')$$

by the formulas:

$$\varphi_\alpha(y) = \alpha \otimes_{[A]} f_A^*(y), \qquad \varphi'_\alpha(y') = (g')_{A^{\mathrm{op}}}^*(\alpha) \otimes_{[A]} (f')_A^*(y')$$

for each $y \in K_*(Y, A)$, $\quad y' \in K_*(Y', A)$.

3.7. Lemma. 1) $(g')_* \circ \varphi'_\alpha = \varphi_\alpha \circ g_*^A$; $\qquad$ 2) $\varphi'_\alpha \circ g_A^* = (g')^* \circ \varphi_\alpha$.

PROOF. Let $y' \in K_i(Y', A)$. From (3.4.1) we have

$$\begin{aligned}((g')_* \circ \varphi_\alpha)(y') &= (g')_*\left[(g')^*_{A^{\mathrm{op}}} \otimes_{[A]} (f')^*_A(y')\right] \\ &= \alpha \otimes_{[A]} (f^*_A \circ g^A_*)(y') = \varphi_\alpha(g^A_*(y')).\end{aligned}$$

Let $y \in K_i(Y, A)$; then we have

$$\begin{aligned}(\varphi^*_\alpha \circ g^*_A)(y) &= (g')^*_{A^{\mathrm{op}}}(\alpha) \otimes_{[A]} (g')^*_A(f^*_A(y)) \\ &= (g')^*(\alpha \otimes_{[A]} f^*_A(y)) = \left[(g')^* \circ \varphi_\alpha\right](y).\end{aligned}$$ □

3.8. In the following proposition we shall use the notation of 3.5.

PROPOSITION. *Let Y be a quasicompact scheme, $A_1, \dots, A_n$ Azumaya algebras on Y (see 3.1), and $\alpha_i \in K_0(X, A_i^{\mathrm{op}})$ $(i = 1, 2, \dots, n)$. Consider the two homomorphisms*

$$\varphi \colon \bigoplus_{i=1}^n K_*(Y, A_i) \longrightarrow K_*(X), \qquad \varphi' \colon \bigoplus_{i=1}^n K_*(Y', A_i) \longrightarrow K_*(X'),$$

defined by the formulas $\varphi = \sum_i \varphi_{\alpha_i}$, $\varphi' = \sum_i \varphi'_{\alpha_i}$ (see 3.6). Suppose that $g_*([\mathcal{O}_{Y'}]) = [\mathcal{O}_Y]$ and φ' is an isomorphism. Then under the conditions of 3.5 the homomorphism φ is an isomorphism.

PROOF. The following diagrams

$$\begin{array}{ccc} K_*(X') & \xleftarrow{(g')^*} & K_*(X) \\ \uparrow{\scriptstyle \varphi'} & & \uparrow{\scriptstyle \varphi} \\ \bigoplus_{i=1}^n K_*(Y', A_i) & \xleftarrow{g^*_\oplus} & \bigoplus_{i=1}^n K_*(Y, A_i) \end{array} \qquad \begin{array}{ccc} K_*(X') & \xrightarrow{g'_*} & K_*(X) \\ \uparrow{\scriptstyle \varphi'} & & \uparrow{\scriptstyle \varphi} \\ \bigoplus_{i=1}^n K_*(Y', A_i) & \xrightarrow{g^\oplus_*} & \bigoplus_{i=1}^n K_*(Y, A_i) \end{array}$$

where $g^*_\oplus = \bigoplus_{i=1}^n g^*_{A_i}$, $g^\oplus_* = \bigoplus_{i=1}^n g^{A_i}_*$, are commutative by 3.7. It is clear now that the proposition follows from Lemma 3.9. □

3.9. LEMMA. *Under the conditions of Proposition* 3.8 *the following hold*:
1) g'_* *is surjective*,
2) $g^*_\oplus$ *is injective.*

PROOF. It follows from the base change formula (1.2.2) that $(g')_*([\mathcal{O}_{X'}]) = [\mathcal{O}_X]$. By the projection formula (1.2.1) $(g')_*(g')^*(x) = g'_*([\mathcal{O}_{X'}]) \cdot x = [\mathcal{O}_X] \cdot x = x$. Therefore, $(g')_*$ is surjective. By the projection formula (3.4.2) $(g^\oplus_* \circ g^*_\oplus)(y) = g_*([\mathcal{O}_{Y'}]) \cdot y = [\mathcal{O}_Y] \cdot y = y$. Therefore, $g^\oplus_*$ is injective. □

§4. K-theory of Severi-Brauer schemes

In this section Quillen's result about the K-theory of Severi-Brauer schemes from his paper **[10]** will be reproved (see 4.9).

Fix a quasicompact scheme Y up to the end of this section. If T', T'' are schemes over Y, then we shall write $T' \times T''$ for $T' \times_Y T''$ throughout this section.

4.1. Let $f\colon X \to Y$ be a Severi-Brauer scheme over Y of relative dimension $(n-1)$; i.e., X is locally isomorphic to the Y-scheme $\mathbb{P}^{n-1}_Y$ for the étale topology on Y. Such schemes are essentially the same as Azumaya algebras of rank n^2 on Y **[3]**.

4.2. The following lemma is proved by the descent method.

4.3. LEMMA. *For each integer i there is a unique invertible sheaf $\mathcal{L}_i$ on $X \times X$ that satisfies the following conditions:*

1) *the sheaf $\mathcal{L}_i$ is locally isomorphic to the sheaf $\mathcal{O}(-i) \boxtimes \mathcal{O}(i)$ on $\mathbb{P}_Y^{n-1} \times \mathbb{P}_Y^{n-1}$ for the étale topology on Y;*
2) *$\Delta^*(\mathcal{L}_i) \cong \mathcal{O}_X$, where $\Delta\colon X \to X \times X$ is the diagonal embedding.*

4.4. LEMMA-DEFINITION. *Put $\mathcal{L} = \mathcal{L}_1$ and $I = (p_1)_*(\mathcal{L})$, where $p_1, p_2\colon X \times X \to X$ are the natural projections. Then the sheaf is locally free on X and is locally isomorphic to the sheaf $\mathcal{O}(-1)^n$ on $\mathbb{P}_Y^{n-1}$ for the étale topology on Y. The canonical homomorphism $v\colon p_1^* I \to \mathcal{L}$ is an epimorphism.*

PROOF. All statements in the lemma are local for the étale topology on Y. The proof of their local on Y variants follows from property 4.3.1. □

4.5. LEMMA-DEFINITION. *Put $A = f_*(\mathrm{End}_X(I)^{\mathrm{op}})$, where* op *denotes the opposed ring structure. Then A is an Azumaya algebra on Y of rank n^2. Moreover, the canonical homomorphism $f^*A \to \mathrm{End}_X(I)^{\mathrm{op}}$ is an isomorphism.*

PROOF. These statements are local for the étale topology on Y. Their local on Y variants follow from 4.3.1. □

4.6. LEMMA. *One has an isomorphism $p_2^*(I^{\otimes k}) \otimes_{A_k} p_1^*((I^{\otimes k})^*) \xrightarrow{\sim} \mathcal{L}^{\otimes(-k)}$, where $A_k = A^{\otimes k}$ and the structure of right (respectively left) A_k-module on the sheaf $p_2^*(I^{\otimes k})$ (respectively $p_1^*((I^{\otimes k})^*)$) is induced by the natural structure on the sheaf $I^{\otimes k}$ (see* 4.5*).*

PROOF. By 4.3 it is suffiicient to verify properties 4.3.1 and 4.3.2 for our sheaves. Evidently, these properties hold for $\mathcal{L}^{\otimes(-k)}$. First our sheaf is locally isomorphic to the sheaf $\mathcal{O}(k)^{(n^k)} \boxtimes_{M_{(n^k)}} \mathcal{O}(-k)^{(n^k)}$ on $\mathbb{P}_Y^{n-1} \times \mathbb{P}_Y^{n-1}$ (see 4.4, 4.5) for the étale topology on Y; i.e., it is locally isomorphic to the sheaf $\mathcal{O}(k) \boxtimes \mathcal{O}(-k)$ and the properties 4.3.1 hold for it. On the other hand, applying Δ^* to it we obtain the sheaf $I^{\otimes k} \otimes_{A_k} (I^{\otimes k})^*$, which is isomorphic to $\mathcal{O}_X$. □

4.7. PROPOSITION. *The epimorphism $v\colon p_1^* I \to \mathcal{L}$ from* 4.4 *determines a morphism $t\colon X \times X \to \mathbb{P}_X(I^*)$ of schemes over X such that $t^*\mathcal{O}(1) \cong \mathcal{L}$, where the scheme $X \times X$ is considered as a scheme over X by the first projection p_1. Moreover, t is an isomorphism.*

PROOF. The existence of the morphism t of schemes over X with the condition $t^*\mathcal{O}(1) \cong \mathcal{L}$ follows from the universal property of projective bundles (see [**5**, II, 7.1]). The statement that t is an isomorphism is local for the étale topology on Y. Therefore one can put $X = \mathbb{P}_Y^{n-1}$, $\mathcal{L} = \mathcal{O}(-1) \boxtimes \mathcal{O}(1)$, $I = \mathcal{O}(-1)^n$, and one can consider that the homomorphism $v\colon p_1^* I \to \mathcal{L}$ coincides with the natural epimorphism $\mathcal{O}(-1) \boxtimes \mathcal{O}^n \to \mathcal{O}(-1) \boxtimes \mathcal{O}(1)$. Now it is evident that t is an isomorphism. □

4.8. COROLLARY. *If the space Y is quasicompact, then the homomorphism $\varphi' = \sum_{i=0}^{n-1} \varphi_i'\colon \bigoplus_{i=0}^{n-1} K_*(X, A^{\otimes i}) \to K_*(X \times X)$ is an isomorphism, where φ_i' is induced by the exact functor $\mathcal{P}(X, A^{\otimes i}) \to \mathcal{P}(X \times X)$, defined by $\mathcal{F} \mapsto p_2^*(I^{\otimes i}) \otimes_{A_i} p_1^*(\mathcal{F})$, where $A_i = A^{\otimes i}$.*

PROOF. For each i consider the Morita equivalence $\eta_i\colon \mathcal{P}(X) \to \mathcal{P}(X, A^{\otimes i})$, defined by $\mathcal{Y} \mapsto (I^{\otimes i})^* \otimes \mathcal{Y}$. To prove that φ' is an isomorphism it is sufficient to prove that the composition $\varphi' \circ \eta_*$ is isomorphism, where $\eta_* = \bigoplus_{i=0}^{n-1} \eta_{i,*}$. We see that the functor $\varphi' \circ \eta_i$ coincides with tensoring with the sheaf $p_2^*(I^{\otimes i}) \otimes_{A_i} p_1^*((I^{\otimes i})^*)$ (here $A_i = A^{\otimes i}$); i.e., it coincides with tensoring with the sheaf $\mathscr{L}^{\otimes(-i)}$ (see 4.6). Since $t^*\mathscr{O}(-1) \simeq \mathscr{L}^{\otimes(-1)}$, the composition $(t^{-1})^* \circ (\varphi' \circ \eta_*)\colon \bigoplus_{i=0}^{n-1} K_*(X) \to K_*(\mathbb{P}_X(I^*))$ coincides with the isomorphism from Theorem 2.1. Using 4.7, we conclude that $\varphi' \circ \eta_*$ is an isomorphism. □

4.9. THEOREM ([**10**]). *Let Y be a quasicompact scheme, $f\colon X \to Y$ a Severi-Brauer scheme of relative dimension $(n-1)$ over Y, and A the Azumaya algebra on Y associated with f (see 4.5). Then one has an isomorphism*

$$\bigoplus_{i=0}^{n-1} K_*(Y, A^{\otimes i}) \longrightarrow K_*(X).$$

PROOF. Consider the Cartesian square

$$\begin{array}{ccc} X \times X & \xrightarrow{p_2} & X \\ \downarrow{\scriptstyle p_1} & & \downarrow{\scriptstyle f} \\ X & \xrightarrow{f} & Y \end{array}$$

and define two homomorphisms (see 3.6, 3.8)

$$\varphi\colon \bigoplus_{i=0}^{n-1} K_*(Y, A^{\otimes i}) \longrightarrow K_*(X), \qquad \varphi'\colon \bigoplus_{i=0}^{n-1} K_*(X, A^{\otimes i}) \longrightarrow K_*(X \times X)$$

by the formulas $\varphi = \sum \varphi_{[I^{\otimes i}]}$, $\varphi' = \sum \varphi'_{[I^{\otimes i}]}$, where I is the sheaf from 4.4 and $I^{\otimes i}$ is its tensor power equipped with the natural right $A^{\otimes i}$-module structure (see 4.5). We show that φ is an isomorphism. By 3.8 and 4.8 it is sufficient to verify that $f_*([\mathscr{O}_X]) = [\mathscr{O}_Y]$. The latter follows from the definition of the homomorphism $f_*\colon K_0(X) \to K_0(Y)$ (see 1.1) and Lemma 4.10. □

4.10. LEMMA. *The canonical homomorphism $\mathscr{O}_Y \to f_*(\mathscr{O}_X)$ is an isomorphism and $R^i f_*(\mathscr{O}_X) = 0$ for $i > 0$.*

PROOF. These statements are local for the étale topology on Y. Their local on Y variants follow from the same properties of the natural projection $\mathbb{P}_Y^{n-1} \to Y$. □

4.11. REMARK. The sheaf I on X from 4.4 coincides with the sheaf J on X from Quillen's calculation [**10,** (8.4)]. Hence the isomorphism φ from 4.9 coincides with the Quillen isomorphism from [**10,** (8.4)].

§5. Main results

In this section we formulate results that may be proved by the methods of the present paper.

5.1. Let G be one of the following group schemes over $\operatorname{Spec}\mathbb{Z}$: SL_n, SO_{2n+1}, Sp_{2n}, or SO_{2n}. Let $T \subset G$ be a maximal untwisted torus, $P \subset G$ a parabolic subgroup of G containing T, and $W(G)$ and $W(P)$ the Weyl groups of G and P, correspondingly. For each general G-bundle $f : X \to Y$ over an arbitrary scheme Y consider the bundle:

$$g\colon X/P \longrightarrow Y.$$

This is the generalized flag bundle over Y associated with the general G-bundle $f\colon X \to Y$.

THEOREM. *If Y is a quasicompact scheme, then the natural homomorphism*

$$K_0(X/P) \otimes_{K_0(Y)} K_*(Y) \xrightarrow{\sim} K_*(X/P)$$

is an isomorphism, and moreover, $K_0(X/P)$ is a free $K_0(Y)$-module of the rank Card $W(G)/$Card $W(P)$.

5.2. Let $\underset{\sim}{PGL}(n)$ be a group sheaf for the étale topology on Y, defined by $u \mapsto PGL(n, \Gamma(u, \mathscr{O}_Y))$. Let $\gamma \in H^1_{\text{et}}(Y, \underset{\sim}{PGL}(n))$ be a 1-cocycle, and $\mathscr{F}(i_1, i_2, \ldots, i_k; \gamma)$ the twisted form of the flag fibre bundle $\mathscr{F}(i_1, i_2, \ldots, i_k; \mathscr{O}^n_Y)$ corresponding to the 1-cocycle γ and the sequence of integers $\underline{i} = (i_1, i_2, \ldots, i_k)$ $(1 \le i_1 \le i_2 \le \cdots \le i_k \le n-1)$ (see **[3]**). Let A be the Azumaya algebra on Y corresponding to the 1-cocycle γ (see **[3]**).

THEOREM. *Let Y be a quasicompact scheme; then one has an isomorphism*

$$\bigoplus_{(\alpha^1, \alpha^2, \ldots, \alpha^k)} K_*(Y, A^{\otimes|\alpha|}) \xrightarrow{\sim} K_*(\mathscr{F}(i_1, i_2, \ldots, i_k; \gamma)),$$

where the sum is taken over all k-tuples $(\alpha^1, \ldots, \alpha^k)$ such that the nonincreasing sequences α^j satisfy the following conditions:

$$n - i_j \ge \alpha^j_1 \ge \alpha^j_2 \ge \cdots \ge \alpha^j_{i_j - i_{j-1}} \ge \alpha^j_{i_j - i_{j-1}+1} = 0$$

(here take $i_0 = 0$) and where $|\alpha| = \sum_j |\alpha^j|$, $|\alpha^j| = \sum_m \alpha^j_m$.

5.3. All results of this paper still hold if one changes in its formulations the groups $K_*(X)$ and $K_*(Y)$ to the groups $K_*(X, A)$ and $K_*(Y, A)$ for an arbitrary Azumaya algebra A on Y. For example, the following result holds.

THEOREM. *Let Y be a quasicompact scheme, A an Azumaya algebra on Y. Under the conditions of Theorem* 1.8 *the homomorphism*

$$K_0(X) \otimes_{K_0(Y)} K_*(Y, A) \longrightarrow K_*(X, A)$$

is an isomorphism.

References

1. A. A. Beilinson, *Coherent sheaves on $\mathbb{P}^n$ and problems of linear algebra*, Funktsional Anal. i Prilozhen. **12** (1978), no. 3, 68–69; English transl. in Functional Anal. Appl. **12** (1978).
2. I. N. Bernstein, I. M. Gelfand, and S. I. Gelfand, *Algebraic vector bundles on $\mathbb{P}^n$ and problems of linear algebra*, Funktsional Anal. i Prilozhen. **12** (1978), no. 3, 66–67; English transl. in Functional Anal. Appl. **12** (1978).
3. A. Grothendieck, *Le group de Brauer.* I, North-Holland, Amsterdam, 1968.
4. A. Grothendieck, P. Berthelot, and L. Illusie, *Théorie des intersections et théorème de Riemann-Roch*, Lecture Notes in Math., vol. 225, Springer-Verlag, Berlin and New York, 1971.
5. R. Hartshorne, *Algebraic geometry*, Springer-Verlag, Berlin and New York, 1977.

6. M. M. Kapranov, *Derived category of coherent sheaves on Grassmann manifolds*, Izv. Akad. Nauk SSSR Ser. Mat. **48** (1984), no. 1, 192–202; English transl. in Math. USSR-Izv. **24** (1985).
7. ———, *On the derived category of coherent sheaves on some homogeneous spaces*, Invent. Math. **92** (1988), 479–508.
8. Yu. I. Manin, *Correspondences, motifs and monoidal transformations*, Mat. Sb. **77** (1968), no. 4, 475–507; English transl. in Math. USSR-Sb. **6** (1968).
9. D. Mumford, *Abelian varieties*, Oxford Univ. Press, Oxford, 1970.
10. D. Quillen, *Higher algebraic K-theory*. I, Lecture Notes in Math., vol. 341, Springer-Verlag, Berlin and New York, 1973, pp. 85–147.
11. J.-P. Serre, *Groupes de Grothendieck des schemas en groupes reductifs déployés*, Inst. Hautes Études Sci. Publ. Math. **34** (1968), 37–52.
12. R. G. Swan, *K-theory of quadric hypersurfaces*, Ann. of Math. (2) **122** (1985), 113–153.
13. I. A. Panin, *Algebraic K-theory of Grassmannian manifolds and their twisted forms*, Funktsional Anal. i Prilozhen. **23** (1989), no. 2, 71–72; English transl. in Functional Anal. Appl. **23** (1989).
14. ———, *On algebraic K-theory of generalized flag fibre bundles and some of their twisted forms*, Adv. Soviet Math., vol. 4, Amer. Math. Soc., Providence, RI, 1991, pp. 21–46.
15. A. A. Suslin, *K-theory and K-cohomology of certain group varieties*, Adv. Soviet Math., vol. 4, Amer. Math. Soc., Providence, RI, 1991, pp. 53–74.

Amer. Math. Soc. Transl.
(2) Vol. **163**, 1995

On Mappings Preserving Convexity

Anna V. Shaidenko-Künzi

Mappings preserving the convexity of sets in Euclidean spaces have been investigated in many papers. It is interesting to reduce the family of the sets whose images are required to be convex. Theorem 1 and the proposition below relate to this question. On the other hand, it is interesting to consider sets more general than convex ones, namely, unions of convex sets. Theorem 2 gives complete conditions under which the preservation of this property of sets by an injective mapping implies the affinity of this mapping.

We consider n-dimensional Euclidean space E^n. Further we use the following notation: $\mathbb{N}$ and $\mathbb{R}$ are the sets of natural and real numbers, respectively; $]x, y[$ and $[x, y]$ are, respectively, the open and the closed interval with the ends x and y; ρ is the distance; $\operatorname{diam} X$, $\operatorname{int} X$, $\bar{X}$, ∂X, $\operatorname{con} X$, and $A(X)$ denote, respectively, the diameter, the interior, the closure, the boundary, the convex hull, and the affine hull of a set X; $D^n(x, r)$ is the closed n-dimensional ball with center x and radius r. In the proof we use the following

THEOREM K (KUZ′MINYH). *An injective mapping $f: E^n \to E^n$ $(n \geq 2)$ preserving collinearity is affine if $f(E^n)$ does not lie on a line.*

Theorem K follows obviously from a theorem in [**1**].

We call a set in E^n a *cylinder* if it can be represented as the product of a subset of a hyperplane and a convex subset of a line not lying in this hyperplane.

THEOREM 1. *Let $M \subset E^n$ $(n \geq 2)$ be a convex set and suppose that* $\operatorname{int} M$ *is not a cylinder. Then a bijective mapping $f: E^n \to E^n$ such that for any $\kappa \in \mathbb{R} \setminus \{0\}$ and any $O \in E^n$ the set $f(H_O^\kappa(M))$ is convex (where H_O^κ denotes the homothety with coefficient κ and center O) is an affine mapping.*

A natural question is whether it is possible to transfer this theorem to the case of convex surfaces, but the following example shows that this cannot be done automatically. The transformation $f: E^2 \to E^2$ given by the formula $f(x, y) = (x, y + x^2)$ is not affine, though it satisfies the conditions of Theorem 1 and transforms all parabolas of the kind $y = ax^2 + bx + c$ into parabolas (and straight lines) of the same kind. However, if the convex surface M is bounded, the conclusion of the theorem stays valid and, more precisely, the following holds.

1991 *Mathematics Subject Classification*. Primary 52A20.

PROPOSITION. *Let $M \subset E^n$ $(n \geq 2)$ be a bounded convex surface (i.e., M is bounded,* $\operatorname{int}\operatorname{con} M \neq \varnothing$*, and* $M = \partial(\operatorname{con} M)$*) such that* $\operatorname{int}\operatorname{con} M$ *is not a cylinder. Then a bijective mapping $f : E^n \to E^n$ such that for any $\kappa \in \mathbb{R}\backslash\{0\}$ and any $O \in E^n$ the set $f\left(H_O^\kappa(M)\right)$ is the boundary of a convex set, is affine.*

Mappings of finite unions of convex sets are described by the following theorem.

THEOREM 2. *Let p, q, and n be natural numbers, $n \geq 2$. Then every injective mapping of E^n into E^n that maps any union of p convex sets onto a union of q convex sets is affine if and only if $q < 2p$.*

PROOF OF THEOREM 1. Evidently, $\operatorname{int} M \neq \varnothing$ because the empty set is a cylinder too. Further, we need the following lemmas.

LEMMA 1. *The set M has n tangent hyperplanes in general position (i.e., the vectors of the normals to these hyperplanes are linearly independent).*

PROOF. The set of points at which a tangent hyperplane exists is dense in ∂M. We shall prove that among all tangent hyperplanes it is possible to choose n of them in general position. The first one we take arbitrary and denote it P_1. We take a point $A \in \operatorname{int} M$ and a line l_1 so that $A \in l_1$, $l_1 \| P_1$ and also we take a point $B_1 \in l_1 \cap \partial M$. If a hyperplane of support at B_1 is tangent we denote it by P_2, and if it is not unique we take a ball S with center A and radius ε such that $S \subset \operatorname{int} M$. In the ε-neighborhood of B_1 there is a point B_1' at which a tangent hyperplane exists. We denote this hyperplane by P_2. Clearly, $P_1 \nparallel P_2$ since otherwise $P_2 \cap \operatorname{int} M \neq \varnothing$. Now we take a line l_2 with $A \in l_2$ and $l_2 \| P_2 \cap P_1$ and get in the way just described a tangent hyperplane P_3, and so on. Continuing this process we take finally a line l_{n-1} with $A \in l_{n-1}$, $l_{n-1} \| P_{n-1} \cap \cdots \cap P_2 \cap P_1$, and get a hyperplane P_n. So n tangent hyperplanes in general position are constructed, and Lemma 1 is proved. □

Further, without loss of generality, we can assume the hyperplanes $P_1, \ldots, P_n$ of Lemma 1 mutually perpendicular; i.e., $P_i \perp P_j$ for all i, j $(i \neq j,\ 1 \leq i, j \leq n)$. We denote by $N_1, \ldots, N_n$ the basis of E^n consisting of vectors normal to $P_1, \ldots, P_n$, and the coordinate hyperplanes determined by this basis are denoted by $P_1', \ldots, P_n'$, so that $N_i \perp P_i'$ for all i $(1 \leq i \leq n)$; thus $P_i' \| P_i$.

LEMMA 2. *The normal vectors to all tangent hyperplanes of M (except those collinear with the vectors of the basis) do not all lie in any hyperplane P_i'.*

PROOF. Suppose the contrary, i.e., that all these vectors lie in some P_i'. Denote by M' the image of M under the orthogonal projection $p \colon E^n \to P_i$. Consider a point $X' \in \operatorname{int} M'$ (where $\operatorname{int} M'$ is the interior of M' in the topology of the hyperplane P_i) such that there is a point $X \in \partial M$, $X \neq X'$ and $p(X) = X'$. If such an X' does not exist, then $\operatorname{int} M$ is a cylinder, namely, the product of $\operatorname{int} M'$ and the ray $A(N_i)$ lying on the same side of P_i as M. Hence, the point X' exists. Take a neighborhood U' (in the hyperplane P_i) of X' such that $U' \subset \operatorname{int} M'$ and a neighborhood U of X such that $P(U) \subset U'$. There is a point $Y \in U$ at which a tangent hyperplane exists; denote this hyperplane by P. Then $P \| P_i$, since otherwise P contains a line l such that $l \| A(N_i)$, $l \cap U' \neq \varnothing$, i.e., $P \cap \operatorname{int} M \neq \varnothing$, which is impossible. All points of $\partial M \backslash P_i$ that are projected (perpendicularly to P_i) onto $\operatorname{int} M'$ lie on P since a convex set has at most two parallel tangent hyperplanes. Thus, $\operatorname{int} M$ is a cylinder in contradiction to the theorem; so Lemma 2 is proved. □

Now let $A \in \partial M$ and P_A be the tangent hyperplane at the point A. Denote the open half-spaces determined by P_A by P_A^+ and P_A^-, respectively, so that $\operatorname{int} M \subset P_A^+$. Then

$$\bigcup_{1<p<\infty} \bigcap_{p\leq\kappa<\infty} H_A^\kappa(M) = P_A^+ \cup X_A^+,$$

where $X_A^+ \subset P_A$. This implies that $f(P_A^+ \cup X_A^+)$ is convex because for any $1 < p < \infty$ the set $f\left[\bigcap_{p\leq\kappa<\infty} H_A^\kappa(M)\right]$ is convex (since for all κ the sets $f\left(H_A^\kappa(M)\right)$ are convex), and if $1 < p_1 < p_2$, then

$$\bigcap_{p_1\leq\kappa<\infty} H_A^\kappa(M) \subset \bigcap_{p_2\leq\kappa<\infty} H_A^\kappa(M).$$

Analogously,

$$\bigcup_{-\infty<p<-1} \bigcap_{-\infty<\kappa\leq p} H_A^\kappa(M) = P_A^- \cup X_A^-,$$

where $X_A^- \subset P_A$, and $f\left(P_A^- \cup X_A^-\right)$ is convex. For any $\kappa \neq 1$ and for any translation $t\colon E^n \to E^n$ there exists a point O such that $H_O^\kappa(M) = t\left(H_A^\kappa(M)\right)$. Therefore, the images of all sets parallel to $P_A^+ \cup X_A^+$ or to $P_A^- \cup X_A^-$ are convex too. Now take the hyperplanes $P_1, \dots, P_n$. Consider a hyperplane $P \| P_i$ (where $1 \leq i \leq n$). The images of all sets parallel to $P_i^+ \cup X_i^+$ or to $P_i^- \cup X_i^-$ (where X_i^+ and X_i^- are the sets for P_i obtained in analogous way as the sets X_A^+ and X_A^- for P_A) are convex. Take two sequences of hyperplanes $\{Q'_j\}_{j=1}^\infty$ and $\{Q''_j\}_{j=1}^\infty$, where $Q'_j \| P \| Q''_j$, $Q'_j \subset P^-$, and $Q''_j \subset P^+$ for any natural j and $\{Q'_j\}_{j\to\infty} \to P$, $\{Q''_j\}_{j\to\infty} \to P$ (the half-spaces P^+ and P^- are obtained, respectively, from P_i^+ and P_i^- by the translation which maps P_i onto P). Then P is the intersection of all the sets obtained from $P_i^+ \cup X_i^+$ by the translations that map P_i onto every Q'_j and of all the sets obtained from $P_i^- \cup X_i^-$ by the translations that map P_i onto every Q''_j. Hence $f(P)$ is convex as an intersection of convex sets. (Note that the convexity of the sets $f(P^+)$, $f(P^-)$, $f(\overline{P^+})$, and $f(\overline{P^-})$ also follows from this construction.)

All hyperplanes that are parallel to some P_i are called *principal*; all planes of smaller dimensions (including lines) that are intersections of the principal hyperplanes are also called *principal*. The images of all principal planes are convex as intersections of convex sets.

The set $f(P)$ has no interior points. Indeed, if a point fO is interior, $O \in P$, then there is a ball B with center fO, $B \subset f(P)$. On the other hand, there is a principal line l for which $l \cap P = O$. However $f(l) \cap f(P) \supset f(l) \cap B$, so that $f(l) \cap f(P)$ consists of more than one point, in contradiction to the injectivity of f. Thus, $f(P)$ lies in a hyperplane. Continuing further in the same way, we show that any principal plane is mapped onto a plane of the same dimension.

From here we obtain that any ray of a principal line is mapped onto a convex subset of a line, since such a ray is the intersection of a principal line with a half-space bounded by a principal hyperplane nonparallel to this line. Thus, the image of any interval of a principal line is also convex.

It follows from this that f preserves the order of points on a principal line. Indeed, suppose it is not so; i.e., on our line there are points A, B, C with the property $C \in [A, B]$ and $fB \in [fA, fC]$. But $f([A, C])$ is a convex subset of a line, $B \notin [A, C]$, so $fB \notin [fA, fC]$, a contradiction. Thus $f|_l$ (where l is a principal line) is continuous and, hence, a homeomorphism. Therefore $f(l)$ is an open convex subset of a line.

Now we show that $f(l)$ is a whole line. Indeed, suppose there is a point $fX \in A(f(l))$ such that $fX \notin f(l)$. There exists a principal hyperplane P with $X \in P$ and a point $X' \in l$ exists for which $P \cap l = X'$. As $f(P)$ is convex, $f(P) \supset [fX, fX']$ and so $f(P) \cap f(l)$ has more than one point, in contradiction to the injectivity of f.

The image of any principal two-dimensional plane is a two-dimensional plane. Indeed, a principal two-dimensional plane contains two intersecting principal lines. Hence, the image of such a plane is a convex subset of a two-dimensional plane containing two intersecting lines; i.e., it is a whole two-dimensional plane. Continuing further in this way, we can get that any principal plane of any dimension is mapped onto a plane of the same dimension. The injectivity of f also implies that f preserves parallelism of principal planes.

By *good* planes of all dimensions (including lines) we mean the planes that are intersections of hyperplanes parallel to any tangent hyperplanes to M. Clearly, good planes have all properties just proved for principal ones. In addition, we need

LEMMA 3. *If a κ-dimensional $(1 < \kappa \leq n)$ coordinate plane P contains a good $(\kappa - 1)$-dimensional plane Q (not a coordinate one) in general position with any $\kappa - 1$ of the coordinate $(\kappa - 1)$-dimensional planes, then the mapping $f|_{l_i}$ is affine for any coordinate line $l_i \subset P$.*

PROOF. Take a two-dimensional coordinate plane $R \subset P$ and let $l_1, l_2 \subset P$ be coordinate lines. We show that $Q \cap R$ is a line nonparallel to l_1 and l_2. Obviously, $Q \cap R$ cannot be a point. Assume that Q is parallel to one of the lines l_1 or l_2, for instance, $Q \| l_2$. Consider, then, all coordinate $(\kappa - 1)$-dimensional planes lying in P and containing l_1 (there are exactly $\kappa - 1$ of them). Their normals are perpendicular to l_1, but the normal to Q is also perpendicular to l_1, which contradicts the condition that Q is in general position with them. So $l := Q \cap R$ is a line and $l \nparallel l_1$ and $l \nparallel l_2$.

Now we show that $f|_{l_i}$ $(i = 1; 2)$ is affine. Consider a triangle T with its sides lying on the lines l_1, l_2, and l_3 respectively (without loss of generality we can assume that Q does not contain the origin of coordinates). Since the mapping f preserves the parallelism of good lines, we conclude that f preserves the mid-points of the sides of the triangle T. Consequently, on the dyadic rational lattices lying on the lines l_1, l_2, l and having the vertices of triangle T as points 0 and 1, the mapping f is affine. And from the continuity of f on good lines it follows that $f|_{l_1}$, $f|_{l_2}$, and $f|_l$ are affine. This means that $f|_{l_i}$ for any coordinate line l_i is affine, since the plane R is chosen arbitrary. Lemma 3 is proved. □

Consider now the set of normals to all tangent hyperplanes of M. The set $\mathcal{N}$ of those of them that are noncollinear with the vectors $N_1, \dots, N_n$ is not empty, as follows from Lemma 2. For any vector $a \in \mathcal{N}$ denote by $X(a)$ the following coordinate plane: $a \in X(a)$ but a does not lie in any coordinate plane of smaller dimension. For any basis vector N_i there is a vector $a \in \mathcal{N}$ such that $N_i \in X(a)$. Indeed, if for some N_i this is not so, then all $\mathcal{N}$ lies in the coordinate hyperplane P_i' and that contradicts Lemma 2. Consider some vector N_i and the corresponding vector a. Let K be a tangent hyperplane with normal a. Then, taking in Lemma 3 the plane $X(a)$ as P and the plane $X(a) \cap K$ as Q, we obtain that $f|_{A(N_i)}$ is affine. And so it is for all basis vectors.

Take an affine mapping $\varphi\colon E^n \to E^n$ such that for any coordinate line $A(N_i)$ the mapping $\varphi \circ f|_{A(N_i)}$ is the identity. Since $\varphi \circ f$ preserves parallelism of all good planes, for any principal hyperplane Q' we have $\varphi \circ f(Q') = Q'$. And as any point of E^n is the intersection of n principal hyperplanes, the mapping $\varphi \circ f$ is the identity and so the mapping f is affine. The theorem is proved. □

A. D. Aleksandrov has observed that in this theorem it is sufficient to require the convexity of the sets $f(H_O^\kappa(M))$ not for all $\kappa \in \mathbb{R}\backslash\{0\}$ but only for $\kappa \in \{\kappa_i\}_{i\in\mathbb{N}} \cup \{\kappa_i'\}_{i\in\mathbb{N}} \subset \mathbb{R}\backslash\{0\}$, where $\{\kappa_i\}_{i\to\infty} \to \infty$ and $\{\kappa_i'\}_{i\to\infty} \to -\infty$.

The condition that $\operatorname{int} M$ be not a cylinder is essential since the mapping $f: E^n \to E^n$, given by the formula $f(x_1, \dots, x_{n-1}, x_n) = (x_1, \dots, x_{n-1}, x_n^3)$, is not affine but it is bijective and it maps any closed or open convex cylinder, the foundation of which is lying in a hyperplane parallel to the hyperplane $x_n = 0$, onto a convex cylinder.

The example of A. V. Kuz′minyh shows that the injectivity of f cannot be dropped; this is an example of an everywhere discontinuous surjective mapping $\psi: E^2 \to E^2$ preserving the convexity of all convex sets. Indeed, as is known **[2]** there exists a function $\varphi\colon E^1 \to E^1$ that maps any nonempty (open) interval onto E^1. Let $\chi\colon E^1 \to E^2$ be a surjective mapping and $\xi\colon E^2 \to E^1$ the mapping given by the formula $\xi(x, y) = y - x^2$. The mapping $\psi = \chi \circ \varphi \circ \xi$ is the desired one.

That the requirement of surjectivity of f is essential is shown by the following example. Let $f: E^2 \to E^2$ be given by the formula $f(x, y) = (e^x, e^y)$. Then $f(E^2) = \{(x, y) \in E^2 \mid x > 0, y > 0\}$; the vertical (respectively horizontal) lines are mapped onto the vertical (respectively horizontal) open rays and the lines $y = x + c$ onto the open rays coming from the origin of coordinates; the image of any triangle homothetic to the triangle $\operatorname{con}\{(0,0), (1,0), (0,-1)\}$ is a triangle. This mapping f is not surjective and although it satisfies all other conditions of Theorem 1, it is not affine.

PROOF OF THE PROPOSITION. We introduce the notation

$$\mathscr{M} := \{\, H_O^\kappa(M) \mid \kappa \in \mathbb{R}\backslash\{-1, 0, 1\},\ O \in E^n \,\}.$$

The family of surfaces $P \in \mathscr{M}$ for which $f(P)$ is a convex surface (i.e., $f(P)$ is not a proper subset of a hyperplane) is denoted by $\mathscr{M}_1$, $\mathscr{M}_1 \subset \mathscr{M}$. In the proof of the proposition we shall use the lemma that the family $\{\, \operatorname{int}\operatorname{con} P \mid P \in \mathscr{M}_1 \,\}$ is a base of the topology in E^n; more precisely, the following will be proved:

LEMMA. *For any point $X \in E^n$ and for any neighborhood U of it, a surface $\Pi \subset \mathscr{M}_1$ exists such that $X \in \operatorname{int}\operatorname{con}\Pi$ and $\Pi \subset U$.*

Using this lemma we prove the proposition. A convex surface P divides $E^n\backslash P$ into two connected components. If P is not a hyperplane, we call $\operatorname{int}\operatorname{con} P$ the *interior* of the surface P and the other component the *exterior* (if P is a hyperplane, then we call the open half-spaces determined by it the interior and exterior of P).

For any surface $P \in \mathscr{M}$ and for any two points X_1 and X_2 belonging to the same connected component of $E^n\backslash P$ there exists a "chain" of surfaces from $\mathscr{M}$ (i.e., a set of surfaces $P_1, \dots, P_N \in \mathscr{M}$ for some natural N with the property: for any $1 \le i < N$ we have $P_i \cap P_{i+1} \neq \varnothing$) such that $X_1 \in P_1$, $X_2 \in P_N$, and $P \cap (\cup_{i=1}^N P_i) = \varnothing$.

Let $P \in \mathscr{M}_1$; we prove that then a connected component of $E^n\backslash P$ is mapped onto a connected component of $E^n\backslash f(P)$. Consider points X_1, X_2 in one component of $E^n\backslash P$ and the corresponding "chain" $P_1, \dots, P_N$. The only disconnected boundary of a convex set is a join of two parallel hyperplanes. Suppose $\cup_{i=1}^N f(P_i)$ is connected. Then fX_1 and fX_2 lie in one connected component of $E^n\backslash f(P)$ because of the connectedness of $\cup_{i=1}^{N} f(P_i)$ and the injectivity of f. The case when $\cup_{i=1}^{N} f(P_i)$ is disconnected and the points fX_1 and fX_2 lie in different connected components of $E^n \setminus f(P)$ is impossible. Indeed, this would mean that for some $1 \leq \kappa \leq N$ the set $f(P_\kappa)$ is a join of two parallel hyperplanes Q_1 and Q_2 lying in different components of $E^n \setminus f(P)$, and if this is so, then $f(P)$ can only be either a hyperplane or a join of two parallel hyperplanes, as $P \in \mathscr{M}_1$. If Q is a hyperplane, $Q \subset f(P)$, then obviously $Q \| Q_1 \| Q_2$. Let $Y_1 \in f^{-1}(Q_1)$, $Y_2 \in f^{-1}(Q_2)$. There is a surface $P'_\kappa \in \mathscr{M}$ such that $P'_\kappa \neq P_\kappa$, Y_1, $Y_2 \in P'_\kappa$, $P'_\kappa \cap P = \varnothing$; but then $f(P'_\kappa) = Q'_1 \cup Q'_2$, where Q'_1 and Q'_2 are hyperplanes, $Q \| Q'_1 \| Q'_2$, $fY_1 \in Q'_1$, and $fY_2 \in Q'_2$. Thus $f(P'_\kappa) = f(P_\kappa)$ in contradiction to the injectivity of f.

The bijectivity of f implies that any connected component of $E^n \setminus P$ is mapped onto a connected component of $E^n \setminus f(P)$. So, the interior of any surface P from $\mathscr{M}_1$ is mapped onto a connected component of $E^n \setminus f(P)$; then using the lemma we conclude that any open set is mapped onto an open one; i.e., f is an open mapping. Therefore, the mapping f^{-1} is continuous and since f^{-1} (like f) is bijective, it follows that f^{-1} is a homeomorphism and, hence, f is a homeomorphism too. So $\mathscr{M}_1 = \mathscr{M}$ and the interior of any surface from $\mathscr{M}$ is mapped onto the interior of its image, which means that con M satisfies the conditions of Theorem 1. Thus, f is affine.

PROOF OF THE LEMMA. Denote by $\mathscr{M}_X^U$ the family of surfaces $P \in \mathscr{M}$ for which $X \in \operatorname{int}\operatorname{con} P$ and $P \subset U$. We shall prove the lemma by contradiction. Let a point $O \in E^n$ and its neighborhood U be such that $\mathscr{M}_O^U \cap \mathscr{M}_1 = \varnothing$. Choose a surface $P \in \mathscr{M}_O^U$ and points A, $B \in P$ so that $O \in]A, B[$, $D^n(O; 3d) \subset U$, where $d = \operatorname{diam} P$, and $|\kappa_0| < 1$, where $P \in H_X^{\kappa_0}(M)$. If we show that

$$\bigcap_{\substack{Q \in \mathscr{M}_O^U \\ A, B \in Q}} Q = \{A, B\},$$

we will get a contradiction to the injectivity of f, since all $f(Q)$ are convex hyperplane sets and so

$$\bigcap_{\substack{Q \in \mathscr{M}_O^U \\ A, B \in Q}} f(Q) \supset [fA, fB].$$

So. we prove that

$$\bigcap_{\substack{Q \in \mathscr{M}_O^U \\ A, B \in Q}} Q = \{A, B\}.$$

For this we take a point $C \in P$, $C \neq A, B$, and show that there is a surface $P' \in \mathscr{M}_O^U$ for which $C \notin P'$ and A, $B \in P'$. Let p be the intersection of P with the plane $A(\{A, B, C\})$. All further constructions are made in the same two-dimensional plane $A(\{A, B, C\})$. We introduce the notation $\mathscr{K} := \{F(p) \mid F$ is either H_X^κ with $X \in A(\{A, B, C\})$ or a translation in $A(\{A, B, C\})$, $F(P) \in \mathscr{M}_O^U\}$. Consider all intervals that are intersections of con p and lines parallel to $[A, B]$. Let x be the maximal length of those of them that belong to p. If $x < \rho(A, B)$ then construct a curve $p_1 \in \mathscr{K}$ for which A, $B \in p_1$ and there exists exactly one more interval $[A', B']$ different from $[A, B]$ (where A', $B' \in p_1$ and $C \neq A'$, B') with the property $\overrightarrow{A'B'} = \overrightarrow{AB}$. For this either take p itself if it satisfies this condition or enlarge p homothetically with some coefficient $1 < \alpha < 2$ so that $\alpha x < \rho(A, B)$ still holds and translate it so that A, B would lie on the p_1 obtained. If still $C \in p_1$, translate p_1 so that $[A', B']$ coincides with $[A, B]$. If C does not lie on the curve $p_2 \in \mathscr{K}$ obtained, then denote p_2 by p'; if again $C \in p_2$ this means that either for the points A, A' or for the points B, B', say, for definiteness sake, for B, B', the following holds: $[B, B'] \subset p_1$, $C \in A(\{B, B'\})$. In this case $A(\{A, A'\}) \cap p_1 = \{A, A'\}$. Then take a curve homothetic to p_1 with the coefficient -1 and translate it so that the image of $[A', B']$ under this homothety coincides with $[A, B]$; so, we obtain a curve $p' \in \mathscr{K}$, $C \notin p'$.

Suppose $x \geq \rho(A, B)$. If there exists an interval $[D, F] \subset p$, $[D, f] \| [A, B]$, $\rho(D, F) \geq \rho(A, B)$ lying on the same side of $A([A, B])$ as the point C, then applying to p a homothety with a positive (not greater than 1) coefficient and after that a translation such that the points A, B belong to the curve $p' \in \mathscr{K}$ obtained, we can always achieve $C \notin p'$, because the line $A([D', F'])$, where $[D', F']$ is the image of $[D, F]$ under the just described transformations, strictly separates the interval $[A, B]$ and the point C. If on that side of $A([A, B])$ where C lies there is no such interval, then apply the same transformation not to p, but to the curve $H_Y^{-1}(p)$, where Y is the mid-point of $[A, B]$ (certainly, only if $C \in H_Y^{-1}(p)$; otherwise, $p' = H_Y^{-1}(p)$) and again we obtain a curve $p' \in \mathscr{K}$ for which A, $B \in p'$, $C \notin p'$.

So, in all cases we have a curve p', a section of the desired surface P', which is obtained from P under the same transformation as p' from p. The lemma is proved. □

Proof of Theorem 2. In what follows the expression "almost all" (or, what is the same, "almost every") means "all except for at most a countable set".

Let $q < 2p$ and let f be some mapping satisfying the condition of the theorem. A plane $P \subset E^n$ is said to be *thin* if $\max_{i=1,\dots,q} \dim Q_i \leq p$, where Q_i are the convex sets for which $f(P) = \cup_{i=1}^q Q_i$. Let $P \subset E^n$ be a thin two-dimensional plane. We prove that $f|_l$ is affine. For this we shall prove the following

Lemma. *Let L be a pencil of lines in P (i.e., either these lines are all parallel, or they pass through a fixed point). Then for almost all $l \in L$ the mapping $f|_l$ is a homeomorphism preserving collinearity.*

(We shall call such lines *proper*.)

Using the lemma we conclude that $f(P)$ lies in a two-dimensional plane. Indeed, take a point $O \in P$ and two proper lines $l_1, l_2 \subset P$ passing through O. Then $f(l_1 \cup l_2)$ lies in a two-dimensional plane. And any point $O' \in P$ is mapped into the same plane, since there is a proper line l with $O' \in l$ for which $l \cap (l_1 \cup l_2)$ is a two-point set.

It is also clear from the lemma that $f(P)$ does not lie on a line.

We show that $f|_P$ preserves collinearity. Assume that there exists a line $l \subset P$ such that $f(l)$ does not lie on a line. Then there are an interval $[K_1, K_2] \subset f(l)$ and a point $fY \notin A([K_1, K_2])$, where $Y \in l$. Take a point X such that $fX \in]K_1, K_2[$. By the lemma, we can take a proper line s with $X \in s$, $s \subset P$ such that $fY \notin A(f(s))$. Again, according to the lemma, a point $Z \in s$, $Z \neq X$ exists for which $A(\{Y, Z\})$ is proper, but $f(A(\{Y, Z\})) \bigcap f(l) \neq fY$, which contradicts the injectivity of f. Thus, $f|_P$ is injective, preserves collinearity, and $f(P)$ does not belong to a line. So from Theorem K it follows that $f|_P$ is affine.

Now assume that for any thin plane P with $\dim P \leq \kappa$ $(2 \leq \kappa < n)$ the mapping $f|_P$ is affine and prove that then for any plane P' with $\dim P' = \kappa + 1$ the mapping $f|_{P'}$ is also affine. Let $P^0 \subset P'$ be a plane of dimension κ. We shall show that

P^0 is thin. Take $(\kappa - 1)$-dimensional planes $P_1, \dots, P_{\kappa+1} \subset P^0$ that bound some κ-dimensional simplex. By P_i' $(i = 1, \dots, \kappa + 1)$ denote a κ-dimensional plane having the properties: $P_i' \subset P'$, $P_i' \perp P^0$, $P_i' \cap P^0 = P_i$. Since P' is thin, almost all κ-dimensional planes parallel to P_i' are thin for every i (due to the injectivity of f), so the restriction of f to any of them is affine and, hence, also for almost all $(\kappa - 1)$-dimensional planes in P^0 parallel to P_i the restriction of f to any of them is affine. Choose among them $(\kappa - 1)$- dimensional planes $Q_i \| P_i$ $(i = 1, \dots, \kappa + 1)$ on which f is affine so that they contain the faces of some κ-dimensional simplex in P^0. Then $f|_{Q_1 \cup \dots \cup Q_{\kappa+1}}$ is affine. Therefore, for every i and for almost every $(\kappa - 1)$-dimensional plane $Q \subset P^0$, $Q \| P_i$ we have that $f|_{\bigcup_{j=1}^{\kappa+1} Q_j \cup Q}$ is affine. Hence, f is affine on the whole of P^0 except maybe at at most a countable set of points. Thus, the whole set $f(P^0)$, except possibly for at most a countable set of points, lies in a κ-dimensional plane. Hence P^0 is thin and so by the assumption $f|_{P^0}$ is affine and this means that $f|_{P'}$ is affine too.

So, by induction on the dimension we obtain that the mapping f is affine, since E^n is also a thin plane.

Now we have only to prove the lemma.

PROOF OF THE LEMMA. In what follows all objects in a preimage are supposed to be lying in the two-dimensional plane P.

It is clear that almost all lines of L are thin; denote the set of them by L_1, $L_1 \subset L$. This means that for every $l \in L_1$ we have $f(l) = \bigcup_{i=1}^q Q_i$, where every Q_i is a convex subset of a line. Let $A^*(l) := \bigcup_{i=1}^q A(Q_i)$. Denote by L_2 the maximal subfamily of L_1 having the property that for any $l \in L_2$ a line l' exists such that $f(l) \subset l' \cup X$ and the set X is finite. We shall prove that the family $L_1 \backslash L_2$ is at most countable. Suppose the contrary. Assume that for some $j \geq 1$ the lines $l_1, \dots, l_j \in L_1 \backslash L_2$ are distinct and $\bigcup_{i=1}^j A^*(l_i)$ contains $2j$ distinct lines. Denote by G the family of all lines $l \in L_1 \backslash L_2$ for each of which the set $f(l) \cap (\cup_{i=1}^j A^*(l_i))$ is infinite. From the injectivity of f it is clear that G is at most countable. Take $l_{j+1} \in (L_1 \backslash L_2) \backslash G$; then $\cup_{i=1}^j A^*(l_i)$ contains $2(j + 1)$ different lines. This process will not stop if the family $L_1 \backslash L_2$ is more than countable, but then $f(\cup_{\kappa=1}^p l_\kappa)$ cannot be represented as a join of q convex sets in contradiction with the condition of the theorem. So $L_1 \backslash L_2$ is indeed at most countable.

Denote now by L_3 a maximal subfamily of L_2 having the property that for any line $l \in L_3$ and for any segment $r \subset l$ the set $f(r) \cap l'$ is convex (below a "segment" means a closed interval of more than one point or a whole line). We shall prove that $L_2 \backslash L_3$ is at most countable. Again, suppose the contrary.

A segment r of a line l from $L_2 \backslash L_3$ for which $f(r) \cap l'$ is not convex is called a *torn* segment. Notice that if L is a pencil of lines passing through some point $O \in P$, then any set of torn segments r_i having mutually different affine hulls l for which fO is an end of a non-one-point connected component of $f(r) \cap l'$ is finite. Indeed, if in this set there existed an infinite set of segments containing O or an infinite set of segments not containing O then in every one of these cases it would be possible to choose p segments, the image of the union of which is not a union of q convex sets, contradicting the conditions of the theorem. Denote by S' the family of lines containing such segments r_i. Take a line $l_1 \in (L_2 \backslash L_3) \backslash S'$ and a torn segment $r_1 \subset l_1$. We show that there exists a segment r_2 such that $A(r_2) \in L_2 \backslash L_3$ and $f(r_1 \cup r_2)$ is not a union of three convex sets. Suppose the contrary. Then, as before, if $S \subset L_2 \backslash L_3$ consists of all lines l for which the set $f(l) \cap l_1'$ is infinite, then S is at most countable.

Let $f(T)$ be the set of all (belonging to $f(E^n)$) ends of non-one-point connected components of the set $f(r_1)\cap l_1'$. Then T is finite and the family S'' of all lines from $L_2\backslash L_3$ intersecting T is also finite. If $f(r_1)\cap l_1'$ has no one-point components, then take a torn segment r with the property $A(r)\in(L_2\backslash L_3)\backslash(S\cup S'\cup S'')$. If $f(r_1\cup r)$ is the union of three convex sets, then there is a point $X\in r_1$ such that fX "connects" two (non-one-point) components of $f(r)$ lying on one line (this means that the join of these two components and the point fX is connected). The number of such r having mutually different affine hulls is infinite. Take p of them. Then the image of their union is not the union of q convex sets, a contradiction.

And if $f(r_1)\cap l_1'$ has a one-point component X_0, then the only difference with the former case is that there can be infinitely many torn segments with mutually different affine hulls having the property that X_0 belongs to the closure of a non-one-point component of the image of each of them. Then we choose among them p segments in a corresponding way and get a contradiction as in the former case.

So, there exists a segment r_2 such that $f(r_1\cup r_2)$ is not the join of three convex sets. Analogously there is a segment r_3 such that $f(r_1\cup r_2\cup r_3)$ is not the join of five convex sets and so on. Finally, there are segments $r_1,\dots,r_p$ such that $f(\cup_{i=1}^p r_i)$ is not the union of q convex sets, a contradiction. Hence, $L_2\backslash L_3$ is at most countable.

Let $l\in L_3$. Denote by $H\subset l$ the set of points whose images do not lie on l'. The set $l\backslash H$ is decomposed into open (in a line) convex sets. Let U be one of them. Take points $X_1, X_2, X_3\in U$ such that X_2 lies between X_1 and X_3. Then the sets $f([X_1,X_2])$ and $f([X_2,X_3])$ are convex; this means that fX_2 lies between fX_1 and fX_3. From this (taking into account the injectivity of f and the choice of l) we conclude that $f|_U$ is a homeomorphism. Thus, if $H\neq\varnothing$, then the convex set $l'\cap f(l)$ can be represented as the union of not less than two nonempty disjoint open sets, which contradicts the connectedness of a convex set.

So $H=\varnothing$; i.e., $f|_l$ is a homeomorphism preserving collinearity. The lemma is proved. □

If $q\geq 2p$, then there exists an injective mapping $\varphi\colon E^n\to E^n$ that is not affine but it maps any union of p convex sets onto the union of q convex sets. The mapping

$$\varphi(x_1,\dots,x_n)=\begin{cases}(x_1,\dots,x_{n-1},2x_n) & \text{if } x_n>0,\\ (x_1,\dots,x_{n-1},x_n) & \text{if } x_n\leq 0\end{cases}$$

is such. This example completes the proof of Theorem 2. □

We note that in the proof of Theorem 2 only p-unions of convex sets consisting of subspaces of all dimensions and of closed intervals were used. □

The requirement of injectivity of f cannot be omitted; the example of A. V. Kuz′minyh shows this.

References

1. A. V. Kuz′minyh, *Generalization of the Darboux's theorem*, Sibirsk. Mat. Zh. **20** (1979), no. 4, 917–921; English transl. in Siberian Math. J. **20** (1979).
2. B. R. Gelbaum and J. Olmsted, *Counterexamples in analysis*, Holden-Day, San Francisco, CA, 1964.

Amer. Math. Soc. Transl.
(2) Vol. **163**, 1995

Affine Crystallographic Groups

G. A. Soifer

Introduction

Let V be a vector space over the field $\mathbb{R}$ of reals, A an affine space associated with V, and let Aff V be the group of all affine motions of A. Clearly the stabilizer

$$G_{a_0} = \{\gamma \in \operatorname{Aff} V : \gamma(a_0) = a_0\}$$

of a point $a_0 \in A$ is isomorphic to the full group of automorphisms $Gl(V)$ of V and the group Aff V is isomorphic to the semidirect product $V \times Gl(V)$. Therefore, each $\gamma \in \operatorname{Aff} V$ will be considered as a pair $\gamma = (v_\gamma, l_\gamma)$ with $v_\gamma \in V, \quad l_\gamma \in Gl(V)$. We call l_γ the linear part of γ. This identification of the groups Aff V and $V \times Gl(V)$ enables us to assume that $V \subset \operatorname{Aff} V$. The map $\gamma \mapsto l_\gamma$ is a group homomorphism. Therefore, one can associate to every subgroup Γ of Aff V the subgroup $l(\Gamma) = \{l_\gamma : \gamma \in \Gamma\}$ of $Gl(V)$. Clearly, the group Aff V has the faithful linear representation

$$\rho_* : \gamma \mapsto \begin{pmatrix} l_\gamma & v_\gamma \\ 0 & 1 \end{pmatrix}.$$

An arbitrary affine transformation γ of A can be decomposed into a product $\gamma = v_0\gamma_0$, where $v_0 \in V$ and $\gamma_0 \in G_{a_0}$ for some $a_0 \in A$ and $v_0\gamma_0 = \gamma_0 v_0$. The vector v_0 is fixed under the transformation l_{γ_0} and the line $\pi_\gamma = \{a_0 + \lambda v_0 : \lambda \in \mathbb{R}\}$ is γ-invariant. For a topological group Γ acting on a topological space M a set $S \subset M$ is called *Γ-covering*, provided that $\bigcup_{\gamma\in\Gamma} \gamma S = M$. If for any compact set $K \subset M$ the set $\{\gamma \in \Gamma : \gamma K \cap K \neq \varnothing\}$ is finite the group Γ is said to act *discontinuously* on M. A group Γ acting on a topological space M is said to be a *crystallographic group* of motions of M provided that

(i) Γ acts discontinuously on M;
(ii) there exists a compact Γ-covering set.

In this paper we study crystallographic groups of motions of an affine space in connection with the following old conjecture.

CONJECTURE A. *A crystallographic group Γ of motions of an affine space is almost solvable* (*i.e., contains a solvable subgroup of finite index*).

Let G^* be an algebraic group containing the group $l(\Gamma)$ and let $G^* = U^*S^*$ be a decomposition into a product of the unipotent radical U^* and a reductive subgroup

1991 *Mathematics Subject Classification*. Primary 57S30.

S^*. The statement of the *Bieberbach-Schoenflies* theorem is the affirmative answer to Conjecture A in the case where $U^* = 1$ and S^* is a compact group.

Note that if S^* is compact, Conjecture A is true without the assumption that Γ is crystallographic [**3**]. That is perhaps why the following conjecture was formulated [**9**] (we give an equivalent statement).

CONJECTURE B. *A discontinuous group of affine motions is almost solvable.*

G. A. Margulis gave a complete solution of this problem by disproving Conjecture B [**8**].

Further steps in proving Conjecture A were made in [**4**] where Conjecture A was proved in the case $\dim V \le 3$. Then it was established in [**6**] that the conjecture is true for $G^* = O(n, 1)$. In [**7**] Conjecture A was proved under the condition that the group G^* is reductive and $\operatorname{rank}_{\mathbb{R}} G^* \le 1$ (specifically, when G^* is locally isomorphic to $O(n, 1)$). In [**13**] this result was shown to be true for the more general case $\operatorname{rank}_{\mathbb{R}} S^* \le 1$.

We shall show that the following more general assertion is true.

THEOREM A. *Let Γ be a crystallographic group of motions of an affine space A and let $G^* \subset \operatorname{Aff} V$ be an algebraic subgroup containing Γ. If the real rank of any simple subgroup in G^* does not exceed 1, then Γ is almost simple.*

The methods used in proving this theorem enable us to prove the following rigidity theorem for almost solvable crystallographic groups of affine motions.

THEOREM B. *Let Γ_1 and Γ_2 be two crystallographic almost solvable groups, $\Gamma_1 \subset \operatorname{Aff} V$, $\Gamma_2 \subset \operatorname{Aff} V$, and let G_1 and G_2 be their algebraic closures. Then any isomorphism $\varphi\colon \Gamma_1 \to \Gamma_2$ extends to a rational isomorphism $\varphi\colon G_1 \to G_2$.*

In conclusion we express deep gratitude to G. A. Margulis for numerous stimulating discussions, advice, and suggestions.

§1. Preliminary notes and necessary definitions

Let Γ be a group of affine motions, $\Gamma \subset \operatorname{Aff} V$, G the algebraic closure of Γ, $G = U \cdot S$ a decomposition of G into the product of a unipotent and reductive subgroup.

Since it suffices to prove the assertion of Theorem A for a subgroup of finite index in Γ and, on the other hand, by Selberg's theorem [**10**], Γ has a torsion-free subgroup of finite index, we replace, where it is necessary, without mentioning the fact, the group Γ by a torsion-free subgroup of Γ of finite index.

PROPOSITION 1. *If Γ is a crystallographic group of affine motions of a space A, then the group U acts transitively on A.*

PROOF (G. A. Margulis). Let $\operatorname{vcd}(\Gamma)$ be the virtual cohomological dimension of the group Γ. Since Γ is crystallographic, $\operatorname{vcd}\Gamma = \dim A$ [**12**]. It follows from the conjugacy of reductive subgroups in G and the fact that the kernel of the homomorphism $\gamma \mapsto l_\gamma$ is unipotent that S has a fixed point a. Thus, the orbits G_a and U_a coincide and, hence, the set U_a is Γ-invariant. Since U_a is a closed submanifold in A, we have $\operatorname{vcd}\Gamma = \dim U_a$, and, hence, $U_a = A$. □

Note that this assertion was proved earlier in a different way in [**5**].

We recall the following

DEFINITION 1. Let Γ be a polycyclic group and let $\Gamma = \Gamma_0 \supset \cdots \supset \Gamma_n = \{e\}$ be a chain of normal subgroups with abelian quotient groups Γ_i/Γ_{i+1}. We set $r(\Gamma) = \operatorname{rank}(\Gamma) = \sum_i \operatorname{rank}(\Gamma_i/\Gamma_{i+1})$. This number depends only on the group Γ and is called the polycyclic rank of Γ.

Clearly, we have

LEMMA 1. *If Γ is a polycyclic group, Γ' a subgroup of Γ of finite index, then* $r(\Gamma) = r(\Gamma')$.

DEFINITION 2. A subgroup Γ of a Lie group G is called uniform if the quotient space G/Γ is compact.

Since any element $\gamma \in \Gamma$ decomposes into a product $\gamma = u \cdot s$, where $u \in U$, $s \in S$, one can consider the map $\pi\colon \Gamma \to U$ obtained by setting $\pi(\gamma) = u$.

PROPOSITION 2. *If Γ is a solvable subgroup in* Aff V *acting discontinuously on A, then*

1) $\pi(\Gamma)$ *contains a uniform subgroup of U.*
2) $r(\Gamma) = \dim U$.

PROOF. Let $\Gamma_0 = \Gamma \cap G^0$, where G^0 is the connected component of the identity of the algebraic closure G of Γ in Aff V. Since Γ_0 is Zariski dense in G^0, $[\Gamma^0, \Gamma^0]$ is a Zariski dense subgroup of the unipotent group $G_1 = [G^0, G^0]$. Therefore $\Gamma_1 = G_1 \cap \Gamma$ is Zariski dense in G_1 and, hence **[10,** Theorem 2.10], is a uniform subgroup of G_1. Let us consider the factor group $\widetilde{G} = G/G_1$ and the subgroup $\widetilde{\Gamma} = \Gamma/\Gamma_1$ of $\widetilde{G}$. Clearly $\widetilde{U} = U/G_1$ is the unipotent radical of the group $\widetilde{G}$. Choose a subgroup Γ_2 of finite index in Γ such that $\Gamma_1 \subset \Gamma_2$ and such that the semisimple part of each element $\gamma \in \Gamma_2$ is not periodic. Let $\widetilde{\Gamma}_2 = \Gamma_2/\Gamma_1$. Clearly, $r(\widetilde{\Gamma}_2) = r(\widetilde{\Gamma})$, $r(\Gamma_2) = r(\widetilde{\Gamma}_2) + r(\widetilde{\Gamma}_1)$, and $r(\Gamma_1) = \dim G_1$. It suffices to prove that $r(\widetilde{\Gamma}_2) = \dim \widetilde{U}$. To this end we shall show that there is no $\gamma \in \Gamma_2$ such that γG_1 is a nontrivial semisimple element of $\widetilde{G}$. If, on the contrary, there exists such an element $\gamma \in \Gamma_2$, then we consider the Jordan decomposition $\gamma = \gamma_n \cdot \gamma_s$ of γ. By the choice of γ, $\gamma_n \in \Gamma_1$.

Since γ_3 is a reductive element, it has a fixed point $a' \in A$. Let us consider the set $X = \{\gamma^m a' \colon m \in \mathbb{Z}\}$ lying in the orbit $G_1 a'$.

Since Γ_1 is a uniform subgroup of G_1, there exists a compact set $K \ni a'$ in $G_1 a'$ such that for each $x \in G_1 a'$ there is $\gamma \in \Gamma_1$ with $\gamma(x) \in K$. So for each γ^m, $m \in \mathbb{Z}$, there exists $\gamma_m \in \Gamma_1$ such that $\gamma_m \gamma^m a' \in K$. Since Γ acts discontinuously on A, there exists an $m \in \mathbb{Z}$ such that $\gamma_m \gamma^m = e$. Hence $\gamma^m \in G_1$, which contradicts the choice of γ. This implies that (i) $r(\Gamma_2) = \dim U$ and (ii) $\pi(\Gamma_2)$ contains a uniform subgroup of U. □

One can prove that $\pi(\Gamma)$ lies in a uniform subgroup of U.

The assertion 1) of Theorem 2 was established in **[13]** under the additional assumption that Γ is a solvable subgroup of a crystallographic group.

DEFINITION 3. We say that a representation ρ of a semisimple group S has *property* $(*)$ if the subspace corresponding to the identity root is nontrivial.

Let $S = S_1 \times S_2$ and let ρ be an irreducible representation of the semisimple group S. Then ρ is a tensor product of irreducible representations ρ_1 and ρ_2 of the groups S_1 and S_2, respectively. It can easily be verified that the following assertion is true.

LEMMA 2. *The representation ρ has property $(*)$ if and only if ρ_1 and ρ_2 have property $(*)$.*

The following assertion can be verified directly.

LEMMA 3. *Let γ be an element of* Aff V *and $\gamma = v_0\gamma_0$, where $v_0 \in V$ and $\gamma_0 \in G$ with $[v_0, \gamma_0] = 1$. If $\gamma_1 \in$ Aff V and $\gamma' = \gamma_1\gamma\gamma_1^{-1}$ then $\pi_\gamma = \{\gamma_1(a) + \lambda l_{\gamma_1}(v_0), \quad \lambda \in \mathbb{R}\}$.*

§2. Proof of main assertions

We recall that Γ stands for a crystallographic group of motions of an affine space A, G is its algebraic closure in Aff V, and $G = U \cdot S$ is a decomposition of G into the product of a unipotent group U and a reductive group S. Let a_0 be a fixed point of the subgroup S, i.e., $S \subset G_{a_0}$. If Γ^* is a subgroup of G, G^* is its algebraic closure, $G^* \cap \Gamma = \Gamma^*$, U^* is the unipotent radical of G^*, then we choose a reductive subgroup S^* of G^* such that $S^* \subset S$ and $U^*S^* = G^*$.

As in [13], we consider the set of orbits $\widetilde{A} = \{G^*a : a \in A\}$. Clearly, this set is in fact a G-manifold isomorphic, by Proposition 1, to the G-manifold U/U^*. Suppose that Γ^* is a crystallographic group of motions of the manifold G^*a_0. The following assertion is proved, in view of Proposition 1, like Lemma 2.1 in [7] (see also [13]).

LEMMA 4. *If $\widetilde{A}$ is the orbit manifold $\{G{*}a : a \in A\}$, then the quotient group $\widetilde{\Gamma} = \Gamma/\Gamma^*$ is a crystallographic group of $\widetilde{A}$.*

We shall assume, unless stated otherwise, that the connected identity component of S is nontrivial.

Obviously, by Proposition 2, any solvable subgroup of Γ satisfies the condition of Lemma 4.

Let R be the solvable radical of G, $\Gamma_1 = R \cap \Gamma$, and let G_1 be the algebraic closure of Γ_1. Let $\widetilde{a}_0 = G_1 a_0 \in \widetilde{A}$. We consider the natural representation d of the group S in the tangent space $T_{\widetilde{a}_0}$ of a manifold $\widetilde{A}$ at a point $\widetilde{a}_0$, $d : S \to GL(T_{\widetilde{a}_0})$.

Before beginning the proof of Theorem A we recall some facts about simple Lie groups of real rank ≤ 1 [7, pp. 23–24]. Let H be a semisimple connected Lie group. We denote by $\dim X_H$ the dimension of the symmetric space of the group H, where $X_H = H/K$ and K is a maximal compact subgroup in H. Let d_H be the dimension of a minimal nontrivial representation of H.

Table 1 shows the values of $\dim X_H$ and d_H for various H.

LEMMA 5. *Let H be a semisimple connected Lie group and let the real rank of an arbitrary simple subgroup of H be ≤ 1. If d^* is the dimension of a faithful representation ρ of H having property $(*)$, then $\dim X_H < d^*$.*

TABLE 1

H is locally isomorphic to	$\dim X_H$	d_H
$O(n, 1)$	n	$n + 1$
$U(n, 1)$	$2n$	$2n + 1$
$Sp(n, 1)$	$4n$	$4n + 4$
F_4II	16	26
$Sl_2(\mathbb{R})$	2	2

PROOF. First suppose that ρ is an irreducible representation of H. Since H is a product of simple connected groups H_i, $i = 1, \dots, s$, $\dim X_H = \sum \dim X_{H_i}$. On the other hand, the representation ρ is a tensor product of irreducible representations ρ_i. By Lemma 3, each of these representations has property $(*)$. If d_i is the dimension of ρ_i then $d = \prod_{i=1} d_i$. Now we apply Table 1. Obviously, $d_i > \dim X_{H_i}$ for all $i = 1, \dots, s$. Hence $\dim X_H = \sum_i \dim X_H < d$.

An arbitrary representation ρ decomposes into a direct sum of irreducible representations. Let $\rho = \bigoplus_i \rho_i$ and let d_i be the dimension of ρ_i. We denote by $\widetilde{H}_i$ a connected subgroup in H such that the restriction $\rho_i \mid \widetilde{H}_i$ is a faithful representation of H_i. It is easy to verify using Table 1 that $d_i \geq \dim X_{\widetilde{H}_i}$. Since ρ has property $(*)$, some of the ρ_i, say, ρ_1, also has property $(*)$. So $d_1 > \dim X_{\widetilde{H}_i}$. Clearly $\sum \dim X_{\widetilde{H}_i} \geq \dim X_H$ and hence $d^* = \sum d_i > \dim X_H$, which proves the lemma. □

With notation as in the previous sections we give the

PROOF OF THEOREM A. By Lemma 4 and **[12]**, we have $\operatorname{vcd} \widetilde{\Gamma} = \dim \widetilde{A}$, and since by Auslander's theorem **[10]** a subgroup of finite index in $\widetilde{\Gamma}$ is isomorphic to a discontinuous group of transformations of the symmetric space $X_{\widetilde{S}}$, where $\widetilde{S}$ is the semisimple part of G, $\operatorname{vcd} \Gamma \leq \dim X_{\widetilde{S}}$. Hence $\dim \widetilde{A} \leq \dim X_{\widetilde{S}}$. We consider again the representation $d \colon S \to Gl(T_{\widetilde{a}_0})$ of the reductive subgroup S. We recall that $\widetilde{a}_0 \in \widetilde{A}$, $\widetilde{a}_0 = G_1 a_0$, and $S \subset G_{a_0}$. Since $\dim T_{\widetilde{a}_0} = \dim \widetilde{A}$, we have for the dimension d^* of d that $d^* \leq \dim X_{\widetilde{S}}$. We remark that this is impossible if the group $\widetilde{S}$ is not compact. We first observe that d has property $(*)$. To this end we choose $\gamma \in \Gamma$ such that the semisimple part γ_S of the Jordan decomposition, $\gamma = \gamma_n \gamma_s$ of γ is in general position. Let γ_S be an element of G_{a_1}. Since the group G acts transitively on A, the representation $\widetilde{d} \colon G_{a_1} \to Gl(T_{a_1})$ and d are conjugate and hence we can assume that $a_1 = a_0$. We consider the curve $\gamma(t)\widetilde{a}_0$ in the space $\widetilde{A}$. By the choice of $\widetilde{a}_0$ and γ_S and the properties of the Jordan decomposition, $\gamma(t)\widetilde{a}_0 = \gamma_n(t)\widetilde{a}_0$ and the points of this curve are fixed relative to γ_S. We first observe that this curve is not trivial. Indeed, if $\gamma_n(t)\widetilde{a}_0 = \widetilde{a}_0$, then $\gamma_n(t)a_0 \subset G_1 a_0$. By Proposition 2, Γ_1 is a crystallographic group of motions of the space $G_1 a$. Hence, there exists a compact neighborhood K of a_0 in $G_1 a_0$ such that for each $a' \in G_1 a_0$ there exists a $\gamma \in \Gamma_1$ with $\gamma a' \in K$. Then, since for each $n \in \mathbb{Z}$, $\gamma^n a_0 = \gamma_u^n a_0 = \gamma_u(n) a_0 \in G_1 a_0$, there exist elements $\gamma_n \in \Gamma_1$ such that $\gamma_n \gamma^n a_0 \in K$. Since Γ acts discontinuously on A, there exists n such that $\gamma^n \in G_1$. This is impossible, because $\widetilde{S}$ is not compact. Thus, the curve $\gamma_u(t)\widetilde{a}_0$ is not trivial and the points of this curve are fixed relative to $d\gamma_S$, which is an element in general position of the group dS. Hence, the representation d of the group S has property $(*)$. By the restriction imposed on the group $G^* \supset \Gamma$, we can use the assertion of Lemma 5. So $\dim X_{\widetilde{S}} < d^*$, which contradicts the inequality $d^* \leq \dim X_{\widetilde{S}}$. Thus the group $\widetilde{S}$ is compact, and hence by Auslander's theorem $\widetilde{F}$ is almost solvable, which completes the proof. □

NOTE. The conjecture that the representation d has property $(*)$ is due to G. A. Margulis.

Now we turn to the proof of Theorem B.

PROOF OF THEOREM B. Let Γ_1 and Γ_2 be two almost solvable crystallographic groups, G_1 and G_2 their algebraic closures in Aff V_1 and Aff V_2, respectively. We note that G_1 (resp. G_2) is an $\mathbb{R}$-algebraic completion of the group Γ in the sense of [**10**, Definition 4.39]. Let U_1 be the unipotent radical of G_1. It is easy to observe, using Proposition 2, that $r(\Gamma_1) = \dim U_1$. It remains to show that the centralizer $Z_{G_1}(U_1)$ is contained in U_1. Indeed, γ is a semisimple element in $Z_{G_1}(U_1)$ and a_0 is a fixed point for γ. Since the group U_1 acts transitively on the affine space A_1 associated with V_1, $\gamma a = \gamma u a_0 = u \gamma a_0 = u a_0 = a, \quad u \in U_1$, and hence $\gamma = 1$. The remaining part of the proof of Theorem B coincides almost verbatim with the proof of Lemma 4.41 in [**10**]. We consider the subgroup $\Delta = \{(\gamma, \varphi(\gamma)) : \gamma \in \Gamma_1\}$ of $G_1 \times G_2$ and its algebraic closure $\widetilde{G}$. Let $\widetilde{U}$ be the unipotent radical of $\widetilde{G}$. Let us denote by α_i the restriction of the projection $\rho_i : G_1 \times G_2 \to G_i$, $i = 1, 2$, to G. We remark that α_1 is an isomorphism. Indeed, from $\dim \widetilde{U} \le \operatorname{rank} \Delta = \operatorname{rank} \Gamma_i$ and $\alpha_i(\widetilde{U}) = U_i$ we deduce by Proposition 2 that $\ker \alpha_i \cap \widetilde{U} = \{1\}$, $i = 1, 2$. Let x be a semisimple element of $\ker \alpha_1$. Then $\alpha_1(x^{-1} y x) = \alpha_1(y)$ for all $y \in U$. Since $\alpha_1 : U \to U_1$ is an isomorphism, $x \in Z_G(\widetilde{U})$. Hence $\alpha_Z(x) \in Z_{G_2}(U)$. Since $Z_{G_2}(U_2) \subset U_2$, $\alpha_2(x) = 1$. Therefore, $x = 1$. Thus, α_i is an isomorphism, $i = 1, 2$. It follows that $\widetilde{\varphi} = \alpha_2 \alpha_1^{-1}$ is the required isomorphism $G_1 \to G_2$ extending φ. □

References

1. L. Auslander, *The structure of complete locally affine manifolds*, Topology **3** (1964), 131–139.
2. ______, *Simply transitive groups of affine motions*, Amer. J. Math. **99** (1977), 809–826.
3. ______, *Bieberbach's theorems on space groups and discrete uniform subgroups of Lie groups*, Ann. of Math. (2) **71** (1960), 579–590.
4. D. Fried and W. M. Goldman, *Three dimensional affine crystallographic groups*, Adv. Math. **47** (1983), 1–49.
5. D. Fried, W. M. Goldman, and M. W. Hirsch, *Affine manifolds with nilpotent holonomy*, Comment. Math. Helv. **56** (1981), 487–523.
6. W. M. Goldman and Y. Kamishima, *The fundamental group of a compact flat Lorentz space form is virtually polycyclic*, J. Differential Geom. **19** (1984), 233–240.
7. F. Grunewald and G. A. Margulis, *Transitive and quasitransitive actions of affine groups preserving a generalized Lorentz-structure*, J. Geom. Phys. **5** (1988), 493–531.
8. G. A. Margulis, *Free completely discontinuous groups of affine transformations*, Dokl. Akad. Nauk SSSR **272** (1983), no. 4, 785–788; English transl. in Soviet Math. Dokl. **28** (1983).
9. J. Milnor, *On fundamental groups of complete affinely flat manifolds*, Adv. Math. **25** (1977), 178–187.
10. M. S. Raghunathan, *Discrete subgroups of Lie groups*, Springer-Verlag, Berlin and New York, 1972.
11. M. Rosenlicht, *On quotient varieties and the affine embedding of certain homogeneous spaces*, Trans. Amer. Math. Soc. **101** (1961), 211–223.
12. J.-P. Serre, *Cohomologie des groupes discrets*, Ann. of Math. Stud., vol. 70, Princeton Univ. Press, Princeton, NJ, 1971, pp. 77–169.
13. G. Tomanov, *The virtual solvability fundamental group of a generalized Lorentz space form*, J. Differential Geom. **32** (1990), 539–547.
14. J. A. Wolf, *Spaces of constant curvature*, McGraw-Hill, New York, 1967.

KEMEROVO STATE UNIVERSITY, KRASNAJA, 6, KEMEROVO 650043, RUSSIA

Amer. Math. Soc. Transl.
(2) Vol. **163**, 1995

Integrals with Respect to Vector-valued Measures: Theoretical Problems and Applications

A. V. Uglanov

Integrals with respect to vector-valued measures arise in attempts to represent linear operators by integrals, in the theory of random processes (stochastic measures) and in the theory of smooth measures. The more fundamental parts of the theory of vector-valued integrals are easily transferred from the scalar case. But the proofs of Fubini-type theorems (even for scalar, but alternating measures) run up against difficulties which, in comparison with their classical (probability) variants, are fundamentally new[1]. In this article we establish two such theorems; some more "secondary" results are given.

§1. Definitions

Let (Ω, Σ) be a measurable space ($\Omega \in \Sigma$), B a Banach space, $M(\Omega, B)$ a space of B-valued measures (i.e., countably additive functions $\Sigma \to B$), and $M(\Omega, [0, \infty])$ the space of positive (not necessarily finite) measures on Ω. For $\mu \in M(\Omega, B)$ let $|\mu| \in M(\Omega, [0, \infty])$ denote the variation of a measure μ. In what follows $M_0(\Omega, B) = \{\mu \in M(\Omega, B) \colon |\mu|(\Omega) < \infty\}$, and $L(B, F)$ is the space of continuous linear mappings of a Banach space B into a Banach space F endowed with the uniform norm. A function $f \colon \Omega \to L(B, F)$ is said to be:

(a) simple, if $\Omega = \bigcup_{n=1}^{\infty} Q_n$, $Q_n \in \Sigma$, $Q_i \cap Q_j = \varnothing$ for $i \neq j$, and $f(x) = l_n = \text{const}$ on Q_n;

(b) measurable, if there exist a sequence of simple functions $f_n \colon \Omega \to L(B, F)$ that converges uniformly to f;

(c) integrable with respect to a measure $\mu \in M_0(\Omega, B)$, if f is measurable and the integral $\int \|f\| \, d|\mu|$ is finite.

For a simple integrable function f let

$$\int_Q f \, d\mu = \sum_{n=1}^{\infty} l_n \mu(Q \cap Q_n) \qquad (Q \in \Sigma);$$

1991 *Mathematics Subject Classification*. Primary 28B05.

[1] Another such example is the theorem on passing to the limit (under the integral sign) for a sequence of measures: see §4.3, Proposition 1, and §5.7.

for an integrable function f let

$$\int_Q f\,d\mu = \lim_{n\to\infty} \int_Q f_n\,d\mu.$$

Assume now that $\Omega = \bigcup_{n=1}^{\infty} \Omega_n$, $\Omega_1 \subset \Omega_2 \subset \ldots \in \Sigma$, $\Sigma_n = \Sigma \cap \Omega_n$. A function $\mu\colon \bigcup_{n=1}^{\infty} \Sigma_n \to B$ is said to be a σ-bounded measure, if $\mu \in M_0(\Omega_n, B)$ for all $n = 1, 2, \ldots$. A function $f : \Omega \to L(B, F)$ is said to be integrable with respect to a σ-bounded measure μ if:

(a) f is integrable with respect to the measure $\mu \in M_0(\Omega_n, B)$ for all $n = 1, 2, \ldots$;

(b) $\sup_n \int_{\Omega_n} \|f\|\,d|\mu| < \infty$.

For such a function f let

$$\int_Q f\,d\mu = \lim_{n\to\infty} \int_{Q\cap\Omega_n} f\,d\mu \qquad (Q \in \Sigma).$$

The integral introduced has all the usual characteristics; in particular it is linear with respect to f and countably additive with respect to Q.

Let $(\overline{\Omega}, \overline{\Sigma})$ be another measurable space. A function $\nu\colon \overline{\Omega} \times \Sigma \to B$ is said to be a transition measure from $\overline{\Omega}$ to Ω if

(a) for any $\overline{w} \in \overline{\Omega}$ we have $\nu(\overline{w}, \cdot) \in M(\Omega, B)$;

(b) for any $A \in \Sigma$, the function $\nu(\cdot, A)$ is measurable (in the sense of the definition given above, we have $B = L(\mathbb{R}^1, B)$).

Let $M(\overline{\Omega}, \Omega, B)$ stand for the collection of all transition measures. The set $M(\overline{\Omega}, \Omega, [0,\infty])$ is defined similarly. For $\mu \in M(\Omega, B)$, $\nu \in M(\overline{\Omega}, \Omega, B)$, $b \in B'$ $\big(B' = L(B, \mathbb{R}^1)\big)$ we use the notation:

$$b\mu : \Sigma \to \mathbb{R}^1 : A \mapsto \big(b, \mu(A)\big);$$

$$b\nu : \overline{\Omega} \times \Sigma \to \mathbb{R}^1 : (\overline{w}, A) \mapsto \big(b, \nu(\overline{w}, A)\big);$$

$$|\nu| : \overline{\Omega} \times \Sigma \to [0, \infty] : (\overline{w}, A) \mapsto |\nu|(\overline{w}, \cdot)(A).$$

For $A \in \overline{\Omega} \times \Omega$ and $\overline{w} \in \overline{\Omega}$ let $S_{\overline{w}}(A) = \{w \in \Omega : (\overline{w}, \overline{w}) \in A\}$. We note further that the function $\overline{\Omega} \to B : \overline{w} \mapsto \nu\big(\overline{w}, S_{\overline{w}}(A)\big)$ is measurable for all $\nu \in M(\overline{\Omega}, \Omega, B)$, $A \in \overline{\Sigma} \times \Sigma$ (see **[1,** p. 20] and **[2,** Chapter III, §2]).

Everywhere below (X, Σ_X), (Y, Σ_Y) are measurable spaces and Σ_Y is a σ-algebra of countable type (i.e., it is generated by a countable class of subsets: **[2,** p. 32]), $Z = X \times Y$, $\Sigma_Z = \Sigma_X \times \Sigma_Y$.

§2. Vector-valued transition measures

In this section, $\mu \in M(X, \mathbb{R}^1)$, $\nu \in M(X, Y, B)$.

LEMMA 1. $|\nu| \in M\big(X, Y, [0, \infty]\big)$.

PROOF. Suppose that $A \in \Sigma_Y$ is fixed and $G = \{G_n\}$ is a countable algebra that generates Σ_Y and let $G_A = \{G_n \cap A\}$. For $x \in X$, $|\nu|(x, A) = \infty$, let N, ε be arbitrary positive numbers. We take successively: $A_1, \ldots, A_m \in \Sigma_Y$, $A_i \cap A_j = \varnothing$, $A_i \subset A$ such that $\sum_{i=1}^m \|\nu(x, A_i)\| > N$; $b_1, \ldots, b_m \in B'$ such that $\|b_i\| = 1$, $\big(b_i, \nu(x, A_i)\big) > \|\nu(x, A_i)\| - \dfrac{\varepsilon}{m}$; $\overline{A}_1, \ldots, \overline{A}_m \in G_A$, $\overline{A}_i \cap \overline{A}_j = \varnothing$ such that

$(b_i, \nu(x, \overline{A}_i)) > (b_i, \nu(x, A_i)) - \dfrac{\varepsilon}{m}$ $(i = 1, \dots, m)$ (the last is possible because the number measures $(b_i, \nu(x, \cdot))$ are finite and hence they are bounded, and the algebra G_A induces a trackable σ-algebra $\Sigma_x \cap A$).

We have that

$$\sum_{i=1}^{m} \|\nu(x, \overline{A}_i)\| \geq \sum_{i=1}^{m} (b_i, \nu(x, \overline{A}_i)) > N - 2\varepsilon,$$

which implies that

$$(1) \qquad |\nu|(x, A) = \sup\left\{\sum_{j=1}^{n} \|\nu(x, C_j)\| : C_j \in G_A,\ C_j \cap C_k = \varnothing\right\}.$$

If $x \in X$ is such that $|\nu|(x, A) < \infty$ then (1) follows from the fact that the algebra G_A induces $\Sigma_x \cap A$. Thus the equality (1) holds for all $x \in X$, which implies that the function $x \mapsto |\nu|(x, A)$ is measurable. The lemma is proved. □

It is next assumed that for any $A \in \Sigma_Z$ the function $X \to B : x \mapsto \nu(x, S_x(A))$ is weakly (in the sense of Pettis) integrable with respect to the measure μ; thus, the function $\nu\mu : \Sigma_Z \to B : \ \nu\mu(A) = \int \nu(x, S_x(A))\, d\mu(x)$ is defined.

LEMMA 2. $\nu\mu \in M(Z, B)$.

PROOF. We assume that $\mu \geq 0$ (otherwise we must take the Hahn decomposition $X = X^+ \cup X^-$). Suppose that $b \in B'$. In view of Lemma 1 ($B = \mathbb{R}^1$) it follows that the function $v : X \to \mathbb{R}: v(x) = |b\nu|(x, Y)$ is measurable. For $n = 1, 2, \dots$ we set $X_n = \{x \in X : v(x) \leq n\}$, $Z_n = X_n \times Y$. From the inequality $|b\nu(x, A)| \leq v(x)$, $\forall A \in Z_n$, and the Lebesgue Dominated Convergence Theorem (on passing to the limit under the integral sign) we get that the function $(b\nu)\mu$ is countably additive on $\Sigma_Z \cap Z_n$. Let $Z_n = Z_n^+ \cup Z_n^-$ be the Hahn decomposition with respect to the measure $(b\nu)\mu$, $Z^{\pm} = \bigcup_{n=1}^{\infty} Z_n^{\pm}$. It is clear that almost everywhere with respect to the measure μ

$$|b\nu(x, S_x(A))| \leq b\nu(x, S_x(Z^+)) - b\nu(x, S_x(Z^-))$$

for any $A \in \Sigma_Z$. We get from this and from the Lebesgue theorem that the function $(b\nu)\mu = (b, \nu\mu)$ is countably additive on Σ_Z. So the function $\nu\mu$ is weakly countably additive, and hence (Pettis theorem: **[2,** p. 172]) it is countably additive. The lemma is proved. □

LEMMA 3. *Suppose that X is a separable topological space, Σ_X is the Borel σ-algebra, and μ is a Radon measure. Then $|\nu\mu| = |\nu||\mu|$.*

PROOF. The inequality $|\nu\mu| \leq |\nu||\mu|$ is trivial, so we have to prove the reverse inequality. We assume, as before, that $\mu \geq 0$.

Assume first that $|\nu\mu|(Z) < \infty$. Fix $A \in \Sigma_Y$ and let G_A be the algebra introduced earlier. Let us consider the measure $\rho : \Sigma_X \to \mathbb{R}^1 : \rho(C) = |\nu\mu|(C \times A)$.

It is clear that ρ is absolutely continuous relatively to μ ($\rho \ll \mu$), so that the Radon-Nikodym density $\frac{d\rho}{d\mu} : X \to \mathbb{R}^1$ is defined. We take $\varepsilon > 0$, set $D_\varepsilon = \left\{x \in X : \frac{d\rho}{d\mu}(x) + \varepsilon < |\nu|(x, A)\right\}$, and assume that $\mu(D_\varepsilon) = \alpha > 0$. We consider further that $\alpha > \varepsilon$. Using Luzin's theorem **[1,** p.179] and the fact that G_A is countable and that μ is a Radon measure, we form a compactum $K_\varepsilon \subset D_\varepsilon$ such that $\mu(D_\varepsilon \setminus K_\varepsilon) < \varepsilon$ and the restrictions of all the functions $\nu(\cdot, C)$ $(C \in G_A)$, $\frac{d\rho}{d\mu}$ to K_ε are continuous. For $x_0 \in K_\varepsilon$ we take sets $C_1, \dots, C_n \in G_A$, $C_i \cap C_j = \varnothing$ in order to have the inequality

$$\sum_{i=1}^{n} \|\nu(x_0, C_i)\| > \frac{d\rho}{d\mu}(x_0) + \frac{\varepsilon}{2}$$

(see (1)), and then we take a neighborhood $U(x_o) \subset K_\varepsilon$ of the point x_0 such that the inequalities

$$\text{(2)} \qquad \begin{aligned} &\sum_{i=1}^{n} \|\nu(x_0, C_i)\| > \frac{d\rho}{d\mu}(x) + \frac{\varepsilon}{4}, \\ &\sum_{i=1}^{n} \|\nu(x, C_i) - \nu(x_0, C_i)\| < \frac{\varepsilon}{4} \end{aligned}$$

hold for all $x \in U(x_0)$. Since $\mu(K_\varepsilon) > 0$, we can choose a set $U = U(x_0)$ of positive measure: $\mu(U) > 0$ from the covering $\{U(x_0) \colon x_0 \in K_\varepsilon\}$ of the compactum K_ε. If we now set $Z_i = U \times C_i$ $(i = 1, \dots, n)$ and use the inequalities (2), then we have that

$$\begin{aligned} |\nu\mu|(U \times A) &\geq |\nu\mu|\Big(\bigcup_{i=1}^{n} Z_i\Big) \geq \sum_{i=1}^{n} \|\nu\mu(Z_i)\| \\ &= \sum_{i=1}^{n} \left\| \int_U \nu(x, C_i)\, d\mu(x) \right\| > \sum_{i=1}^{n} \left\| \int_U \nu(x_0, C_i)\, d\mu(x) \right\| - \frac{\varepsilon}{4}\mu(U) \\ &= \sum_{i=1}^{n} \|\nu(x_0, C_i)\|\, \mu(U) - \frac{\varepsilon}{4}\mu(U) > \int_U \frac{d\rho}{d\mu}\, d\mu = \rho(U) = |\nu\mu|(U \times A), \end{aligned}$$

which is impossible. Thus $\mu(D_\varepsilon) = 0$; i.e., $\frac{d\rho}{d\mu}(x) \geq |\nu|(x, A)$ for μ-almost all x. Integrating this inequality with respect to a set $C \in \Sigma_X$ we get that $|\nu\mu|(C \times A) \geq |\nu|\mu(C \times A)$ which, by the finiteness of the measure $|\nu\mu|$, implies the inequality $|\nu\mu| \geq |\nu|\mu$.

Now suppose that $|\nu\mu|(Z) = \infty$. If $A \in \Sigma_Z$ is such that $|\nu\mu|(A) = \infty$, then so much more $|\nu||\mu|(A) = \infty$. If $|\nu\mu|(A) < \infty$, then for the transition measure $\nu_A \colon X \times \Sigma_Y \to B \colon \nu_A(x, C) = \nu(x, C \cap S_x(A))$ we have $|\nu_A\mu|(D) = |\nu\mu|(A \cap D)$ $(D \in \Sigma_Z)$ and in particular, $|\nu_A\mu|(Z) = |\nu\mu|(A) < \infty$. Using what has been proved above with ν replaced by ν_A, we get $|\nu_A\mu| = |\nu_A||\mu|$, which implies $|\nu\mu|(A) = |\nu||\mu|(A)$. The lemma is proved. □

THEOREM 1. *Assume that the conditions of Lemma* 3 *are satisfied, that the measure* $\nu\mu$ *is* σ*-bounded, and that* $f \colon Z \to L(B, F)$ *is a* $\nu\mu$*-integrable function. Then for* μ*-almost all* x *the function* $y \mapsto f(x, y)$ *is* $\nu(x, \cdot)$*-integrable and the function* $x \mapsto \int_Y f(x, y)\nu(x, dy)$ *is equivalent* (mod μ) *to a* μ*-integrable one and*

$$\text{(3)} \qquad \int_Z f\, d\nu\mu = \int_X \int_Y f(x, y)\nu(x, dy)\, d\mu(x).$$

PROOF. Since f is $\nu\mu$-integrable, then $\int_Z \|f\|\, d|\nu\mu| < \infty$. This, together with Lemma 3 and the Fubini theorem for positive measures **[2,** p.110], gives us

$$\text{(4)} \qquad \int_Z \|f\|\, d|\nu\mu| = \int_X \int_Y \|f(x, y)\| |\nu|(x, dy)\, d|\mu|(x) < \infty.$$

Now we prove the theorem for $|\nu\mu|(Z) < \infty$. Let $X_1 = \{x \in X : |\nu|(x, Y) = \infty\}$, $X_2 = \{x \in X : \int_Y \|f(x,y)\| |\nu|(x,dy) = \infty\}$. The condition $|\nu\mu|(Z) < \infty$, Lemma 3, and inequality (4) give us $|\mu|(X_1) = |\mu|(X_2) = 0$. If $f = l \cdot \mathcal{I}_Q$ ($l \in L(B,F), \mathcal{I}_Q$ is the indicator of the set $Q \in \Sigma_Z$), then the equality (3) follows from the definition of the measure $\nu\mu$ and the continuity of the operator l. By the linearity of both sides with respect to f, the equality (3) is valid also for functions of the form $f = \sum_{i=1}^n l_i \mathcal{I}_{Q_i}$ ($Q_i \cap Q_j = \varnothing$). Suppose that f is a simple function: $f = \sum_{i=1}^\infty l_i \mathcal{I}_{Q_i}$ ($\{Q_i\}$ is a decomposition of Z). Let $f_n = \sum_{i=1}^n l_i \mathcal{I}_{Q_i}$. For any $x \in X \setminus (X_1 \cup X_2)$, the function $y \mapsto f(x,y)$ is simple, $\nu(x,\cdot)$-integrable, and by the definition of the integral of such functions,

$$(5) \qquad \int_Y f(x,y)\nu(x,dy) = \lim_{n\to\infty} \int_Y f_n(x,y)\nu(x,dy).$$

This, together with the equality $|\mu|(X_1 \cup X_2) = 0$ gives us that the function $x \mapsto \int_Y f(x,y)\nu(x,dy)$ is equivalent to a measurable one. Then by the inequalities

$$\left\| \int_Y f_n(x,y)\nu(x,dy) \right\| \le \int_Y \|f_n(x,y)\| |\nu|(x,dy) \le \int_Y \|f(x,y)\| |\nu|(x,dy),$$

relations (4) and (5), and the Lebesgue theorem we have

$$\lim_{n\to\infty} \int_X \int_Y f_n(x,y)\nu(x,dy)\, d\mu(x) = \int_X \int_Y f(x,y)\nu(x,dy)\, d\mu(x).$$

But from the validity of (3) for the functions f_n and the definition of the integral of simple functions it turns out that

$$\lim_{n\to\infty} \int_X \int_Y f_n(x,y)\nu(x,dy)\, d\mu(x) = \lim_{n\to\infty} \int_Z f_n\, d\nu\mu = \int_Z f\, d\nu\mu.$$

Thus, in the case of a simple function all assertions of the theorem are proved. Now, if f is an arbitrary $\nu\mu$-integrable function, then representing it as a uniform limit of a sequence of simple functions, using (4) and the μ-integrability of the function $x \mapsto |\nu|(x, Y)$ (Lemma 3), and applying twice the Lebesgue theorem we get all assertions of Theorem 1.

Now suppose that $|\nu\mu|(Z) = \infty$, $Z = \bigcup_{n=1}^\infty Z_n$, $Z_1 \subset Z_2 \subset \dots \subset Z_n \subset \dots$, $Z_n \in \Sigma_Z$, $|\nu\mu|(Z_n) < \infty$. Let us consider a transition measure $\nu_n : X \times \Sigma_Y \to B$: $\nu_n(x, A) = \nu(x, A \cap S_x(Z_n))$. It is clear that for $D \in \Sigma_Z$

$$(6) \qquad (\nu_n\mu)(D) = (\nu\mu)(D \cap Z_n),$$

so that $|\nu_n\mu|(Z) < \infty$ and hence, the equality (3) is valid when ν is replaced by ν_n; from this and the equality (6) we have

$$(7) \qquad \int_{Z_n} f\, d\nu\mu = \int_X \int_{Y_n} f(x,y)\nu(x,dy)\, d\mu(x)$$

($Y_n = S_x(Z_n)$). We remark that for almost all x (modμ), $|\nu|(x, Y_n) = |\nu_n|(x, Y) < \infty$, so that the measure $\nu(x,\cdot)$ is σ-bounded. From (4) and the definition of integral with respect to a σ-bounded measure, we now get that for almost all x

$$\lim_{n\to\infty}\int_{Y_n} f(x,y)\nu(x,dy)=\int_Y f(x,y)\nu(x,dy).$$

From this, the estimate $\|\int_{Y_n} f(x,y)\nu(x,dy)\|\le\int_Y\|f(x,y)\||\nu|(x,dy)$, inequality (4), and the Lebesgue theorem we obtain the passage to the limit under the outer integral sign as $n\to\infty$ on the right-hand side of (7) and the proof of the theorem is complete. □

§3. Vector-valued initial distribution

In the present section $\mu\in M(X,B)$, $\nu\in M(X,Y,\mathbb{R}^1)$.

DEFINITION. A measurable function $f:X\to\mathbb{R}^1$ is said to be weakly μ-integrable if $\forall A\in\Sigma_X\ \exists I\in B\ \forall b\in B'$ the function $f\mathcal{I}_A$ is $b\mu$-integrable and $(b,I)=\int f\mathcal{I}_A\,db\mu$; in addition $\int_A f\,d\mu\overset{\text{def}}{=}I$.

For weakly μ-integrable functions $f: X\to\mathbb{R}^1$ we define $f\mu\colon\Sigma_X\to B$: $f\mu(A)=\int_A f\,d\mu$; it is easy to see that $f\mu\in M(X,B)$.

LEMMA 4. $|f\mu|=|f||\mu|$.

PROOF. We consider $f\ge 0$ and suppose that $X_n=\{x:\frac{1}{n}\le f(x)\le n\}$. For any $A\in\Sigma_x\cap X_n, b\in B', \|b\|=1$ we have $|f\mu|(A)\ge|b(f\mu)|(A)=f|b\mu|(A)\ge\frac{1}{n}|b\mu|(A)$; then it follows from $|f||\mu|(A)=\infty$ (and $|\mu|(A)=\infty$) that $|f\mu|(A)=\infty$. The implication $|f\mu|(A)=\infty\Longrightarrow|f||\mu|(A)=\infty$ is evident. If $A\subset X_n$ is such that $|f||\mu|(A)<\infty$, then $|\mu|(A)<\infty$ and the equality $|f\mu|(A)=|f||\mu|(A)$ arises from (by passing to the limit) its validity for simple functions and the estimate $|f\mu|(A)\le(\sup|f|)|\mu|(A)$. Thus $|f\mu|=|f||\mu|$ on $\Sigma_X\cap X_n$, and hence (by the countable additivity of $|f\mu|$, $|f||\mu|$), also on $\Sigma_X\cap\{0<f<\infty\}$. It is evident that $|f\mu|=|f||\mu|=0$ on $\Sigma_X\cap\{f=0\}$. The lemma is proved. □

It is further assumed that for any $A\in\Sigma_Z$ the function $X\to\mathbb{R}^1$, $x\mapsto\nu(x,S_x(A))$ is weakly μ-integrable; so, a function $\nu\mu\colon\ \Sigma_Z\to B$, $\nu\mu(A)=\int\nu(x,S_x(A))\,d\mu(x)$ is defined.

LEMMA 5. $\nu\mu\in M(Z,B)$.

PROOF. Assume that $b\in B'$. From the definition of the weak μ-integral we obtain that $b(\nu\mu)=\nu(b,\mu)$. By Lemma 2 (with $\nu\in M(X,Y,B)$ replaced by $\nu\in M(X,Y,\mathbb{R}^1)$ and $\mu\in M(X,\mathbb{R}^1)$ replaced by $b\mu\in M(X,\mathbb{R}^1)$) $\nu(b\mu)\in M(Z,\mathbb{R}^1)$; so $\nu\mu$ is weakly countably additive, and hence, it is countably additive. The lemma is proved. □

LEMMA 6. *If for any $A\in\Sigma_Z$ the function $f_A\colon X\to\mathbb{R}^1$, $x\mapsto\nu(x,S_x(A))$ is $|\mu|$-integrable, then the function $\nu|\mu|\colon\ \Sigma_Z\to\mathbb{R}^1$, $A\mapsto\int f_A\,d|\mu|$ is countably additive.*

PROOF. If $|\mu|(X)<\infty$, then the statement follows from Lemma 5 (with μ replaced by $|\mu|$).

Suppose that μ is σ-bounded and $X = \bigcup_{n=1}^{\infty} X_n$, $X_1 \subset X_2 \subset \cdots$, $X_n \in \Sigma_X$, $|\mu|(X_n) < \infty$. Then $\nu|\mu|$ is countably additive on $Z_n = X_n \times Y$. Let $Z_n = Z_n^+ \cup Z_n^-$ be the Hahn decomposition for $\nu|\mu|$, $Z^\pm = \bigcup_{n=1}^{\infty} Z_n^\pm$. It is clear that for any $A \in \Sigma_Z$ $|\nu(x, S_x(A))| \leq \nu(x, S_x(Z^+)) - \nu(x, S_x(Z^-))$ for $|\mu|$-almost all $x \in X$, and the Lebesgue theorem implies the countable additivity of $\nu|\mu|$ on Z.

Finally, consider the case of an arbitrary (i.e., not σ-bounded) measure μ. Assume that $A_1, \dots, A_n, \dots$, $A_n \in \Sigma_Z$, $A_1 \subset A_2 \subset \dots$, $A = \bigcup_{n=1}^{\infty} A_n$. Let $X_{n,m} = \{x \in X \colon |f_{A_n}|(x) > \frac{1}{m}\}$, $X_0 = \bigcup_{n,m=1}^{\infty} X_{n,m}$. The condition of the lemma implies that $|\mu|(X_{n,m}) < \infty$; so $|\mu|$ is σ-bounded on X_0 which by what we have already proved, implies that $\nu|\mu|$ is countably additive on $X_0 \times Y$. Further, we have

$$\begin{aligned}
\lim_{n\to\infty} \nu|\mu|(A_n) &= \lim_{n\to\infty} \int_X f_{A_n}\, d|\mu| = \lim_{n\to\infty} \int_{X_0} f_{A_n}\, d|\mu| \\
&= \lim_{n\to\infty} \nu|\mu|(A_n \cap (X_0 \times Y)) = \nu|\mu|(A \cap (X_0 \times Y)) \\
&= \int_{X_0} f_A\, d|\mu| = \int_X f_A\, d|\mu| = \nu|\mu|(A),
\end{aligned}$$

and the lemma is proved. □

Before the following statement we remark that by the boundedness of a numerical measure and Lemma 1 ($B = \mathbb{R}^1$) $|\nu| \in M(X, Y, [0, \infty))$, so that the measure $|\nu||\mu|$ is well defined.

LEMMA 7. $|\nu\mu| = |\nu||\mu|$.

PROOF. The inequality $|\nu\mu| \leq |\nu||\mu|$ is trivial, so we have to prove the reverse inequality. Assume first that $|\nu\mu|(Z) < \infty$. For any $A \in \Sigma_Z$ the function $f_A \colon X \to \mathbb{R}^1 \colon x \mapsto \nu(x, S_x(A))$ is weakly μ-integrable, so that by Lemma 4

$$|f_A\mu| = |f_A||\mu|. \tag{8}$$

But it is clear that

$$|f_A\mu| \leq |\nu\mu|(A) \leq |\nu\mu|(Z) < \infty, \tag{9}$$

which implies that $|f_A||\mu|(X) < \infty$. By Lemma 6, $\nu|\mu| \in M(Z, \mathbb{R}^1)$. Let $Z^\pm$ be the Hahn decomposition for $\nu|\mu|$. We introduce the transition measures $\rho^\pm \colon X \times \Sigma_Y \to \mathbb{R}^1$, $\rho^\pm(x, A) = \pm\nu(x, S_x(A \cap Z^\pm))$. Because $\pm\nu|\mu| \geq 0$ on $Z^\pm$ and $\rho^\pm|\mu|(A) = \pm\nu|\mu|(A \cap Z^\pm)$ $(A \in \Sigma_Z)$, we have $\rho^\pm|\mu| \geq 0$. It follows from this that for any $A \in \Sigma_Y$, $\rho^\pm(x, A) \geq 0$ for $|\mu|$-almost all x. In particular, we may take a set $X_0 \in \Sigma_X$ such that $|\mu|(X \backslash X_0) = 0$ and for all $x \in X_0$, $n = 1, 2, \dots$, the inequalities $\rho^\pm(x, G_n) \geq 0$ hold ($\{G_n\}$ is a countable algebra, inducing Σ_Y). It follows from the countable additivity of the functions $\rho^\pm(x, \cdot)$ that $\rho^\pm \geq 0$ on $X_0 \times \Sigma_Y$. From this and the minimal Jordan decomposition $\nu(x, \cdot) = \nu^+(x, \cdot) - \nu^-(x, \cdot)$ we get that $\rho^\pm(x, \cdot) \geq \nu^\pm(x, \cdot)$ for $x \in X_0$ (in fact, these become equalities for $x \in X_1$, $|\mu|(X \backslash X_1) = 0$), which implies that

$$\rho^+|\mu| + \rho^-|\mu| \geq |\nu||\mu|. \tag{10}$$

By (9) and (8) with A replaced by $A^\pm \subset Z^\pm$, we see that $|\nu\mu|(A^\pm) \geq |f_{A^\pm}||\mu|(X) = \rho^\pm|\mu|(A^\pm)$.

Summing these inequalities and using (10), we find that $|\nu\mu| \geq |\nu||\mu|$ which is what was required.

The case $|\nu\mu|(Z) = \infty$ is treated in a way analogous to that in Theorem 1. The lemma is proved. □

Before we formulate the following theorem, let us recall that we do not require that the measure μ is σ-bounded. In §1, however, the integrals were defined with respect to σ-bounded measures only; the introduction of the set $\overline{X}$ below is occasioned by this circumstance.

THEOREM 2. *We assume that the measure $\nu\mu$ is σ-bounded and the function $f: Z \to L(B, F)$ is $\nu\mu$-integrable. Then for $|\mu|$-almost all x the function $y \mapsto f(x, y)$ is $\nu(x, \cdot)$-integrable, there exists a set $\overline{X} \in \Sigma_X$ such that $\nu(x, \cdot) \equiv 0$ for $x \in X \backslash \overline{X}$, the measure μ is σ-bounded on $\overline{X}$, and the function $x \mapsto \int f(x, y)\nu(x, dy)$ is equivalent* (mod $|\mu|$) *to a μ-integrable one on $\overline{X}$. Then*

$$\int_Z f \, d\nu\mu = \int_{\overline{X}} \int_Y f(x, y)\nu(x, dy) \, d\mu(x).$$

PROOF. Suppose that $Z = \bigcup_{n=1}^{\infty} Z_n$, $Z_1 \subset Z_2 \subset \ldots$, $Z_n \in \Sigma_Z$, $|\nu\mu|(Z_n) < \infty$. Let $X_{n,m} = \{x \in X : |\nu|(x, Z_n) > \frac{1}{m}\}$. By Lemmas 1 and 7, $|\mu|(X_{n,m}) < \infty$, so that μ is σ-bounded on the set $\overline{X} = \bigcup_{n,m=1}^{\infty} X_{n,m}$. From $\int \|f\| \, d|\nu\mu| < \infty$ (by the condition of the theorem), Lemma 7, and the Fubini theorem for the positive measures, we get inequality (4). Arguments similar to those in the proof of Theorem 1 (with X replaced by $\overline{X}$) complete the proof of Theorem 2. □

§4. Some applications

1. Conditional expectations with respect to "vector-valued probability". Let X, Y be complete separable metric spaces with their Borel σ-algebras Σ_X, Σ_Y, $(Z, \Sigma_Z) = (X \times Y, \Sigma_X \times \Sigma_Y)$, $P \in M_0(Z, B)$. For $A \in \Sigma_Y$, let $P_X^A : \Sigma_X \to B : C \mapsto P(C \times A)$, $|P|_X^A : \Sigma_X \to [0, \infty) : C \mapsto |P|(C \times A)$, $\mu = |P|_X^Y$. It is clear that $P_X^A \ll \mu$ and we may assume that for any $A \in \Sigma_Y$ there exists a μ-integrable function $\nu^A : X \to B$, satisfying the equality

$$\int_C \nu^A \, d\mu = P_X^A(C) \tag{11}$$

for any $C \in \Sigma_X$. (Here ν^A is the Radon-Nikodym density $\dfrac{dP_X^A}{d\mu}$. For the existence of such a density see [**3**, Chapter 8, §19].) We take a transition measure $\rho: X \times \Sigma_Y \to [0, \infty]$ such that for any $C \in \Sigma_X$, $A \in \Sigma_Y$

$$\int_C \rho(x, A) \, d\mu(x) = |P|_X^A(C). \tag{12}$$

($\rho(\cdot, A)$ is a regular representative of the conditional probability $|P|(X \times A \mid \Sigma_X$.) In view of the assumptions about Y such a representative always exists: [**4**, p. 53]). It follows from (11) and (12) that for any $A \in \Sigma_Y$ $\|\nu^A(x)\| \leq \rho(x, A)$ for μ-almost all x. We take a countable algebra $\{G_n\}$ inducing Σ_Y and, after removing from X a set of μ-measure zero, we can assume that the inequalities

$$\left\| v^{G_n}(x) \right\| \leq \rho(x, G_n) \tag{13}$$

are valid everywhere on X.

It follows from (11) and the additivity of P that the set of x such that even one of the equalities $v^A(x) = v^{A_1}(x) + v^{A_2}(x)$ fails has μ-measure zero; removing this set from X, we can assume that the function $A \mapsto v^A(x)$ is additive on the algebra G for all x. But then (13) implies that the function $A \mapsto \left(b, v^A(x)\right)$ is countably additive on G for any $b \in B'$, and hence, it is extended, with preservation of the countable additivity, to σ-algebra Σ_Y. Thus the mapping $v(x)\colon \Sigma_Y \to B''$: $\left(b, v(x, A)\right) = \left(b, v^A(x)\right)$ is defined. By (13) for all $A \in \Sigma_Y$ the vector $v(x, A) \in B''$ is approximated by vectors $v^{G_n}(x) \in B$ with respect to the norm of B'', so that $v(x, A) \in B$. Using the Pettis theorem we get that the function $v(x, \cdot)\colon \Sigma_Y \to B$, $A \mapsto v(x, A)$ is countably additive. It is evident that for any $A \in \Sigma_Y$ the function $v(\cdot, A)\colon X \to B$, $x \mapsto v(x, A)$ is measurable, so that $v \in M(X, Y, B)$. From (11) and the equality $v(x, A) = v^A(x)$ for $A \in G$ we get the equality $P = v\mu$ (the formula of complete "vector-valued probability": $v(\cdot, A)$ is a conditional "probability" of the event $X \times A$ with respect to the σ-algebra $\Sigma_X \times Y$). If $f\colon Z \to L(B, F)$ is a P-integrable function, then using Theorem 1 we obtain the equality

$$\int f\, dP = \int_X \int_Y f(x, y) v(x, dy)\, d\mu(x), \tag{14}$$

which is the vector analogue of the classical theoretical probability relation (the inner integral is the conditional expectation of the "random variable" f with respect to the σ-algebra $\Sigma_X \times Y$).

As in the proof of Theorem 1, it is easy to see that the formula (14) is valid also in the case when P is a σ-bounded vector-valued measure.

2. Newton-Leibniz formula on a Banach space. Suppose that $X = [\alpha, \beta]$; Y is a separable Banach space; Σ_X, Σ_Y are the Borel σ-algebras; μ is Lebesgue measure on X; $\varphi\colon Y \to \mathbb{R}^1$ is a Fréchet continuously differentiable function; $\Omega_x = \{y \in Y : \varphi(y) = x\}$ ($x \in [\alpha, \beta]$); $V = \{y \in Y : \alpha \leq \varphi(y) \leq \beta\}$; $\rho \in M(Y, B)$; $|\rho|\{y \in V : \varphi'(y) = 0\} = 0$.

Assume that the measure ρ is τ_s-differentiable in the directions of a dense linear subset $Y_0 \subset Y$ with respect to the topology $\sigma(B, B')$ **[5**, Chapter IV, §2] we denote by $D_y\rho \in M(Y, B)$ the corresponding differential.

For any $b \in B'$ the measures $b\rho$, $b\rho^{\pm}$ ($b\rho = b\rho^+ - b\rho^-$ is the Jordan decomposition of the measure $b\rho$) are τ_s-differentiable with respect to Y_0 and $D_y b\rho = bD_y\rho = D_y(b\rho^+) - D_y(b\rho^-)$. Thus, according to the construction in **[6]** if Ω_x is a surface, then the σ-bounded surface measures $(b\rho)_{\Omega_x}$, $(b\rho)^{\pm}_{\Omega_x}$ on the surface Ω_x are defined, where $(b\rho)^{\pm}_{\Omega_x} \geq 0$ and

$$(b\rho)_{\Omega_x} = (b\rho)^+_{\Omega_x} - (b\rho)^-_{\Omega_x}. \tag{15}$$

As is proved in **[7]** (see Theorem 1 there; in the special situation under consideration

$F(t,\omega) = t - \varphi(\omega)$, $(b\rho)^{\pm} \geq 0$, and the second condition of the theorem is unnecessary), the functions

$$\nu_b^{\pm}\colon [\alpha,\beta] \times \Sigma_Y \to [0,\infty] : \nu_b^{\pm}(x,A) = \int_{A\cap\Omega_x} \|\varphi'\|^{-1}\, d(b\rho)^{\pm}_{\Omega_x}$$

are well defined ($\varphi' \in Y'$ is the Fréchet derivative of φ), they are transition measures from $[\alpha,\beta]$ to Y, and for any $A \in \Sigma_Y$

$$(b\rho)^{\pm}(A\cap V) = \int_\alpha^\beta \nu_b^{\pm}(x,A)\, dx \qquad (dx = d\mu(x)).$$

This and (15) give us

$$b\rho(A\cap V) = \int_\alpha^\beta \nu_b(x,A)\, dx, \tag{16}$$

where $\nu_b = \nu_b^+ - \nu_b^- \in M(X, Y, \mathbb{R}^1)$. For any x, A the function $b \mapsto \nu_b(x,A)$ is linear and continuous; i.e., the vector $\nu(x,A) \in B''$: $(\nu(x,A), b) = \nu_b(x,A)$ is defined. From the construction of the surface measures, the definition of $\nu(x,A)$, and the inclusion $D_y\rho \in M(Y,B)$ it follows that there exists a sequence $V_1 \subset \dots \subset V_n \subset \dots$, $V_n \in \Sigma_Y$, $\bigcup_{n=1}^{\infty} V_n = V$, such that $\nu(x,A) \in B$ for any $A \subset V_n$. This and the Pettis theorem imply that the function $\nu(x,\cdot) : A \mapsto \nu(x,A)$ is countably additive on $V_n \cap \Sigma_Y$ (the measure $\rho_{\Omega_x} = \|\varphi'\|\nu(x,\cdot)$ is a B-valued surface measure). As it is similarly done in [7] for the scalar case, it is proved that the function $\nu(\cdot,A)\colon x \mapsto \nu(x,A)$ is measurable ($A \in V_n$). So $\nu \in M(X, V_n, B)$. By (16)

$$\nu\mu(X\times A) = \rho(A), \qquad |\nu\mu|(X\times A) = |\rho|(A) \tag{17}$$

($A \in V_n$). Thus, if the measure ρ is σ-bounded on V, then the measure $\nu\mu$ is also σ-bounded. Now if $f\colon V \mapsto L(B,F)$ is a ρ-integrable function, then by (17) the function $f_n\colon X\times V_n \to L(B,F)$, $f_n(x,y) = f(y)$ is $\nu\mu$-integrable, so that Theorem 1 and (17) imply

$$\int_{V_n} f\, d\rho = \int_{X\times V_n} f_n\, d\nu\mu = \int_X \int_{V_n} f_n(x,y)\nu(x,dy)\, d\mu(x)$$
$$= \int_\alpha^\beta \int_{V_n\cap\Omega_x} f\|\varphi'\|^{-1}\, d\rho_{\Omega_x}\, dx.$$

Passing here to the limit as $n\to\infty$ (which is valid by the condition $\int_V \|f\|\, d|\rho| < \infty$ and Lemma 3) we have

$$\int_V f\, d\rho = \int_\alpha^\beta \int_{\Omega_x} f\|\varphi'\|^{-1}\, d\rho_{\Omega_x} dx$$

(the Newton-Leibniz formula or the formula of iterated integration in curvilinear coordinates; in [7] this formula is established in the case when f, ρ are numerical functions).

3. Formulas for vector analysis on Banach spaces[2]**.** The propositions in the present subsection are given without proofs in view of their complexity and volume.

The following proposition, unlike all other results of §4, is only indirectly connected with the Fubini theorem (see below, the text preceding Proposition 2).

PROPOSITION 1. *Suppose that a Banach space B_0 is imbedded in B, where the imbedding operator $I\colon B_0 \to B$ is absolutely summing* **[8]**. *Assume that the sequence $\mu_n \in M(X, B_0)$ is such that for any $A \in \Sigma_x$ there exists a limit (with respect to norm in B_0) $\lim_{n\to\infty} \mu_n(A) = \mu(A)$.*

Then

(1) $\mu \in M(X, B_0)$;

(2) $\sup_n |I\mu_n|(X) < \infty$;

(3) *for any measurable bounded function $f\colon X \to L(B, F)$ we have*

$$\lim_{n\to\infty} \int f\, dI\mu_n = \int f\, dI\mu$$

(here $M(X, B) \ni I\mu_n(I\mu)\colon A \mapsto I\mu_n(A)(I\mu(A))$).

Below we use the following notation: H is a separable Hilbert space densely imbedded in B and the imbedding operator $I\colon H \to B$ is absolutely summing; I^* is the adjoint operator giving a natural imbedding $I^*\colon B^* \to H$; $V^0, \overline{V}$, and ∂V are the interior, closure, and boundary, respectively, of a set $V \subset B$; $C^m(V^0, Y)$ is the space of functions defined on V^0, taking values in a Banach space Y and having derivatives (in the Fréchet sense) of orders $0, 1, \dots, m$ that are bounded continuous functions on V^0; $C^m(\overline{V}, Y)$ is the subspace of $C^m(V^0, Y)$ consisting of the functions whose derivatives of orders $0, 1, \dots, m-1$ admit continuous extension to the set $\overline{V}$.

A point $\omega \in \partial V$ is said to be an ordinary point if there exist an open set $U(\omega) \ni \omega$ and a function $f \in C^1(U(\omega), \mathbb{R}^1)$ such that $f'(\omega) \neq 0$ and $\partial V \cap U(\omega) = f^{-1}(0)$; the collection of all ordinary points is denoted by ∂V_0. Suppose that the measure $\mu \in M(B, \mathbb{R}^1)$ (Σ_B is the Borel σ-algebra) is differentiable in all directions of H. Then a σ-bounded surface measure can be defined according to the construction in **[6]** for any surface $\Omega \subset B$. A set $V \in \Sigma_B$ is said to be regular if ∂V is a surface and $|\mu_\Omega|(\partial V \backslash \partial V_0) = 0$. For $\omega \in \partial V$ we let $n(\omega)$ stand for the outward (with respect to V) unit normal to ∂V if it exists and is unique; we suppose that $n(\omega) = 0$ if the outward normal does not exist or is two-valued.

Everywhere below, V is a regular set. We note that all encountered integrands are defined almost everywhere with respect to the corresponding measures. The following Propositions 2–5 are proved with the help of the theorems from §§2 and 3 and Proposition 1.

PROPOSITION 2. *Suppose that $U \in C^1(\overline{V}, \mathbb{R}^1)$, $a \in H$, and the function $u(n, a)\colon \partial V \to \mathbb{R}^1$ is $\mu_{\partial V}$-integrable. Then*

$$\int_V D_a u\, d\mu + \int_V u\, dD_a\mu = \int_{\partial V} u(n, a)\, d\mu_{\partial V}$$

(integration by parts formula).

[2] The results of this subsection were obtained jointly with E. I. Efimova.

We remark further that by the results of [**5**, Chapter IV, §3, p. 3)] for any $A \in \Sigma_B$ the function

$$D\mu(A)\colon H \to \mathbb{R}^1\colon h \mapsto D_h\mu(A)$$

is linear and continuous; thus the formula $(\mu'(A), h) = D\mu(A)h$ determines a measure $\mu' \in M(B, H)$ unambiguously. Since the imbedding $I\colon H \to B$ is absolutely summing, for $b \in C^1(V^0, B^*)$, $x \in V^0$ the operator $I^*b'(x)I\colon H \to H$ is nuclear, so that the function

$$\div b\colon V^0 \to \mathbb{R}^1\colon x \mapsto \operatorname{Tr} I^*b'(x)I$$

is well defined.

PROPOSITION 3. *Suppose that* $b \in C^1(\overline{V}, B^*)$ *and that the function* $\|b\|_H \cdot \|n\|_H\colon \partial V \to \mathbb{R}^1$ *is* $\mu_{\partial V}$*-integrable. Then all the integrals in* (18) *exist, and*

$$\int_V \div b\,d\mu + \int_V b\,d\mu' = \int_{\partial V} (b, n)\,d\mu_{\partial V} \tag{18}$$

(*Gauss-Ostrogradskiĭ formula*).

Below, $A \in C^1(\overline{V}, L(B, B))$, $b \in C^1(\overline{V}, B^*)$, $u \in C^2(\overline{V}, \mathbb{R}^1)$, and $\alpha \in C^0(\overline{V}, \mathbb{R}^1)$. Let

$$\mathscr{L}^*u\colon V^0 \to \mathbb{R}^1\colon x \mapsto \div[A^*(x)u'(x)] - \div u(x)b(x) + \alpha(x)u(x).$$

PROPOSITION 4. *Suppose that the function* $\|A^*u'\|_H \cdot \|n\|_H\colon \partial V \to \mathbb{R}^1$ *is* $\mu_{\partial V}$*-integrable. Then all the integrals in* (19) *exist, and*

$$\int_V [\mathscr{L}^*u + \div ub - \alpha u]\,d\mu + \int_{V.} (A^*u')\,d\mu' = \int_{\partial V} \frac{\partial u}{\partial A_n}\,d\mu_{\partial V} \tag{19}$$

(*the first Green formula*).

Assume now that $A(x)IH \subset H$ for any $x \in V^0$ and that $AI \in C^1(\overline{V}, L(H, H))$. Then for any orthonormal basis $\{e_n\}$ in H, the functions $a_{ij}\colon \overline{V} \to \mathbb{R}^1$, $x \mapsto (e_i, A(x)e_j)$ belong to $C^1(\overline{V}, \mathbb{R}^1)$ and, consequently, the iterated series on the left-hand side of the formal equality

$$\sum_{j=1}^{\infty}\sum_{i=1}^{\infty}\left[\int_V u\frac{\partial a_{ij}}{\partial e_i}\,d\frac{\partial \mu}{\partial e_i} + \int_V ua_{ij}\,d\frac{\partial^2 \mu}{\partial e_i \partial e_j}\right] = \int_V u\,d(\div A\mu')$$

is defined unambiguously. If this series converges for any orthonormal basis $\{e_i\}$ and its sum does not depend on the choice of basis, then we define

$$\int_V u dZ\mu = \int_V u\,d(\div A\mu') + \int_V ub\,d\mu' + \int_V u\alpha\,d\mu$$

and say that the integral on the left-hand side of the last equality exists.

Further, assume that for any $h \in H$ the measure (h, μ') is differentiable in the directions of H. A function $\varphi\colon \partial V \to \mathbb{R}^1$ is said to be $\mu'_{\partial V}$-integrable if φ is (h, μ')-integrable for any $h \in H$; a surface $\Omega \subset \partial V$ is $\mu'_{\partial V}$-integrable if the function $\mathscr{I}_\Omega\colon \partial V \to \mathbb{R}^1$ is $\mu'_{\partial V}$-integrable. In the last case the equality

$$(h, \mu'_{\partial V}(Q)) = (h\mu')_{\partial V}(Q) \quad \text{for any } h \in H,\ Q \in \Sigma_\Omega$$

(Σ_Ω is the Borel σ-algebra on Ω) determines a surface measure $\mu'_{\partial V} \in M(\Omega, H)$ unambiguously. According to the results in [**6**], for the surface ∂V (and also for any surface) there exists a decomposition $\partial V = \bigcup_{n=1}^{\infty} \Omega_n$, $\Omega_n \in \Sigma_B$, $\Omega_1 \subset \dots \subset \Omega_n \subset \dots$, Ω_n are $\mu'_{\partial V}$-integrable. Then $I\mu'_{\partial V} \in M_0(\Omega_n, B)$ for all $n = 1, 2, \dots$; i.e., $I\mu'_{\partial V}$ is a σ-bounded measure on ∂V.

PROPOSITION 5. *Suppose that the functions* $u\|b\|_H \cdot \|n\|_H$, $\|A^*u'\|_H \cdot \|n\|_H : \partial V \to \mathbb{R}^1$ *are* $\mu_{\partial V}$*-integrable, and that the functions* $u\|A^*n\|_{B^*}$, $u\|A\|_{L(H,H)} : \partial V \to \mathbb{R}^1$ *are* $\mu'_{\partial V}$*-integrable. Then all the integrals in* (20) *exist, and*

$$(20)\qquad \int_V Z^* u\, d\mu - \int_V u\, dZ\mu = \int_{\partial V} \left[\frac{\partial u}{\partial A n} - u(b, n) \right] d\mu_{\partial V} - \int_{\partial V} u A^* n\, dI\mu'_{\partial V}$$

(*the second Green formula*).

§5. Remarks and comments

1. Occasionally, in the definition of a transition measure the measurability of the function (strong measurability) $x \mapsto v(x, A)$ is not required: if this function is weakly measurable, then all the results of §1 become invalid (corresponding counterexamples exist).

2. The topological restrictions imposed on the space (X, Σ_x, μ) in Lemma 3 and Theorem 1 seem alien to our considerations. These restrictions can be "transferred" to the spaces (Y, Σ_y), B. Whether the Fubini theorem (for vector-valued transition measures) remains valid without the topological conditions is an open question (besides the trivial case $v(x, A) \equiv v(A)$).

3. It is interesting to note that the "mirror-like" analogy of Lemma 4 (i.e., when the measure μ is numerical and the function f is vector-valued and weakly μ-integrable) is false: see **[1,** p. 105].

4. From Theorems 1 and 2, it is easy to get their analogues in the case when the measurability of f is understood in some extended case (Borel-, Bochner-, $(Z, \overline{\Sigma}_z, \overline{|v_\mu|})$-measurability, and so on).

5. Attempts to synthesize Theorems 1 and 2, i.e., to give a proof of the Fubini theorem in the case when both measures μ, v are vector-valued, are vain (it is enough to take $B = \mathbb{R}^2$, with μ and v taking values in orthogonal spaces).

6. A very weak variant (comparatively) of Theorems 1 and 2 is given in **[9]**.

7. Suppose that, in addition to the conditions of Proposition 1, $\dim B < \infty$, while the imbedding I is not required to be absolute summing. Then, as is well known (and follows from Proposition 1) all the conclusions of the proposition remain valid. The situation changes sharply in the case when $\dim B = \infty$: variations of the measures $I\mu_n$ (even when they are bounded) may not be bounded in total, and if this is the case, then, as a rule, the passage to the limit under the integral sign is impossible (the author knows some corresponding examples). Thus, the requirement that I be absolute summing in Proposition 1 is essential.

8. The author has used special cases of the Newton-Leibnitz and Green formulas in solving infinite-dimensional differential equations **[10,11]** and the analysis of random process characteristics **[12]**.

References

1. N. N. Vakhaniya, V. I. Tarieladze, and S. A. Chobanyan, *Probability distributions on Banach spaces*, "Nauka", Moscow, 1985; English transl., Reidel, New York, 1987.
2. J. Neveu, *Mathematical foundations of the calculus of probability*, Holden-Bay, San Francisco, CA, 1965.
3. R. E. Edwards, *Functional analysis. Theory and applications*, Holt, Rinehart and Winston, New York, 1965.
4. I. I. Gihman and A. V. Skorohod, *Theory of stochastic processes*, Vol. I, "Nauka", Moscow, 1971; English transl., Springer-Verlag, Berlin and New York, 1974.
5. Yu. L. Daletskiĭ and S. V. Fomin, *Measures and differential equations in infinite-dimensional spaces*, "Nauka", Moscow, 1983; English transl., Kluwer, Dordrecht, 1991.
6. A. V. Uglanov, *Surface integrals in a Banach space*, Mat. Sb. **110** (1979), no. 2, 189–217; English transl. in Math. USSR-Sb. **38** (1981).
7. ———, *The Newton-Leibniz formula on Banach spaces and the approximations of functions of an infinite-dimensional argument*, Izv. Akad. Nauk SSSR Ser. Mat. **51** (1987), no. 1, 152–170; English transl. in Math. USSR-Izv. **30** (1988).
8. A. Pietsch, *Nuclear locally convex spaces*, Academie-Verlag, Berlin, 1965.
9. H. Watanabe, *Path integral for some systems of partial differential equations*, Proc. Japan Acad. Ser. A Math. Sci. **60** (1984), 86–89.
10. A. V. Uglanov, *Division of generalized functions of an infinite number of variables by polynomials*, Dokl. Akad. Nauk SSSR **264** (1982), no. 5, 1096–1099; English transl. in Soviet Math. Dokl. **25** (1982).
11. ———, *Integral representation of functions in a Banach space*, Dokl. Akad. Nauk SSSR **282** (1985), no. 4, 800–804; English transl. in Soviet Math. Dokl. **31** (1985).
12. ———, *On the smoothness of distribution for functionals of random processes*, Teor. Veroyatnost. i Primenen. **33** (1988), no. 3, 535–544; English transl. in Theory Probab. Appl. **33** (1988).

YAROSLAVL′ STATE UNIVERSITY, SOVETSKAYA 14, YAROSLAVL′, RUSSIA

Amer. Math. Soc. Transl.
(2) Vol. **163**, 1995

Generalized Derivations of Algebras

E. B. Vinberg

In this paper we discuss a generalization of the notion of a derivation of an algebra that possesses a number of interesting properties.

We shall suppose that the base field K is of characteristic 0.

The structure of an algebra on a vector space V may be considered as an element of the space $\Pi_2^1(V) = V \otimes V^* \otimes V^*$ of tensors of type $(1,2)$ on V.

Let ρ be the natural linear representation of the Lie algebra $\mathfrak{gl}(V)$ in the space $\Pi_2^1(V)$. In these terms, a derivation of the algebra determined by a tensor $T \in \Pi_2^1(V)$ is a linear transformation $D \in \mathfrak{gl}(V)$ such that

$$\rho(D)T = 0. \tag{1}$$

Let us call a linear transformation D such that

$$\rho(D)^2 T = 0 \tag{2}$$

a generalized derivation, or a quasiderivation. (We might take any other exponent, but then we would not have those interesting properties that are given below.)

For any linear transformation D, the tensor $\rho(D)T$ determines some new operation on V. We shall call it the derived operation (with respect to D) of the original operation determined by T.

The transformation D is a quasiderivation if and only if it is a derivation of the derived operation.

Let the original operation be denoted by the usual multiplication, while the derived one will be denoted as $x * y$. According to the definition we have

$$x * y = D(xy) - (Dx)y - x(Dy). \tag{3}$$

This enables us to get the following explicit definition of a quasiderivation in terms of the space V itself:

$$D^2(xy) - 2D\big((Dx)y + x(Dy)\big) + \big((D^2x)y + 2(Dx)(Dy) + x(D^2y)\big) = 0. \tag{4}$$

In an analogous manner, we can define quasiautomorphisms. Namely, let R be the natural linear representation of the group $\mathfrak{GL}(V)$ in the space $\Pi_2^1(V)$. An automorphism of the algebra determined by a tensor $T \in \Pi_2^1(V)$ is a linear transformation $A \in \mathfrak{GL}(V)$ such that

$$\big(R(A) - E\big)T = 0. \tag{5}$$

1991 *Mathematics Subject Classification*. Primary 16W25.

(Here E is the identity transformation of V.)

Let us call a linear transformation A a quasiautomorphism if

$$(R(A)-E)^2T=0 \tag{6}$$

or, in an explicit form,

$$A^2(xy)-2A(Ax)(Ay)+(A^2x)(A^2y)=0. \tag{7}$$

Quasiautomorphisms are related to quasiderivations as automorphisms are to derivations. Namely, whenever exponentials exist (for example, if the base field K is $\mathbb{R}$ or $\mathbb{C}$ and D is locally finite dimensional) a linear transformation D is a quasiderivation if and only if $A(t)=\exp tD$ is a quasiautomorphism for any $t\in K$.

Indeed, if D is a quasiderivation, then the 2-dimensional subspace $U=\langle T,\rho(D)T\rangle \subset \Pi_2^1(V)$ is invariant under $\rho(D)$ and $\rho(D)$ induces on U the linear transformation with the matrix $\left(\begin{smallmatrix}0&1\\1&0\end{smallmatrix}\right)$. Hence, U is invariant under $\exp t\rho(D)=R(A(t))$ and $R(A(t))$ induces on U the linear transformation with the matrix

$$\exp t\begin{pmatrix}0&0\\1&0\end{pmatrix}=\begin{pmatrix}1&0\\t&1\end{pmatrix}.$$

So we have

$$(R(A(t))-E)^2T=0;$$

i.e., $A(t)$ is a quasiautomorphism. The inverse implication is proved by the reverse motion.

It should be noted that, unlike derivations, the quasiderivations of a given algebra in general do not form a Lie algebra, or even a vector space. Correspondingly, the quasiautomorphisms in general do not form a group. But the sum of commuting derivation and quasiderivation is still a quasiderivation.

EXAMPLE. In any associative algebra the left multiplication $L_c\colon x\mapsto cx$ is a quasiderivation if $c^2=0$. Indeed, the derived operation has the form

$$x*y=c(xy)-(cx)y-x(cy)=-xcy;$$

so, we have

$$(L_cx)*y+x*(L_cy)=-cxcy-xc^2y=-cxcy=L_c(x*y).$$

More generally, if

$$c^2=ac-ca \tag{8}$$

for some a, then the sum L_c+D_a is also a quasiderivation (with the same derived operation), where D_a denotes the derivation $x\mapsto ax-xa$. It is not so hard to show that in the matrix algebra the equation (8) for a has a solution if and only if c is nilpotent.

CONJECTURE. *All quasiderivations of a semisimple finite-dimensional associative algebra are of the form L_c+D_a, where c and a satisfy* (8).

In the case of a commutative algebra this conjecture means that there are no non-zero quasiderivations at all. This has been proved by A.S. Anan'in (see his paper in the present volume).

Now let us proceed to properties of quasiderivations.

THEOREM 1. *If D is a quasiderivation of some algebra, then the derived operation satisfies all the identities that the original one satisfies.*

It is not difficult to prove this theorem in a purely algebraic manner, but in the case when exponentials exist there is the following very transparent proof. Let T be the tensor determining the original operation. Then

$$R(\exp tD)T = (\exp t\rho(D))T = T + t\rho(D)T,$$

so all the operations of the form $\alpha T + \beta\rho(D)T$, where $\alpha \neq 0$, are isomorphic to T, hence, satisfy all the identities that the original one satisfies. Passing to the limit, we see that the operation $\rho(D)T$ satisfies all these identities as well.

This theorem is trivially generalized to algebras with arbitrarily many operations under the assumption that D is a quasiderivation of each of them. Moreover it is valid for algebras with operations with arbitrary many arguments. (The notion a quasiderivation is naturally generalized to such operations.)

The next property of quasiderivations seems to be more technical, but it is important for the application that will be discussed below.

THEOREM 2. *Let D be a quasiderivation and $x, y \in V$ elements such that $(D^n x)y = x(D^n y) = 0$ for $n = 0, 1, 2, \ldots$. Then $(D^k x)(D^l y) = 0$ for all k, l.*

The proof is by induction on $k + l$. Let $(D^k x)(D^l y) = 0$ for $k + l \leq n$. Put $D^{k-1}x$ and $D^{l-1}y$ for x and y in (4). For $k + l = n + 1$ we get

$$(D^{k+1}x)(D^{l-1}y) + 2(D^k x)(D^l y) + (D^{k-1}x)(D^{l+1}y) = 0.$$

It follows that the elements

$$z_k = (-1)^k (D^k x)(D^{n+1-k}y) \qquad (k = 0, 1, \ldots, n+1)$$

form an arithmetic progression. Since its extreme members $z_0 = x(D^{n+1}y)$ and $z_{n+1} = (-1)^{n+1}(D^{n+1}x)y$ are equal to 0 by assumption, then all its members are equal to 0 and the theorem is proved.

Now consider the main example giving rise to the notion of a quasiderivation.

Let $\mathfrak{g}$ be a Lie algebra and U its universal enveloping algebra. There is an increasing filtration

$$U = \bigcup_{k=0}^{\infty} U^k \tag{9}$$

of U in which U^k is the linear span of the products of $\leq k$ elements of $\mathfrak{g}$. According to the Poincare-Birkhoff-Witt theorem the corresponding graded algebra

$$P = \bigoplus_{k=0}^{\infty} P_k, \qquad P_k = U^k / U^{k-1}, \tag{10}$$

is canonically isomorphic to the symmetric algebra of the vector space $\mathfrak{g}$ (identified with P_1). But one can define in P another natural operation—the Poisson-Berezin bracket, or, simply, the Poisson bracket—arising from the commutator in U. This operation is denoted by braces. It satisfies the Jacobi identity and, together with the multiplication, the Leibniz identity

$$\{xy, z\} = \{x, y\}z + y\{x, z\}. \tag{11}$$

It is homogeneous of degree -1, i.e.,

$$\{P_k, P_l\} \subset P_{k+l-1}. \tag{12}$$

Now let δ be a linear function on $\mathfrak{g}$ and D the derivation of the multiplication in P, defined on generators by the formula

$$Dx = \delta(x) \qquad (x \in P_1 = \mathfrak{g}). \tag{13}$$

It is evident that

$$DP_k \subset P_{k-1}. \tag{14}$$

We prove that D is a quasiderivation of the Poisson bracket.

We have to prove that the second derived operation of the Poisson bracket with respect to D is zero. It is easy to see that the derived operation of any operation satisfying the Leibniz identity satisfies it as well. So the second derived operation of the Poisson bracket satisfies the Leibniz identity. It follows from (12) and (14) that it is homogeneous of degree -3.

In particular, it is zero on $P_1 \times P_1$. Making use of the Leibniz identity we conclude that it is zero everywhere.

Elements $x, y \in P$ are said to be commuting if $\{x, y\} = 0$. In accordance with this definition we shall use terms such as "a commutative subalgebra" and "the center" of P. It follows from the Leibniz identity that if some elements $x_1, \dots,$ $x_m \in P$ mutually commute, then the subalgebra (under the multiplication) generated by them is commutative.

Let D be the transformation defined above and Z the center of P. Then according to Theorem 2 the subalgebra generated by the elements of the form $D^n z$, where $z \in Z$, $n = 0, 1, 2, \dots,$ is commutative.

This may be considered as the underlying algebraic reason of the method of translating invariants, by means of which A. S. Mishchenko and A. T. Fomenko have proved the existence of a complete involutory system of left-invariant functions on the cotangent bundle over an arbitrary semisimple Lie group.

The notion of a quasiderivation is naturally extended to superalgebras. It turns out that a linear transformation D is a quasiderivation if and only if

$$\begin{aligned} D^2(xy) - (1 + (-1)^{pD})D((Dx)y + (-1)^{pDpx}x(Dy)) \\ + (-1)^{pD}\Big((D^2x)y + (1 + (-1)^{pD})(Dx)(Dy) + x(D^2y)\Big) = 0 \end{aligned} \tag{15}$$

(where p denotes the parity).

For odd D this means that D^2 is an (even) derivation. So odd quasiderivations are nothing else than square roots of derivations!

Department of Mathematics, Moscow State University, Moscow, Russia

Recent Titles in This Series

(*Continued from the front of this publication*)

(See the AMS catalog for earlier titles)